Animal Models of Neurological Disorders

Puneet Kumar Bansal
Rahul Deshmukh
Editors

Animal Models of Neurological Disorders

Principle and Working Procedure for Animal Models of Neurological Disorders

Springer

Editors
Puneet Kumar Bansal
Department of Pharmaceutical Sciences and Technology
Maharaja Ranjit Singh Punjab Technical University (State Govt. University)
Bathinda, Punjab
India

and

Department of Pharmacology
ISF College of Pharmacy
Moga, Punjab
India

Rahul Deshmukh
Department of Pharmacology
ISF College of Pharmacy
Moga, Punjab
India

ISBN 978-981-10-5980-3 ISBN 978-981-10-5981-0 (eBook)
https://doi.org/10.1007/978-981-10-5981-0

Library of Congress Control Number: 2017947848

Printed on acid-free paper

This Springer imprint is published by Springer Nature
The registered company is Springer Nature Singapore Pte Ltd.
The registered company address is: 152 Beach Road, #21-01/04 Gateway East, Singapore 189721, Singapore

This Book is Dedicated to Small Animals who Sacrifice their Life for the Betterment of Humankind

Preface and Acknowledgements

This "**Animal Models of Neurological Disorders—Principle and Working Procedure for Animal Models of Neurological Disorders**" book is introduced for the first time and contains detailed information on different types of animal models of neurological disorders along with working principle, experimental procedure, different neurotoxins with doses, and route of administration. During last decade, substantial progress has been made in understanding the pathophysiological basis of various neurological disorders, but still the development of drug therapy for neurological disorders is still an uphill task to the world of neuroscientists.

Therefore, this book is an attempt to help the researchers with updated information in designing new therapeutic strategies. The animal models described in this book for inducing neurological disorders reflect the similar sign and symptoms which are clinically relevant to the human neurological disorders. To provide all details regarding the disorders related to the CNS at a single platform, we have provided different procedures to induce a single disease with the help of chemicals on the basis of various literature surveys. Readers/audience will get to know that what type of pathological changes occurs in a particular neurological disorder that reflects specific signs and symptoms of any particular disease, and how it affects the normal life of a person. Researchers can use this book to select a single protocol to induce the disease of interest according to the availability of resources and can give a try to eradicate these neurological disorders or at least to improve the present status in society by testing different test drugs for their beneficial effects. It is an excellent practical source for all pharmaceutical professionals involved in neurological disorders and for researchers seeking a comprehensive, in-depth overview of animal models of neurological disorders.

This book will be helpful to the undergraduate, postgraduate students, researchers by knowing that how an animal model can be developed with the help of chemicals, what is the mode of action of particular chemical, and how we can design a research work on the CNS. As the prevalence of neurological disorders is increasing day by day, there are urgent need to develop new therapeutic strategies. This is possible only by complete understanding of disease pathology and developing suitable animal models for designing new therapeutic strategies. This book contains specific procedures for the neurological disorders, so that anyone can induce the specific disease in laboratory according to the availability of resources, as this book contains many protocols to induce a single neurological disease.

The editors wishes to thank Prof. M.P.S. Ishar (Vice Chancellor, Maharaja Ranjit Singh Punjab Technical University, Bathinda), Mr. Parveen Garg (Chairman, ISF College of Pharmacy, Moga), and colleagues for their valuable assistance and help in the preparation of this book. The author also appreciates the cooperation of Ph.D. scholars and M. Pharmacy students of Department of Pharmacology, ISF College of Pharmacy, Moga, in preparation of this book. Criticism and suggestions on this book from anyone are welcomed and shall be duly acknowledged by the author.

Bathinda, India Puneet Kumar Bansal
Moga, India Rahul Deshmukh

Contents

Editors and Contributors

About the Editors

Dr. Puneet Kumar Bansal is an Associate Professor of Pharmacology at the Department of Pharmaceutical Sciences and Technology, Maharaja Ranjit Singh Punjab Technical University, Bathinda, Punjab, India. Previously, he worked as a Professor at the ISF College of Pharmacy, Moga, Punjab. Dr. Bansal completed his Ph.D from University Institute of Pharmaceutical Sciences, Panjab University, Chandigarh. His research interests encompass the pharmacological screening of herbal and semisynthetic and synthetic drugs in animal models of various diseases such as movement disorders. He has published more than 100 peer-reviewed research papers/chapters and has more than 10 years of teaching and research experience. Dr. Bansal completed two major research projects for government-funded agencies. He received the International Brain Research Organization (IBRO) fellowship for attending IBRO neuroscience schools in Hong Kong and Malaysia and presented his research work at various national and international conferences. Recently, he was selected as Fellow of *The Linnean Society of London,* which is the world's oldest active biological society, and is also a member of various highly respected national and international professional societies.

Dr. Rahul Deshmukh is a Professor and head of the Department of Pharmacology, ISF College of Pharmacy, Moga, Punjab. He received his doctoral degree from Punjab Technical University, Jalandhar. He has 14 years of teaching and research experience in the field of pharmacology. His research area is focused on understanding the neurobiology of cognitive and movement disorders. He has published over 50 research and review articles in international peer-reviewed journals as well as books and chapters. He received several research grants from funding agencies of the Government of India. He also delivered various invited talks at a number of conferences and is a member of numerous highly respected national and international professional societies.

Contributors

Kiran Kumar Akula Department of Biology and Biochemistry, University of Houston, Houston, TX, USA

Onkar Bedi Department of Pharmacology, ISF College of Pharmacy, Moga, Punjab, India

Ganesh Singh Bhakuni Department of Pharmacology, ISF College of Pharmacy, Moga, Punjab, India

Rajat Bhardwaj Department of Pharmacology, ISF College of Pharmacy, Moga, Punjab, India

Swati Datta Department of Pharmacology, ISF College of Pharmacy, Moga, Punjab, India

Harsimran Kaur Gill Insectary Lab, Cornell University, Ithaca, NY, USA

Shubhangi Gupta Department of Pharmacology, ISF College of Pharmacy, Moga, Punjab, India

Sumit Jamwal Department of Pharmacology, ISF College of Pharmacy, Moga, Punjab, India

Priya Jaswal Department of Pharmacology, ISF College of Pharmacy, Moga, Punjab, India

Kashish Department of Pharmacology, ISF College of Pharmacy, Moga, Punjab, India

Harmanpreet Kaur Department of Pharmacology, ISF College of Pharmacy, Moga, Punjab, India

Karamjeet Kaur Department of Pharmacology, ISF College of Pharmacy, Moga, Punjab, India

Navneet Kaur Department of Pharmacology, ISF College of Pharmacy, Moga, Punjab, India

Tavleen Kaur Department of Pharmacology, ISF College of Pharmacy, Moga, Punjab, India

Prashant Koshal Department of Pharmacology, ISF College of Pharmacy, Moga, Punjab, India

Sanpreet Kour Department of Pharmacology, ISF College of Pharmacy, Moga, India

Mandeep Kumar Department of Pharmacology, ISF College of Pharmacy, Moga, Punjab, India

Vanita Rani Department of Pharmacology, ISF College of Pharmacy, Moga, Punjab, India

Nidhika Sharma Department of Pharmacology, ISF College of Pharmacy, Moga, Punjab, India

Vir Vikram Sharma Dayanand Medical College and Hospital, Ludhiana, India

Sneha Shree Department of Pharmacology, ISF College of Pharmacy, Moga, Punjab, India

Shamsher Singh Department of Pharmacology, ISF College of Pharmacy, Moga, Punjab, India

Vandana Department of Pharmacology, ISF College of Pharmacy, Moga, Punjab, India

Abbreviations

3-HAO	3-hydroxyanthranilic acid oxygenase
3-NP	3-Nitropropionic Acid
5-HT	Serotonin
5-HTP	5-Hydroxytryptophan
6-OHDA	6-Hydroxy Dopamine
AD	Alzheimer's disease
ADHD	Attention Deficit/Hyperactivity Disorder
AE	Acquired Epilepsy
AEDs	Antiepileptic drugs
ALS	Amyotrophic Lateral Sclerosis
AMPA	α-amino-3-hydroxy-5-methyl-4-isoxazole propionic acid
ANS	Autonomic Nervous System
AP	Anteroposterior
APOE4	Apolipoprotein E4 isoform
APP	Amyloid Precursor Protein
ASD	Alzheimer Senile Dementia
ASD	Autism Spectrum disorder
BA	Brucella Abortus
BBB	Blood-Brain Barrier
BDZs	Benzodiazepines
BK	Bradykinin
BP	Blood Pressure
BS	Basal Ganglia
BZD	Benzodiazepines
CBF	Cerebral Blood Flow
CCI	Chronic Constriction Injury
CCI	Controlled Cortical Impact
CER	Conditioned Emotional Response
CFS	Chronic Fatigue Syndrome
CGRP	Calcitonin Gene-Related Peptide
$CHCl_3$	Carbon tetrachloride
CHI	Closed Head Injury
CNS	Central Nervous System
CO_2	Carbon Dioxide

COX	Cyclooxygenase
CPZ	Cuprizone
CRI	Constant Rate Infusions
CS	Conditional Stimulus
CSD	Cortical Spreading Depression
CSF	Cerebrospinal Fluid
CUS	Chronic Unpredictable mild Stress
DA	Dopamine
DAT	Dopaminergic Transporter
DBS	Deep Brain Stimulation
DEP	Diesel exhaust particle
DG	Dentate Gyrus
DHA	Dehydroxyascorbate
DIP	Drug-Induced Parkinsonism
DNA	Deoxyribose Nucleic Acid
DOI	2,5-dimethoxy-4-iodophenyl-2-aminopropane
DRG	Dorsal root of ganglia
DTBI	Diffuse Traumatic Brain Injury
DV	Dorsoventral
EAAs	Excitatory Amino Acids
EAAT	Excitatory Amino Acid Transporter-2
EAE	Experimental Autoimmune Encephalomyelitis
EC	Entorhinal cortex
EEG	Electroencephalogram
ELISA	Enzyme-Linked Immunosorbent Assay
ELS	Early Life Stress
EMG	Electromyogram
eNOS	Endothelial NOS
FJ	Facial Jerking
FPI	Fluid Percussion Injury
FPS	Fear Potential Startle
FST	Forced Swimming Test
GA	General Anesthetics
GABA	Gamma-aminobutyric acid
GAD	Glutamate Decarboxylase
GAHBS	Group A-β-hemolytic Streptococcal
GCS	Glasgow Coma Scale
GD	Gestation Day
GluR	Glutamate Receptor
GP	Globus Pallidus
GSH	Glutathione
GT	Gastrointestinal
GTN	Glyceryl Trinitrate
HD	Huntington's Disease
HDAC	Histone Deacetylase

HMV	Murine Hepatitis Virus
HPA	Hypophyseal Pituitary Axis
HTT	Huntingtin
i.c.v.	Intracerebroventrical
i.m.	Intramuscular
i.p.	Intraperitoneal
i.v.	Intravenous
ICD	Impulse Control Disorder
IDE	Insulin Degrading Enzyme
IDPN	Iminodipropionitrile
IEGs	Immediate Early Genes
IFN-γ	Interferon Gamma
IL	Interleukins
IL-6	Interleukin-6
iNOS	Inducible NOS
ISI	Interstimulus Interval
ITI	Intertrial Interval
JEV	Japanese Encephalitis Virus
JNK	c-Jun N-terminal kinase
KA	Kainic Acid
KCl	Potassium Chloride
LA	Local Anesthetics
L-BMAA	β-N-Methylamino-1-alanine
LD	Levodopa
LFB	Luxol-Fast Blue
LFS	Load Force Swimming
LH	Learned Helplessness
LPS	Lipopolysaccharide
MABP	Mean Arterial Blood Pressure
MAM	Methylazoxymethanol
MAO	Monoamine Oxidase
MAO-B	Monoamine Oxidase-B
MBP	Myelin Basic Protein
MCD	Malformations of Cortical Development
MCP	Monocyte Chemoattractant Protein
MeHg	Methyl Mercury
MFB	Median Forebrain Bundle
$MgSO_4$	Magnesium Sulfate
ML	Mediolateral
MMP	Matrix Metalloproteinase
MMP-9	Matrixmettalloprotease-9
mpPVH	medial-parvicellular para-ventricular hypothalamic nucleus
MPTP	1-methyl-4-phenyl-1,2,3,6-tetrahydropyridine
MRI	Magnetic Resonance Imaging
MS	Multiple Sclerosis

MSN	Medium Spiny Neurons
MTLE	Mesial Temporal Lobe Epilepsy
N_2O	Nitrous Oxide
nAcc	Nucleus accumbens
NADPH	Nicotinamide Adenine Dinucleotide Phosphate
NAT	Noradrenaline Transporter
NCP	Nucleus Coudatus Putamen
NE	Norepinephrine
NFT	Neurofibrillary tangles
NIH	Novelty-induced Hypophagia
NIPS	Neuroleptics-Induced Parkinsonism
NMDA	N-methyl-D-aspartate
NMSS	National Multiple Sclerosis Society
nNOS	Neuronal NOS
NO	Nitric Oxide
NOS	Nitric Oxide Synthase
NPCs	Neural Precursor Cells
NREM	Non-rapid Eye Movement
NSAIDS	Non-Steroidal Anti-Inflammatory Drugs
NSF	Novelty-Suppressed Feeding
NSF	Novelty-suppressed Feeding
OB	Olfactory Bulbectomy
OCD	Obsessive Compulsive Disorder
$ONOO^-$	Nitronium Ion
p.o.	Per Oral
PACAP	Pituitary Adenylate Cyclase-Activating Peptide
PANDAS	Pediatric Autoimmune Neuropsychiatric Disorder Associated With Streptococci
PARP	Poly (ADP-ribose) polymerase
PBBI	Penetrating Ballistic Brain Injury
PCB	Polychlorinated Biphenyls
PCPA	p-chlorophenylalanine
PD	Parkinson's Disease
PDN	Peripheral Diabetic neuropathy
PEG	Polyethylene Glycol
PETN	Pentaerythritol tetranitrate
Pfc	Prefrontal cortex
PGE_2	Prostaglandin E2
PGG_2	Prostaglandin G2
PHFs	Paired Helical Filaments
PHN	Post-herpetic Neuralgia
PKc	Protein Kinase
PLP	Proteolipid Protein
PND	Postnatal Day
PNS	Peripheral Nervous System

PPE	Plasma Protein Extravasation
PPI	Prepulse Inhibition
PSD	Paradoxical Sleep Deprivation
PSNL	Partial sciatic nerve ligation
PTA	Post-traumatic Amnesia
PTSD	Post-traumatic Stress Disorder
QA	Quinolinic Acid
RAGE	Receptor for advanced glycation end products
RAS	Reticular Activating System
REM	Rapid Eye Movement
ROS	Reactive Oxygen Species
RTN	Reticular thalamic nucleus
RWA	Running Wheel Activity
s.c.	Subcutaneous
SCI	Spinal Cord Injuries
SCT	Staircase Test
SD	Sleeping Deprived
SDH	Succinate Dehydrogenase
SEM	Standard Error of Mean
SETX	Senataxin
SIF	Synthetic Interstitial Fluid
SIN	Sciatic Inflammatory Neuritis
SIP	Schedule-Induced Polydipsia
SNI	Spared nerve injury
SNpc	Substantia Nigra Pars Compacta
SNRI	Selective Norepinephrine Reuptake Inhibitor
SOD	Superoxide Dismutase
SSRIs	Selective Serotonin Reuptake Inhibitors
STZ	Streptozotocin
SWD	Spike-wave discharge
SWS	Slow Wave Sleep
TBI	Traumatic Brain Injury
TBZ	Tetrabenazine
TC	Thalamocortical nucleus
TD	Tardive Dyskinesia
TGF	Transforming Growth Factor
THAL	Thalidomide
TIPN	Taxane-induced peripheral neuropathy
TMEV	Theiler's Murine Encephalomyelitis Virus
TMEV-IDD	TMEV-Induced Demyelinating Disease
TNF-α	Tumor necrosis Factor-alpha
TP	Tongue Protrusions
TS	Tourette's Syndrome
TST	Tail Suspension Test
TT	Tetanus Toxin

UCS	Unconditioned Stimuli
UV	Ultrasonic Vocalization
VAMP	Vesicle-Associated Membrane Protein
VCMs	Vacuous Chewing Movement
VIP	Vasoactive Intestinal Polypeptide
VMAT	Vesicular Monoamine Transporter
VPA	Valproic Acid
WDI	Weight-Drop Injury

Introduction

Puneet Kumar Bansal

The human brain is the command center for the human nervous system. It contains 86 billion neurons along with other cells that make more than 100 trillion connections. Neurons are the building blocks of the nervous system which includes the brain and spinal cord. Neurons normally do not reproduce or replace themselves, so when they are damaged or die they cannot be replaced by the body. This unique anatomy of brain makes it more complex to understand. Brain is divided into different parts to perform specific functions, i.e., memory formation, movement, emotions, thought, speech, thinking, and many more. The brain is considered as the most complex part of the human body. This three-pound wonderful organ is the center of intelligence, interpreter of the senses, initiator of body movement, and controller of behavior. Contained in its bony shell and bathed by protective fluid, the brain is the originator of all the qualities that define humanity. For centuries, the brain has attracted and fascinated scientists and philosophers, and recently, it has been viewed the brain as nearly incomprehensible. Now, however, the brain is started to relinquish its secrets. Scientists have unfolded many secrets about the brain during 10 years than in past centuries due to the accelerating pace of research in neurological and behavioral science by virtue of new research techniques.

Damage or alteration in any part of brain results in impaired brain function. Brain disorders are mainly classified into two categories: neurodegenerative disorders and neuropsychiatric disorders. Neurodegenerative disorder is an umbrella term for varying conditions that primarily affect the neurons in the human brain. Neurodegenerative diseases such as Alzheimer's disease, Parkinson' disease, and

P.K. Bansal (✉)
Department of Pharmaceutical Sciences and Technology,
Maharaja Ranjit Singh Punjab Technical University (State Govt. University),
Bathinda, Punjab, India
e-mail: punnubansal79@gmail.com

P.K. Bansal
Department of Pharmacology, ISF College of Pharmacy, Moga 142001, Punjab, India

P.K. Bansal and R. Deshmukh (eds.), *Animal Models of Neurological Disorders*, https://doi.org/10.1007/978-981-10-5981-0_1

Huntington's disease are incurable and debilitating conditions and result in degeneration and/or death of nerve cells. Neuropsychiatric disorders are the illness of a psychological origin manifested either as symptom of emotional distress or in abnormal behavior. Neuropsychiatric disorders (psychiatric disorders) have been classified on the basis of criteria laid down in American Psychiatric Association: Diagnostic and Statistical Manual of Mental Disorders—DSM-IV-TR, 2000, into major clinical depression, bipolar disorder, schizophrenia, anxiety disorders, attention deficit/hyperactivity disorder, migraine, and subsequent subcategories.

The development of drug therapy for neurological disorders is still an uphill task to the medical world. This may be due to their complex pathologies, inadequate understanding, and treatment. The development of novel drugs relies on the understanding of these complex mechanisms. Testing and evaluating new molecule directly on human subjects is impractical because of ethical issues. Thus, for drug discovery research, animal models have always played a crucial role. Several animal models have been designed and utilized in drug performance and evaluation studies. These animal models have played crucial role in developing new drug entities as well as understanding the pathophysiology and etiology of diseases. Neurological disorders represent major threat to human race due to increasing prevalence day by day. During past two decades, substantial progress was made in the understanding the pathophysiology of neurological disorders, but exact mechanisms is still to be established in most of the cases. A better understanding of these mechanisms is essential for the development of new animal models and design of novel therapeutic approaches for mitigation of neurological disorders. Animal models represent an attempt to mimic the pathologies associated with human disease in a preclinical setting. The creation of an animal model that assists in understanding the basic mechanism of pathology in a systematic way in order to establish the biological basis is thus a prerequisite in such studies. In this context, it is important to establish validity of correspondence of such model to clinical state in human being. However, it has become clear that introducing human disorders in an animal does not necessarily trigger pathogenetic cascades identical to those observed in the human disease. Thus, diseases need to be studied simultaneously with the animal models to ensure that they simulate some pathogenesis, against which the new therapeutics may be tested.

With the above object in view, we have introduced a new subject namely **"Animal Models for Neurological disorder."** This completion could very useful in the meant for the university/college students especially those who are engaged in research at any level. This book describes various animal models for neurological disorders with a special emphasis on working principle and the procedure. The data which have been provided in this book are on the basis of validated experimental procedures. The model included in this book for each disease successfully simulates sign and symptoms clinically relevant to neurological disorder prevalent in human beings. This book will provide certainly a coherent platform for effective learning of a wide variety of new validated research techniques that can be learnt.

The basic purpose of this book is to familiarize the students and researchers with basic concepts of neurological disorders and their validated animal models with the working procedures so as to hone their skill in the field of neuroscience for carrying useful research.

Non-transgenic Animal Models of Alzheimer's Disease

Sneha Shree, Rajat Bhardwaj, Kashish and Rahul Deshmukh

1 Introduction

Alzheimer's disease (AD) is a leading cause of progressive dementia and most common neurodegenerative disease worldwide. The prevalence rate of AD is expected to get triple by 2050. Pathological hallmarks of AD are amyloid-β containing plaques and neurofibrillary tangles (NFT's), which are composed of hyperphosphorylated forms of the microtubule-associated protein tau. AD may be classified as early-onset (familial AD) and late-onset (sporadic AD). Familial AD occurs mostly in individuals of age 30–60 years and associated with mutations in amyloid precursor protein (APP) or presenilin (PS1 and PS2) genes, while sporadic AD mainly affects persons after 65 years of age and associated with mutations in apolipoprotein E4 isoform (apoE4) IR dysfunction, etc. Clinical interpretation of AD basically involves progressive deterioration in capabilities of memory, language, calculation, judgment, and behavior. AD is associated with disruption of mitochondrial function, calcium homeostasis, hormonal balance, and increased oxidative stress and neuroinflammation. Animal model has played a major role in

S. Shree · R. Bhardwaj · Kashish · R. Deshmukh (✉)
Department of Pharmacology, ISF College of Pharmacy,
Moga 142001, Punjab, India
e-mail: login2rd@gmail.com

S. Shree
e-mail: snehashree19@gmail.com

R. Bhardwaj
e-mail: 9rajatbhardwaj9@gmail.com

Kashish
e-mail: kashish331993@gmail.com

P.K. Bansal and R. Deshmukh (eds.), *Animal Models of Neurological Disorders*, https://doi.org/10.1007/978-981-10-5981-0_2

defining critical disease-related mechanisms and evaluating novel therapeutic approaches in research. The sporadic form of AD itself probably involves several different etiopathogenic mechanisms. Neuroinflammation, head trauma, brain ischemia, and diabetes have been implicated as risk factors for AD.

Molecular genetic studies have proved the central role of β-amyloid in the pathogenesis of AD. Approximately 5% of the total number of AD cases are because of the mutations in amyloid precursor protein on chromosome 21q21 (APP) and presenilin 1 on chromosome 14q24 (PSEN1), and presenilin 2 on chromosome 1q42 (PSEN2), which are involved in the cleavage of APP. Alzheimer's disease is associated with two major pathological traits. First, there is an accumulation of β-amyloid (Aβ) peptide. Second, hyperphosphorylation of tau protein occurs, which leads to neurofibrillary tangle formation and loses its microtubule binding and stabilizing role, contributing to neuronal degeneration. Aβ peptides originate from the amyloid precursor protein (APP) by amyloidogenic processing when acted upon by the alternate enzyme b-secretase instead of a-secretase followed by g-secretase. Aβ oligomers are neurotoxic and rapidly blocked long-term potentiation, a classic experimental paradigm for synaptic plasticity, which cause local structural disruption of synapses. Aβ triggers mitochondrial oxidative stress and dysregulation of calcium ion (Ca^{2+}) homeostasis, resulting in impairment of the electron transport chain, increased production of superoxide anion radical (O_2), and decreased production of ATP (Fig. 1).

Moreover, the levels of enzymes involved in energy metabolism (cytochrome c oxidase, pyruvate dehydrogenase complex, and a ketoglutarate dehydrogenase complex) are found to be decreased in brain cells of AD patient. Increased deposition of Aβ results in increased mRNA and protein expression of iNOS and

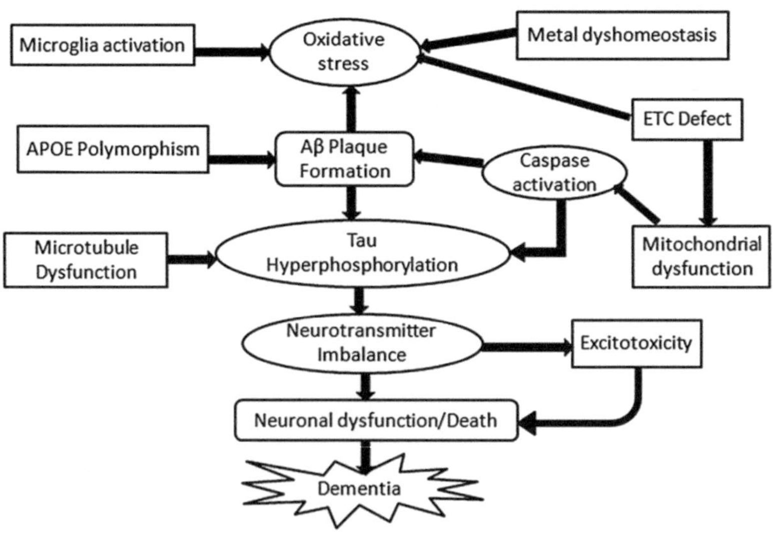

Fig. 1 Pathophysiology of Alzheimer's disease

generation of NO. NO can interact rapidly with superoxide anion O_2 forming more reactive peroxynitrite (ONOO) and induce lipid peroxidation and functional alterations in proteins and DNA, eventually leading to neuronal death. Aβ deposition activates inflammatory reactions by activating microglia and astrocytes, which leads to release of chemokines such as interleukin 8, interferon-g–inducible protein, macrophage inflammatory protein-1a, macrophage inflammatory protein-1b, and monocyte chemoattractant protein-1 (MCP-1) and cytokines such as IL-1, IL-6, TNF-a, and transforming growth factor-b (TGF-β).

2 Classification of Models AD

Animal models are mainly used to study the mechanisms underlying AD pathogenesis, genetic interactions with genes of interest, and environmental risk factors that cause sporadic AD, as well as to test the therapeutic effects of drugs on neuropathology and cognitive function (Fig. 2).

2.1 Chemical-Induced AD Models

2.1.1 STZ-Induced AD

The intracerebroventricular (i.c.v.) Streptozotocin (STZ) is a glucosamine–nitrosourea compound that, when metabolized, produces a cytotoxic product that preferentially destroys pancreatic β cells. The alkylating properties of STZ metabolites produce reactive oxygen species and finally resulting in oxidative stress. In moderate-to-low dosage in short-term experiments, STZ systemic administration caused insulin resistance by a decreasing phosphorylation of IRS-1 in rats. IRS-1 plays an important role in transferring signals from insulin receptors to intracellular pathways. Intracerebroventricular STZ (2-deoxy-2-(3-methy1-3-

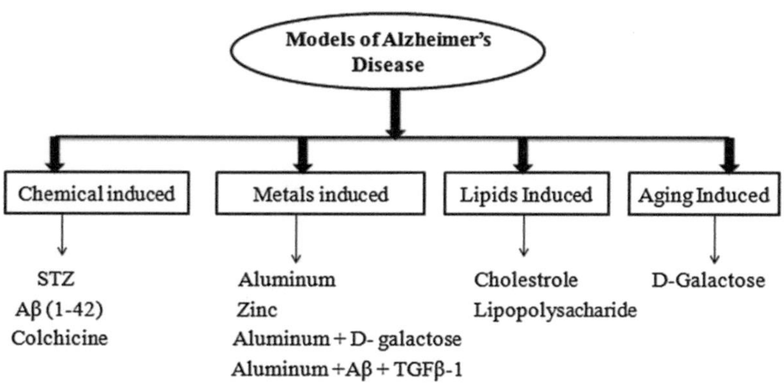

Fig. 2 Classification of animal models of Alzheimer's disease

Table 1 Different doses of STZ-induced AD model

S. no.	Dose/days	Route	Species	References
1.	STZ (1.25 mg/kg) bilateral	I.C.V.	Male Han: Wistar rats	Nitsch and Hoyer (1991)
2.	STZ (1.5 mg/kg) bilateral	I.C.V.	Male Lewis rats	Blokland and Jolles (1993)
3.	STZ (3 mg/kg) bilateral (1st and 3rd day)	I.C.V.	Male Wistar rats	Kumar and Gupta (2003)
4.	STZ (1 mg/kg) bilateral	I.C.V.	Male Wistar rats	Salkovic et al. (2006)

nitrosoureido)) animal model 233 Q19 was developed by Lannert and Hoyer in 1998. It has been reported that there is a decrease in glucose utilization in the brains of AD patients. This leads to the hypothesis that the cognitive dysfunction of AD is related to a reduction in central glucose metabolism. It has been hypothesized that the cholinergic deficit and amyloid accumulation in the brain are caused by a significant decrease in glucose metabolism in the AD brain. Various studies have shown that I.C.V. administration of subdiabetogenic doses of STZ reduce central glucose in rats. STZ administration in non-diabetogenic doses was shown to induce memory impairment, insulin receptor dysfunction in hippocampus and leads to progressive cholinergic impairment, glucose hypo-metabolism, oxidative stress, and neurodegeneration. STZ significantly improves BACE-1, p-p38 MAPK expression, and NF-kB (p65) translocation in astrocytes. Together, these effects significantly result in increase in amyloidogenic processing of APP and eventually cause increased activation of various deleterious pathways leading to neuronal degeneration (Table 1).

Procedure

1. Anesthetize the rats and positioned on digital stereotaxic apparatus.
2. Remove hairs and do sagittal incision on scalp and set the coordinates.
3. Drill the holes of skull which are identified and place cannula into the lateral cerebral ventricles.
4. Animals that take the delivery of either I.C.V. artificial cerebrospinal fluid (ACSF; 10 L/site) or ICV STZ (3 mg/kg bilaterally using a micro syringe). The composition of the ACSF was (in mmol/L): NaCl 147; KCl 2.9; $MgCl_2$ 1.6; $CaCl_2$ 1.7; dextrose 2.2.
5. To ensure diffusion of the administered drug, the cannula is left in place for a period of 2 min following the injection.
6. The stereotaxic coordinates for ICV injection were 0.8 mm posterior to bregma, 1.8 mm lateral to the sagittal suture, and 3.6 mm beneath the cortical surface.

Advantages

- ICV application of the STZ has been shown to cause oxidative stress, glucose/energy metabolism alteration, and cholinergic hypo-function accompanied by memory deficit by impairment of the neuronal insulin receptor transduction cascade (including insulin, IR, IRS-1, PI3 K, Akt, and GSK-3).
- Free radicals generation after ICV STZ injection is an important factor in causing cognitive impairment in rats.

Disadvantages

- Histological appearance of Aβ and Tau neuropathology takes long time to develop. Technical expertise is required for ICV infusion.

Clinical Relevance

- It has been demonstrated that AD in postmortem human brains is associated with significantly decreased expression of insulin/IGF trophic factors, their receptors, and IRS proteins, and further enhances the severity with progression of dementia and neurodegeneration (Fig. 3).

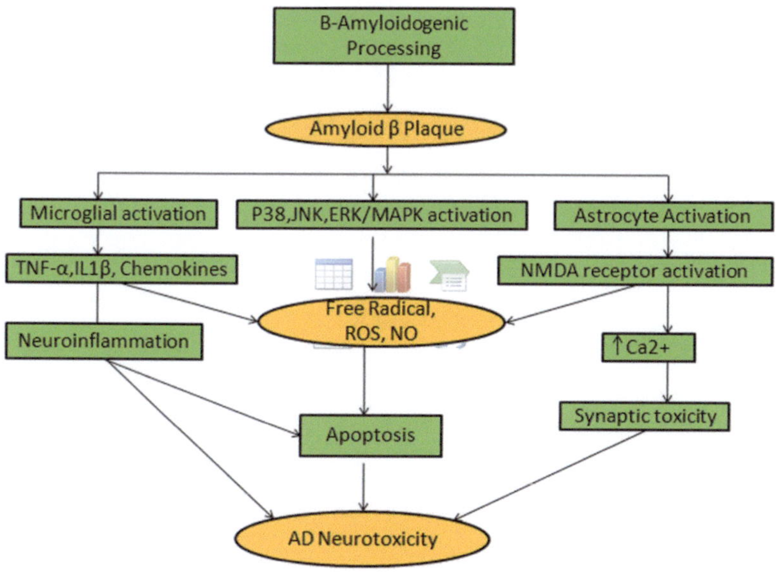

Fig. 3 Amyloid-β-induced neurotoxicity. ROS-reactive oxygen species, NO-nitric oxide, JNK-c-Jun N-terminal kinase, MAPK-mitogen-activated protein kinase, NMDA-, TNF-tumor necrosis factor

- In postmortem cases of advanced AD, it has been explored and resoluted if the neurodegeneration was correlated with significant abnormalities in the expression of genes encoding insulin, IGF-1 and IGF-2 peptides, their receptors, and downstream signaling mechanisms. In that study, we indicated advanced AD to be associated with remarkably diminished levels of insulin, IGF-1 polypeptide, and receptor genes in the brain.

2.1.2 Amyloid Beta-Induced AD

Aβ oligomers are small assemblies or aggregates of this so-called amyloid beta protein. It was discovered in 1992. Aβ peptide is 39-43 amino acids in length, and these are formed by the cleavage of transmembrane protein amyloid precursor protein (APP) by enzyme secretase. α-secretase is the primary enzyme acts APP under physiological considerations followed by γ-secretase. APP undergoes amyloidogenic processing when acted upon by the alternate enzyme β-secretase instead of α-secretase. The extent of amyloid-β deposits correlates with the degree of neuronal damage, cognitive impairment, and memory loss. The level of cholinergic neurons found to be decreased in amyloid-β-induced animal model of AD. Aβ (1-42) activates microglia as well as astrocytes, initiates a chemotactic inflammatory action, and releases cytokines including superoxide, proinflammatory cytokines such as tumor necrosis factor (TNF) and interleukins (IL-1, IL-6) and excitatory amino acids including glutamate. These inflammatory markers are associated with neurodegenerative processes in AD. The amyloid-β_{1-42} (Aβ42) peptide rapidly aggregates to form oligomers, protofibrils, and fibrils en route to the deposition of amyloid plaques associated with Alzheimer's disease. In this model, rats were infused ICV with Aβ (1-42) oligomer (3 nmol/3μl) dissolved in ACSF. It has been demonstrated that oligomeric Aβ alteres the expression of Bcl-2, Bim and Bax and contributes in neuronal apoptosis. On the other hand, Aβ induces NO production either by disrupting calcium homeostasis and subsequent increase in intracellular Ca2+ (nNOS and eNOS-mediated NO release) or by interactions with glial cells (iNOS-mediated NO release). These reactive oxygen species induce a variety of

Table 2 Different doses of amyloid beta-induced AD

S. no.	Dose/days	Route	Species	References
1.	Continuous infusion of the β-amyloid (3, 30, 300 pmol/day) for 14 days	I.C.V.	Male Kbl Wistar rats	Nitta et al. (1994)
2.	3 nmol/mouse bilaterally	I.C.V.	Male Swiss mice	Maurice et al. (1996)
3.	3 mM dose of β-amyloid (25-35) bilaterally	Intra hippocampal	Male Sprague–Dawley rats	Chen et al. (2006)

neurotoxic mechanisms, including DNA/ protein alterations, poly ADP-ribose polymerase (PARP) over activation, mitochondrial dysfunction, lipid peroxidation, neuroinflammation, and apoptosis. Also it has been found that there is disruption in the various intracellular signaling pathways by activation of stress-related kinases c-Jun N-terminal kinase (JNK), and p38 is associated with neuronal death in AD models. Amyloid beta itself can generate the free radicals upon interaction with metals, such as iron and aluminum, which cause neuronal cell death and AD (Table 2).

Procedure

1. Anesthetize the rats and positioned on digital stereotaxic apparatus.
2. Remove hairs and do sagittal incision on scalp and set the coordinates.
3. Drill the holes of skull which are identified and place cannula into the lateral cerebral ventricles.
4. Animals receive either I.C.V. artificial cerebrospinal fluid (ACSF; 10 L/site) or ICV STZ (3 mg/kg bilaterally using a microsyringe).
5. Infused I.C.V. in rats with either artificial cerebrospinal fluid (ACSF; in nmol/L: 147 Nacl, 2.9 Kcl, 1.6 $Mgcl_2$, 1.7 $CaCl_2$, and 2.2 dextrose) or amyloid-$\beta_{1\text{-}42}$ oligomer (3 nmol/3μl) dissolved in ACSF according to the method described by Maurice.
6. Drug solution (3μl) should be injected twice by using Hamilton microsyringe positioned in the injection cannula.
7. The stereotaxic coordinates for ICV injection were 0.8 mm posterior to bregma, 1.8 mm lateral to the sagittal suture, and 3.6 mm beneath the cortical surface.

Advantages

- The 'amyloid cascade hypothesis' emphasizes a central role for Aβ in the pathogenesis of AD. Thus, Aβ has become a major therapeutic target, with various anti-Aβ strategies being pursued.
- Aβ proteinosis is an important structure of senile plaques and is thought to be the main reason for the loss of neurons and the resulting memory disability.

Disadvantages

- This model has high mortality rate. Aβ can be infused into animal brain stereotaxically, which requires more care after surgery.

Clinical Relevance

- AD in humans results in significant atrophy of the brain, particularly in the entorhinal cortex, hippocampus, and amygdale.

- In human, Aβ deposits begin exclusively in the neocortex, then extends into allocortical brain regions, such as the hippocampus. Areas of the diencephalon, where the thalamus and hypothalamus are situated, as well as the striatum and cholinergic nuclei of the basal forebrain are influenced later. Finally, in the last stages of disease, Aβ pathology can be detected in areas of the brain stem and the cerebellum.

2.1.3 Colchicine-Induced AD

Colchicine, an alkaloid extracted from some plants of the lily family, has been used for centuries to treat acute gouty arthritis. It was first isolated in 1820 by the French chemists P.S. Pelletier and J.B. Caventou. In 1883, P.L. Geiger purified an active ingredient, which he named colchicines. Since 1973, it has been recognized as an effective remedy for prophylaxis of attacks of familial Mediterranean fever. Colchicine is a well-known neurotoxin, a cytotoxicant that binds irreversibly to tubulin dimers and induces neurofibrillary degeneration, thereby blocks mitosis and axonal transport, leading to disruption in the microtubule polymerization, death of cerebral granule cells, olfactory bulb neurons, cells of subventricular zone, dentate gyrus cells, and basal forebrain cholinergic neurons, thus causing cognitive impairment. Central administration of colchicine exaggerates free radical generation and oxidative DNA damage that eventually produces marked destruction of hippocampal granule cells and septohippocampal pathways resulting in loss of cholinergic neurons and decreased activities of acetylcholinesterase and choline acetyltransferase, thereby resulting in decreased ability to learn and in loss of memory. Colchicine has been found to be significantly induce free radical generation and deplete antioxidant defense system in rat brains. 15μg/15μl colchicine administered intracerebroventricularly (i.c.v.) in rats to induced memory impairment and oxidative damage. Colchicine is neurotoxic toward various neuronal populations including cerebral granule cells (CGC's) and causes high molecular weight DNA fragmentation and nucleus condensation and causes caspase-2-induced (induces caspases 3) proteolytic degradation of PARP (poly (ADP-ribose) polymerase, DNA repair enzymes, cytoskeletal proteins, and fodrin (Table 3).

Table 3 Different doses of colchicine-induced AD

S. no.	Dose/days	Route	Species	References
1.	Colchicine (15, 30, 60, and 120 pg/kg) for 10 days, 90 min. before tests	I.P.	Male Wistar AF rats	Bensimon and Chermat (1991)
2.	15 μg/5 μL dissolved in 15 μl of artificial cerebrospinal fluid	I.C. V.	Male Wistar rats	Kumar et al. (2009)
3.	Colchicine (15 μg/rat), dissolved in 15 μl of artificial cerebrospinal fluid	I.C. V.	Wistar Albino rats	Pitchaimani et al. (2012)

Procedure

1. Anesthetize the rats and positioned on digital stereotaxic apparatus.
2. Remove hairs and do sagittal incision on scalp and set the coordinates.
3. Drill the holes of skull which are identified and place cannula into the lateral cerebral ventricles.
4. Close the scalp with suture. In sham-operated rats, the surgery was identical except for drilling of holes and placement of the cannula.
5. After surgery, gentamicin (5 mg/kg, ip) is given to all animals to prevent sepsis. Intracerebroventricularly (icv) infused with either artificial cerebrospinal fluid (ACSF; in mM: 147 NaCl, 2.9 KCl, 1.6 $MgCl_2$, 1.7 $CaCl_2$, and 2.2 dextrose) or 15 μg colchicine dissolved in ACSF in rats.
6. Solutions were injected using a Hamilton microsyringe positioned in the injection cannula, and the injection cannula was left in place for 2–3 min following infusion.
7. Special care was taken during the postoperative period to provide food and water inside the cage of the rat.
8. The stereotaxic coordinates for ICV injection were 0.8 mm posterior to bregma, 1.8 mm lateral to the sagittal suture, and 3.6 mm beneath the cortical surface.

Advantages

- Colchicine induces hippocampal lesions resulting in learning and memory impairment, reduction in choline acetyltransferase, suggesting that it could be used as a suitable model for studying Alzheimer's disease.

Disadvantages

- The neurotoxic mechanism of colchicine is not fully understood.
- This model is time-consuming and requires a large number of animals due to high mortality rate.
- Rats treated with colchicine produce a decrease in appetite and transient diarrhea, adipsia, and aphasia after 7–10 days of its administration. Also there are results in myoclonic twitches, aggressive behavior, decreased body weight, which exhibits acoustic startle behavior and decreased threshold to pain.

Clinical Relevance

- I.C.V. colchicine model can be further used to understand better SDAT pathogenesis in AMT humans. It has several features that are consonant with SDAT in humans, including an insidious onset, time-dependent changes in behavioral and biochemical patterns.

2.2 Metal-Induced AD

In the AD brain, variation in metal levels may indicate insufficiencies or excesses of specific metalloproteins or defective metal transporters. Indeed, the levels of various Cu and Fe regulatory and storage proteins are changed in AD brain. There is an accumulating indication that interactions between β-amyloid and copper, iron, and zinc are integrated with the pathophysiology of Alzheimer's disease (AD). Iron and Cu form crucial components of various enzymes required for vital brain functions including energy production, neurotransmitter synthesis, and antioxidant function. Change in metal homeostasis may be a key factor which resulting in AD pathogenesis. It has been found to have role in accumulation of Aβ. Which is mediated by interaction with metals, in particular copper (Cu), zinc (Zn), and iron (Fe). Metal induces oxidative DNA damage by the disruption of calcium homeostasis and intracellular signal transduction pathways. Aβ catalyzes the process of formation of reactive oxygen species (ROS), which in turn contributes to Aβ accumulation by generating modified Aβ species. Due to redox-active nature of Cu and Fe, inappropriate regulation of these metals can lead to reaction with O2 and the formation of ROS, causes cellular toxicity. AD brain indicates marked oxidative damage of proteins, lipids, and nucleic acids. Currently, the copper- and zinc-chelating agent clioquinol depict a potential therapeutic route that may not only obstruct β-amyloid neurotoxicity, but may also reverse the aggregation of neocortical β-amyloid.

2.2.1 Aluminum-Induced AD

The contributions of neurotoxicity of Al in experimental animals were first reported in 1897 by Dollken. It has been also shown that Al salts administered intracerebrally or peripherally in rabbit, cat, mice, rat, and monkey induce the formation of neurofibrillary tangles. Aluminum is the third most common ubiquitous element in the environment. Main sources of aluminum exposure are water, cooking utensils, food additives, grain products, processed cheese, and drugs like antacids and deodorants. Aluminum levels increase in the brain with advancing age. Aluminum (300 mg/kg daily, p.o.) has been shown to abrogate neurotransmission, cognitive behavior. Thus, it has been found to be associated with several neurodegenerative diseases including AD. The neurodegeneration effects of aluminum in experimental animals are dependent on the route of administration, type of aluminum salt, species of animal, dose of treatment, and time of exposure. In most studies, the duration of aluminum exposure ranges from 8 weeks to 6 months. Aluminum has been found to be involved in the formation of paired helical filaments (PHFs) of tau proteins by inhibiting PP2A (protein phosphatase 2A) activity. Moreover, PP2A accounts for approximately 70% of the total tau phosphatase activity in human brain. The mechanism of aluminum also involves accumulation of aluminum in the neurons which leads to cell depolarization and disruption in the Na+/Ca2+ exchange pump. Ultimately cause excessive accumulation of intra-mitochondrial Ca2+ levels leading to opening of the MPT with subsequent release of cytochrome c, activation of caspases and apoptosis. Increase in intra-mitochondrial Ca2+ levels also cause increased production of toxic free radicals. All together, these events cause an

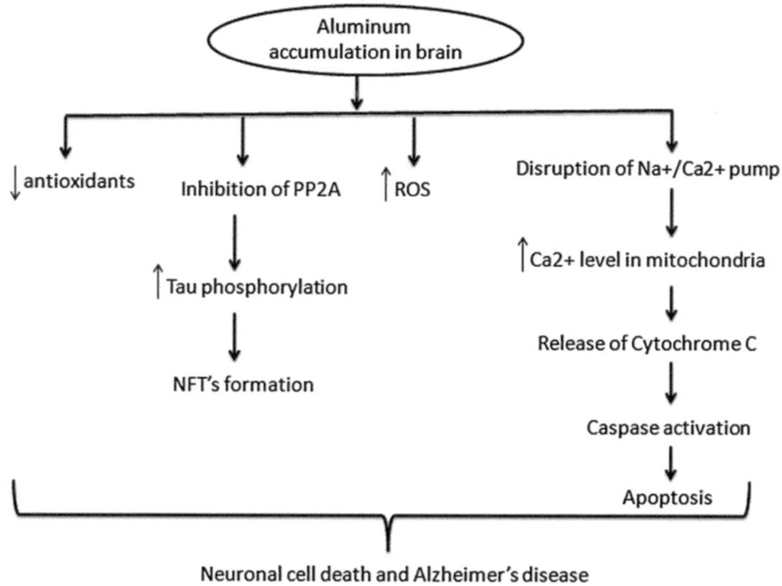

Fig. 4 Mechanism of aluminum-induced AD

Table 4 Different doses of aluminum-induced AD

S. no.	Dose/days	Route	Species	References
1.	300 mg/kg daily	Oral	Wistar rats	Lu et al. (2013)
2.	Single dose of Al maltolate, usually ~ 100 μL of a 50 mM solution in 100 μl of sterile saline	Intra-cisternal	New Zealand (white rabbit)	Savory et al. (2006)

increase in amyloid plaques and neurofibrillary tangles in brain. Mechanism of aluminium-induced neurotoxicity is described in Fig. 4 (Table 4).

2.2.2 Aluminum-, Aβ- and Transforming Growth Factor Beta-1(Tgfβ-1)-Induced AD

The aim was to design an animal model with multifactorial etiologies in animal that mimic the complicated pathological features and biochemical changes of AD. However, most of the classic AD models addressing or targeting single etiologies issues. Therefore, this new animal model may provide a successful multifactorial AD model. Aluminum and Aβ can interact and enhance each other, thereby intensify the neuronal deterioration. Also it has been found that TGF-1 can accelerate the deposition of Aβ in brain and damage capillary vessel to promote the occurrence of AD. Moreover, the impairment of cholinergic system impairment lasts for 3 months in this model. So this could be used for fairly long-term

Table 5 Different doses of aluminum, Aβ and transforming growth factor beta-1(TGFβ-1)

S. no.	Dose/days	Route	Species	Reference
1.	**Low dose**-Aβ1-40 2μg/day (14 days) + 1% $AlCl_3$ solution 2μl/day (5 days) **High dose**-Aβ1-40μg/day (14 days) + 1% AlC13 solution 3μl/day (5 days)	Lateral cerebral	Male Sprague–Dawley rats	Fang et al. (2013)

pathogenesis and drug research. It is more close to the process of etiopathology and pathological and biochemical changes of AD than other classic models of AD. IN early-onset familial AD (FAD), highly aggregative Aβ42 is over-produced by mutant presenilin 1/2 genes. Furthermore, Aβ aggregation may be promoted by apolipoprotein E and by environmental factors, such as oxidative stress and diabetes. Oxidative stress-inducing metal ions, such as iron (Fe) Copper (Cu) zinc (Zn) and aluminum (Al), may contribute to the formation of SPs/NFTs and subsequent neuronal damage in the AD brain (Table 5).

Procedure

1. Anesthetize the rats and positioned on digital stereotaxic apparatus.
2. Remove hairs and do sagittal incision on scalp and set the coordinates.
3. Drill the holes of skull which are identified and place cannula into the lateral cerebral ventricles.
4. Coordinates for the injection lateral cerebral ventricle (behind the bregma 1.2 mm, meta 2 mm, deep 4 mm), and the other point was the anterodorsal nucleus of thalamus (at the opposite side of the cerebrum, behind the bregma 2.1 mm, meta 1.4 mm, deep 4.6 mm).
5. The catheter shall be inserted into the lateral cerebral ventricle point, fixed at the cranium with dental base acrylic resin powder and 502 glue mixture, and at the same time, the catheter was blocked by the nylon line core which was a little less than the catheter inner diameter in order to prevent from infection and leakage of cerebrospinal fluid.

2.2.3 $AlCl_3$- and D-Galactose-Induced AD

$AlCl_3$ and D-galactose induce Alzheimer's disease by alterations in β-amyloid peptide metabolism-related molecules in the early stages of predementia. This model shows alteration in the expression of BACE1 (β–secretase, an enzyme involved in the cleavage of APP), neprilysin (NEP) (Aβ degrading enzymes), insulin degrading enzyme (IDE), and receptor for advanced glycation end products (RAGE), which is strongly involved in Aβ influx back into the brain earlier than Aβ processing. These pathways eventually cause memory impairment and high Aβ levels in the cortex (Co) and hippocampus (Hi), which shows pathological features similar to Alzheimer's disease. This model showed that the combination of Aland D-gal caused more mouse cognitive impairment as compared to treating with Al

Table 6 Different doses of aluminum and D-galactose-induced AD

S. no.	Dose/days	Route	Species	Reference
1.	$AlCl_3$ 40 mg/kg/day, or D-gal 90 mg/kg once (90 days)	i.p.	Kunming strain mice (20–25 g)	Luo et al. (2009)

and D-gal individually. This model could be employed for the early diagnosis of AD hence helpful in investigating future mechanism and drug screening for AD (Table 6).

Advantages

- This model could be employed for the early diagnosis of AD hence helpful in investigating future mechanism and drug screening for AD.
- Most of the classic AD models are mono-factor models, but this model shows multifactor mechanism that mimics the complicated etiology, pathological and biochemical feature of an ideal animal model of AD.
- The cholinergic system impairment lasts for 3 months in this model.

Disadvantages

- Aluminum itself is not that sufficient to cause Alzheimer's disease.

Clinical Relevance

- However, it was found that in brain tissue from dialysis patients, only 1 case out of 50 exhibited the accumulation of Aβ42 and p-tau. Phosphorylated neurofilaments and cytoplasmic argyrophilic inclusions have been shown to be characteristics of Al-induced pathology in DAE, rather than the paired helical filaments that are consistently observed in AD.

2.2.4 Zinc-Induced AD

Zinc plays an important role in normal growth and development. In physiologic conditions, zinc is also associated in variety of neuronal functions viz neurogenesis, neuronal migration, synaptogenesis, and neurotransmission, thus is highly involved in cognitive functions. A total of 100–200 mg/kg zinc content is required for regulating learning and memory function in rats. Zn is a key modulator for synaptic neural transmission, reaching 150–300 μM during synaptic activity, thus possibly causes Aβ accumulation. It has been indicated that zinc also plays an essential role in cellular homeostasis and various biochemical functions, such as protein synthesis and nucleic acid metabolism. Majority (80–90%) of the zinc present in the brain bounded with proteins, and the remaining percentage is packaged within synaptic vesicles of a large subpopulation of excitatory neurons. The synaptic or vesicular zinc cause activation of several neurotransmitter receptors, including NMDA, AMPA, GABA, and glycine receptors as well as voltage-dependent ion channels in an activity-dependent manner. Zinc is highly enriched in the hippocampus, amygdala, cerebral cortex, thalamus, and olfactory cortex. Recent studies showed that secreted zinc plays vital role in information processing, synaptic plasticity, learning, and memory. In fact, zinc is found to be vital in the hippocampus for the stimulation of long-term potentiation, a form of synaptic information storage thus making it a well-known model for the mechanisms underlying memory formation. In pathologic condition, a significant amount of zinc is released into synapse due to membrane depolarization where it triggers a number of prejudicial signaling processes including those that lead to further ROS generation, marking the start of a positive feedback loop including intracellular zinc release and ROS generation. Synaptic zinc is also connected with neuronal dysfunction by its transport from overactive presynaptic zinc-containing neurons to postsynaptic cells via calcium-permeable channels, including but not restricted to a subclass of AMPA receptors. Zinc has been found to diminish several enzymes, mitochondrial respiration, thus resulting in energy depletion and ROS generation. Excessive glutamate release causes an increase in intracellular Ca2+ levels, which trigger several apoptotic pathways, such as calpain or caspases activation ultimately, resulting in neuronal death. Thus, disruption in zinc homeostasis has been implicated in several neurodegenerative diseases, including Alzheimer's disease. Both zinc depletion and excess zinc can cause severe damage to neurons.

Both extracellular and intracellular zinc cause the Aβ-induced neurotoxicity during AD. It has been found that plaques themselves are rich in copper and zinc and aggregation of Aβ peptide resulting in increasing the state of oxidative stress during AD through direct formation of oxidants as well as through microglia activation and subsequent production of ONOO− . The ONOO– generation from both neurons and microglia appears to be a lead trigger of zinc-dependent neuronal apoptosis. Moreover, hydrogen peroxide, an oxidant, can cause the release of zinc from MT III, subsequently resulting in accumulation of Aβ. Zinc secreted into the synapse, through zinc-containing glutamatergic vesicles facilitates Aβ aggregation following cleavage of peptide from membrane-bound APP. With age, the concentration of vesicular zinc and the expression of ZnT3 (transporter responsible for packaging zinc into synaptic vesicles) are also found to be diminished in brain. Further, both Aβ aggregates and zinc have been found in mitochondria, lysosomes, and the ER. Similar to Aβ, zinc can also directly bind to tau to facilitate tau hyperphosphorylation, and Aβ can mediate calcium release from the ER, which cause free radical generation in brain.

Advantages

- Zinc can directly bind to Tau protein and thus contribute in microtubule destabilization, which is one of the major pathologic hallmarks of AD.

Disadvantages

- The neurotoxic mechanisms of zinc are not clear.

2.3 Lipids-Induced AD

2.3.1 Cholesterol-Induced AD

Cholesterol is important for cell structure, repair and signaling, hormone production, and bile acid synthesis and individuals with enhanced cholesterol levels during midlife have risk to develop AD. Cholesterol may be directly involved in Aβ aggregation: Abnormal oxidative metabolites like cholesterol-derived aldehydes can modify Aβ, which is associated with AD and Aβ, NFT's pathology. Cholesterol flux is elevated in AD brain, and cholesterol dyshomeostasis is closely associated with Alzheimer's synaptic loss and cognitive impairment. Individuals with high cholesterol in their early 40s are more likely to develop AD than those with low cholesterol. Alteration in the total cholesterol, HDL and LDL levels, in serum correlate with a disturbed cholesterol metabolism and Aβ load in the AD brain. The cholesterol metabolite 24S-hydroxycholesterol is more soluble than cholesterol and is more easily exported from the brain, and high levels of 24S-hydroxycholesterol have been reported in AD brain. Cholesterol and apoE are involved in fibrillar plaque formation. Lipid rafts are heterogenous, cholesterol- and sphingolipid-rich membrane and responsible for development of nerve axon growth, and they have a major role in APP amyloidogenic processing at γ-secretase level. However, increased cholesterol level (more than 5.8 mmol/L) is correlated to the amyloid plaques associated with AD. Apolipoprotein E (apoE), apoJ, ATP-binding cassette subfamily member 7, and sortilin-related receptor are AD susceptibility genes, which are involved in cholesterol metabolism or transport. ApoE is the major cholesterol carrier in the brain, and it regulates the brain cholesterol metabolism and triglycerides in the body. There are three common alleles of the ApoE genes ε2, ε3, and ε4. ApoE has been demonstrated to play a role in Aβ generation by increasing cellular cholesterol.

2.3.2 Lipopolysaccharides-Induced AD

Lipopolysaccharide (LPS) is a component of the cell wall of gram-negative bacteria, and it has been used experimentally to induce the production of the endogenous IL-1 and b-amyloid precursor protein (β-APP). It is an endotoxin and potent activator of microglia and astrocyte. It plays a substantial role in

Table 7 Different doses of lipopolysaccharides-induced AD

S. no.	Dose/days	Route	Species	References
1.	LPS 50 Ig/mouse for 7 days	i.p.	Swiss mice of either sex	Patil et al. (2003)
2.	LPS 0.25 μg/h for 4 weeks	i.c.v.	Male Harlan Sprague–Dawley	Hauss-Wegrzyniak et al. (1998)

neuroinflammation and brings on phosphorylation of proteins, protein tyrosine kinase, mitogen-activated protein kinases C and A, G protein, and ceramide-activated protein kinase. LPS are reported to induce various cytokines (interleukin (IL) 1 ß, IL-6, interferon alpha, tumor necrosis factor alpha), cyclooxygenase-2 (COX-2), and amyloid precursor protein mRNA levels within the basal forebrain. These proinflammatory mediators successively activate astrocytes and microglia in hippocampus, pituitary gland, and hypothalamus and produce degeneration of CA3 pyramidal neurons and eventually leading to impairment in spatial memory. Moreover, LPS causes increase in the Ca2+ -independent iNOS activity and protein kinase C activation, which in turn leading NADPH oxidation, free radicals generation and activates protein tyrosine kinase. Further leads to neuronal death and spatial memory impairment. Moreover, intraperitoneal injection of LPS (250 Ìg/kg) has been reported to impair cognition by depressing the social exploration of rats which was thought to be mediated by the central inflammatory cytokine IL-1 (Table 7).

Procedure

1. An (Alzet Palo Alto), CA osmotic minipump (model 2002; 0.5 μl/h) containing LPS (Sigma; E. coli, serotype 055:B5, TCA extraction; 1.0 μg/ml) was implanted into the dorsal abdomen and attached via Tygon tubing (0.060 o.d) to a chronic indwelling cannula (Model 3280P, Osmotic pump connect, 28 gauge, Plastics One, Roanoke, VA) that had been positioned stereotaxically so that the cannula tip extended to these coordinates: 2.5 mm posterior to Lambda, on the midline, and 7 mm ventral to the dura.
2. Controls were infused with artificial cerebrospinal fluid (aCSF): (in mM) 140 NaCl; 3.0 KCl; 2.5 CaCl; 1.0 MgCl; 1.2. 2 2 Na HPO, pH 7.4. The LPS was dissolved in aCSF.
3. The rats were infused with either aCSF or LPS for a total of 4 weeks.
4. A volume overload to the CSF space was discounted because the 0.5 μl/h administered contributed to only about 0.4% of the CSF volume produced by the rat each hour and was only 0.25% of the rat's total CSF volume.

Advantages

- LPS can impair cognition by depressing the social exploration of rats which was thought to be mediated by the central inflammatory cytokine IL-1.

Clinical Relevance

- In AD, brain cholesterol flux is elevated: When compared to controls, higher levels of the more soluble form of cholesterol, 24S-hydroxycholesterol, are found in both CSF and plasma of AD patients, even in early stages of dementia.
- Levels of total cholesterol and LDL in serum have been found to correlate with Aβ load in the brains of patients with AD. Epidemiological evidence suggests that elevated cholesterol levels during midlife increase the risk of developing AD.

2.4 Aging-Induced AD

2.4.1 D-Galactose-Induced AD

Aging is considered as a major risk factor in AD and mitochondrial dysfunction, and oxidative DNA damage has been demonstrated to have major role in aging. D-gal is a physiological nutrient and a reducing sugar. Generally, galactose is metabolized by D-galactokinase and galactose-1-phosphate uridyltransferase in animals, but its high levels result in abnormality of metabolism. D-galactose is converted into galactitol and started accumulating in the cell, which in turn causes activation of receptor for advanced glycation end products (RAGE), thereby results an increase in oxidative stress and cellular damage. However, high levels of D-galactose cause oxidative metabolism of D-galactose into aldose and hydroperoxide in presence of enzyme galactose oxidase and finally result in excessive ROS and free radical generation in the brain. Mitochondrial dysfunction induces an increased reactive oxygen species and free radical generation and decreased oxidative phosphorylation. It also causes microglia activation and thereby leads to inflammation and causes oxidation of nucleotide acids, especially DNA. Hippocampus plays a major role in formation of memory. Recently, it has been reported that the chronic administration of D-galactose (D-gal) significantly can lead to neuronal damage and symptoms like deterioration of cognitive abilities same as in aging. Also it has been demonstrated that continuous administration of D-gal (s.c.) in mice induced an increase in cell karyopyknosis, apoptosis, and caspase-3 protein levels in hippocampal neurons. Excessive ROS generation coupled with disrupted antioxidant defenses leads to oxidative damage to mitochondria, and chronic administration of D-galactose causes hippocampal dysfunction in rodents, ultimately causes decline in spatial learning and memory function by

Table 8 Different doses of D-galactose-induced AD

S. no.	Dose/days	Route	Species	References
1.	100 mg/kg, 6 weeks	s.c.	Male Laca mice	Kumar et al. (2011)
2.	Ovariectomy + D-gal 20 mg daily once + 17-β estradiol 0.2 mg (i.m.)	i.p.	Female Sprague–Dawley rats	Hua (2007)

impaired oxidative defense and mitochondrial complex (I, II, and III) enzyme activities (Table 8).

Advantages

- It has been reported that the chronic administration of D-galactose (D-gal) can lead to neuronal damage and symptoms like deterioration of cognitive abilities same as in aging.

Clinical Relevance

- Age-related memory loss starts in the dentate gyrus. A recent gene expression study in the human dentate gyrus postmortem, normalized for gene expression in the entorhinal cortex that is not affected by aging, identified 17 genes showing reliable age-related changes.

3 Conclusion

As Alzheimer's disease has multiple etiologies, so in this review we focus on different animal models of Alzheimer's disease that has been identified and discussed the potential of pharmacologically induced rat models of AD, which are more relevant to the sporadic form of AD. Chemical-induced animal models of memory deficits have been more commonly employed for understanding the pathogenesis and for management of dementia and other cognitive deficits. We believe that investigation of currently available animal models for AD will help to clarify the pathogenic mechanism and allow assessment of the effects of new treatment strategies.

Ethical Statement All institutional guidelines, national guidelines, state and local laws, and regulations with professional standards for the care and use of laboratory animals should be followed. Studies involving animals must state that the institutional animal ethical committee has approved the protocol. For authors using experimental animals, a statement should be made that the animals' care is in accordance with institutional guidelines and animals used have been treated humanely and with regard for the alleviation of suffering. Researchers should treat animals as sentient and must consider their proper care and use and the avoidance or minimization of discomfort, distress, or pain as imperatives. Animal experiments should be designed only after due consideration of animal health. It should be ensured that all researchers who are using animals have received instruction in research methods and in the care, maintenance and handling of the species being used. All the surgical procedures should be performed under appropriate anesthesia and follow only those procedures which avoid infection and minimize pain during and after surgery.

References

Bensimon G, Chermat (1991) Microtubule disruption and cognitive defects: effect of colchicineon learning behavior in rats. Pharmacol Biochem Behav 38:141–145

Blokland A, Jolles J (1993) Spatial learning deficit and reduced hippocampal ChAT activity in rats after an ICV injection of streptozotocin. Pharmacol Biochem Behav 4:491–494

Chen CF, Lang SY, Zuo PP et al (2006) Effects of D-galactose on the expression of hippocampal peripheral type benzodiazepine receptor and spatial memory performances in rats. Psychoneuroendocrinology 31:805–811

Fang F, Yan N, Feng Z et al (2013) Alzheimer's disease animal model by aluminum, beta-amyloid and transforming growth factor beta-1. Aging Neurodegeneration 1:15–19

Hauss-Wegrzyniak B, Dobrzanski P, James D et al (1998) Chronic neuroinflammation in rats reproduces components of the neurobiology of Alzheimer's disease. Brain Res 780:294–303

Hua X (2007) Long-term D-galactose injection combined with ovariectomy serves as a new rodent model for Alzheimer's disease. Life Sci 80:1897–1905

Kumar MHV, Gupta YK (2003) Effect of centellaasiaticaon cognition and oxidative stress in an intracerebroventricularstreptozotocin model of Alzheimer's disease in rats. Clin Exp Pharmacol Physiol 30:336–342

Kumar A, Prakash A, Dogra S (2009) Neuroprotective effects of centellaasiatica against intracerebroventricular colchicine-induced cognitive impairment and oxidative stress. Int J Alzheimer's Dis 1–8

Kumar A, Prakash A, Dogra S (2011) Centellaasiatica attenuates D-galactose-induced cognitive impairment, oxidative and mitochondrial dysfunction in mice. SAGE-Hindawi Access Res Int J Alzheimer's Dis 1–9

Lu H, Hu J, Jiang Y (2013) Optimal dose of zinc supplementation for preventing aluminum-induced neurotoxicity in rats. Neural Regen 8(29):2754–2762

Luo Y, Niu F, Sun Z (2009) Altered expression of Ab metabolism-associated molecules from D-galactose/AlCl3 induced mouse brain. Mech Ageing Dev 130:248–252

Maurice T, Lockhart BP, Privat A (1996) Amnesia induced in mice by centrally administered beta amyloid peptide involved cholinergic dysfunction. Brain Res 706:181–193

Nitsch R, Hoyer S (1991) Local action of the diabetogenic drug, streptozotocin, on glucose and energy metabolism in rat brain cortex. Neurosci Lett 128:199–202

Nitta A, Itoh A, Hasegawa T, Nabeshima T (1994) β-Amyloid protein-induced Alzheimer's disease animal model. Neurosci Lett 170:63–66

Patil CS, Singh VP, Satyanarayan PSV et al (2003) Protective effect of flavonoids against aging- and lipopolysaccharide-induced cognitive impairment in mice. Pharmacology 69:59–67

Pitchaimani V, Arumugam S, Thandavarayan RA (2012) Nootropic activity of acetaminophen against colchicine induced cognitive impairment in rats. J Clin Biochem 50:241–244
Salkovic-Petrisic M, Tribl F, Schmidt M (2006) Alzheimer-like changes in protein kinase B and glycogen synthase kinase-3 in rat frontal cortex and hippocampus after damage to the Insulin signalling pathway. J Neurochem 96:1005–1015
Savorya J, Hermanb MM, Ghribi O (2006) Mechanisms of aluminum-induced neurodegeneration in animals. Implications Alzheimer's Dis 10:135–144

Animal Models of Parkinson's Disease

Nidhika Sharma, Sumit Jamwal, Shamsher Singh,
Harsimran Kaur Gill and Puneet Kumar Bansal

1 Introduction

Parkinson's disease (PD) is late-onset, progressive neurodegenerative, and hypokinetic movement disorder characterized by relatively selective degeneration of nigrostriatal dopaminergic neurons and presence of fibrillar cytoplasmic inclusions containing α-synuclein and ubiquitin. Major pathological features of PD include degeneration of dopaminergic neurons coupled with fibrillar intracytoplasmic inclusions known as Lewy bodies, and these Lewy bodies are also found in the hypothalamus, cranial nerve motor nuclei, locus coeruleus, nucleus basalis, cerebral cortex, and central and peripheral components of the ANS. PD is presented

N. Sharma · S. Jamwal · S. Singh · P.K. Bansal (✉)
Department of Pharmacology, ISF College of Pharmacy, Moga 142001, Punjab, India
e-mail: punnubansal79@gmail.com

N. Sharma
e-mail: nidhika444@gmail.com

S. Jamwal
e-mail: jamwalisf@gmail.com

S. Singh
e-mail: shamshersinghbajwa@gmail.com

H.K. Gill
Insectary Lab, Cornell University, 169, Helios Circle, Ithaca, NY 14853, USA
e-mail: harsimrangill.pau@gmail.com

P.K. Bansal
Department of Pharmaceutical Sciences and Technology,
Maharaja Ranjit Singh Punjab Technical University (State Govt. University),
Bathinda, Punjab, India

P.K. Bansal and R. Deshmukh (eds.), *Animal Models of Neurological Disorders*, https://doi.org/10.1007/978-981-10-5981-0_3

with four primary motor manifestations: tremor at rest, rigidity, bradykinesia (or slowing of movement), and postural instability. Initially, not all patients present with all of the classic signs of PD, there may be only one or two and non-motor symptoms includes neuropsychiatric disturbances (i.e., depression, anxiety, and dementia), cognitive impairment, sleep disturbances or hallucinations, autonomic dysfunctions, fatigue, apathy, and orthostatic hypotension. It is estimated that approximately 5–10% of cases are occur or happened due to inheritable genetic mutation. The remaining 90% of newly diagnosed PD cases are of idiopathic origin. PD is a second most common age related neurodegenerative disease after Alzheimer's disease. PD is thought to affect more than 1 million people in the USA alone, 1 of every 100 individuals, beyond the age of 55 suffered with disease. The prevalence of PD is also increases with age, affecting about 1–2% adults above the age of 60 years and 4% of above the age of 80 years.

Pathophysiology

Pathophysiology of PD includes 70–80% dopaminergic neurons cell's death in the substantia nigra pars compacta (SNPc) and specifically the ventral part of the pars compacta. Without treatment, PD progresses over 5–10 years to rigid, akinetic state in which patients are incapable of caring for themselves. Loss of dopaminergic neurons of SNPc leads to functional changes in complex circuit of basal ganglia which further resulting into enormous motor inhibition, and there is development of Parkinsonism like symptoms. A reduced activity of the mitochondrial complexes I has been observed in patients with PD in the SNPc. Mitochondrial complex I toxic derangement promotes alpha-synuclein aggregation and formation of Lewy body-like inclusions. There is possibility that protein aggregation may play a role in Parkinsonism had long been suggested by the presence of Lewy bodies in brain. Increasing evidences indicate that the ubiquitin–proteasome pathway dysfunction occurs in PD. A dysfunction in proteasome might contribute to the accumulation and aggregation of alpha-synuclein and other neurotoxic proteins because the finding suggested that Lewy bodies are ubiquitin-positive aggregates. The most important metal involved in the pathogenesis of PD is iron because of its ability to generate free radicals. It accumulates in different brain regions in high concentration and forms selectively complexes with neuromelanin that may induce oxidative stress. During the metabolism of dopamine, reactive oxygen species (ROS) are formed in very high concentrations. Then, there is formation of hydrogen peroxide, which further converted to either hydroxyl radical via a Fenton reaction or to superoxide anion and peroxynitrite. The reactive oxygen species (ROS) further activates mitogen-activated protein kinase and c-Jun N-terminal kinase, which further leads to the liberation of cytochrome C from mitochondria and sequential activation of caspase 9 and caspase 3, which results in activation of apoptotic factors. Elevated levels of glutamate in the synaptic cleft also contribute to the neurodegeneration of dopaminergic neurons in PD.

After more than 40 years of clinical use, levodopa (LD) remains the gold standard treatment for PD. On comparison with other available dopaminergic therapies, dopamine substitution with levodopa (LD) is associated with significantly reducing disability with the greatest improvement in motor function and mortality and increasing patient quality of life. Long-term conventional levodopa therapy is associated with wearing-off and dyskinesia which can result in motor dysfunction.

Need of animal model

The animal models are extremely useful to understand the origin or pathophysiology of the disease and discovery of novel treatments. Models based on specific pathogenic mechanisms may subsequently lead to the development of neuroprotective agents for PD. So there is a need for development of new animal models by activating pathogenic mechanism involved in PD so that we can develop target specific drugs.

2 Classification of Animal Models of PD (Fig. 1)

2.1 Toxin-Induced Models

2.1.1 6-Hydroxy Dopamine (6-OHDA)-Induced Parkinsonism

6-OHDA is a classic animal model for PD. 6-OHDA was first isolated in 1960 and used by Ungerstedt to produce animal model of akinesia. 6-OHDA is toxic both at a peripheral and central level; however, 6-OHDA is unable to cross the blood–brain barrier (BBB), toxicity in the CNS is accomplish only when 6-OHDA directly injected into the brain (SNpc or the striatum) by means of stereotaxic surgery. Neurotoxic effects of 6-OHDA occur by accumulation of the toxin into catecholaminergic neurons. It uses DAT and NAT transporters to gain access to the cytosol where it can auto-oxidize, hence generating an intracellular hydrogen peroxide, superoxide, and hydroxyl radicals. These free radical cause neuronal cell death by oxidative stress. The intracellular storage of 6-OHDA is mediated by the dopamine or noradrenaline membrane transporters (DAT and NAT respectively), which recognize and uptake 6-OHDA due to its structural similarity with endogenous catecholamines (Fig. 1). The primary endpoint for the unilateral 6-OHDA model is rotation, as assessed in a rotameter. Rotation results from an imbalance in functionality of dopamine systems on the left and right sides of the brain following the unilateral lesion. The chances to rotate and the direction of rotation (i.e., ipsilateral or contralateral to the lesion) is affected by pharmacological agents that either simulate or block dopamine pathways. One week following injection, d-amphetamine can be administered in a dose of (2.5 mg/kg i.p) to subjects and the net rotations (i.e., rotational asymmetry) over a 90-min trial recorded. Alternatively, 3 weeks postlesioning, apomorphine in a dose of (0.25 mg/kg) can be administered, and monitored rotational asymmetry over 40 min. D-amphetamine acts by releasing dopamine stored in the remaining nerve terminals, whereas apomorphine is an agonist for both D1/D2 receptors.

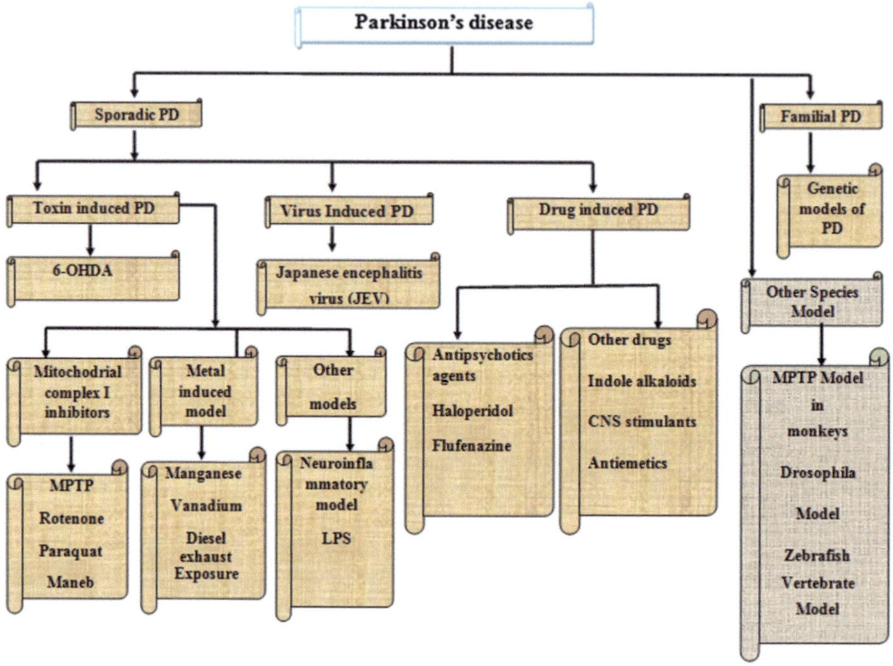

Fig. 1 Classification of animal model of PD

Different doses & routes of administration of 6-OHDA

S. no.	Dose	Route	Species	References
1.	Single-dose administration of 6-OHDA (2 μg/1 μl or 8 μg/μl)	Nucleus caudatus putamen (NCP)	Sprague–Dawley rats	Urban Ungerstedt (1968)
2.	Single-dose administration of 6-OHDA (8 μg/4 μl or 8 μg/2 μl)	Right unilateral SNPc	Male Wistar rats	Hritcu (2008)
3.	6-OHDA (6 μg/2 μl) for 14 days	Intracerebroventricular (I.C.V)	Male Wistar rats	Kahale et al. (2014)

Procedure:

- Brain surgery is performed using a stereotaxic apparatus.
- The rats are anaesthetized using ketamine (80 mg/kg, i.p) and xylazine (5 mg/kg, i.p), then their heads are mounted in a stereotaxic apparatus frame at flat skull position.
- The cannulas are implanted unilaterally above dorsal striatum and 6-OHDA is administered using following coordinates: anteroposterior (AP) −0 mm from the

bregma, mediolateral (ML) ±3.6 mm from the midline, and dorsoventral (DV) −4.5 mm from the skull surface.

- After surgery, immediately the rats are injected with gentamycin (5 mg/kg i.p) to avoid infection and are left in observational boxes until they had recovered from anesthesia.
- Then, they are housed individually in polypropylene cages (30 cm × 19 cm 12 cm) for a week.

Advantages
6-OHDA is the first animal model of PD associated with SNPc dopaminergic neuronal death was introduced more than 30 years ago.

Limitation
The time course and severity of the PD model depend on the number, amount, and location (striatum, substantia nigra, or medial forebrain bundle) of the 6-OHDA injections

Clinical relevance
6-OHDA is mediated its action by DAT and NAT transporters and generating the free radicals. These free radical cause neuronal cell death by oxidative stress, and oxidative stress is the contributing factor in the pathogenesis of PD.

2.1.2 Mitochondrial Complex I (NADPH Dehydrogenase) Inhibitors

MPTP-Induced Neurotoxicity

MPTP (1-methyl-4-phenyl-1,2,3,6-tetrahydropyridine) is a neurotoxin precursor to MPP^+, which causes Parkinson's disease by destroying dopaminergic neurons in the substantia nigra pars compacta (SNPc) of the brain. MPTP as a model of PD was developed after an accidental event that happened in 1980s when several drug addicts from California developed Parkinsonian like syndrome after intravenous use of synthetic meperidine that was contaminated with MPTP.

MPTP is highly lipophilic and after systemic administration crosses the blood–brain barrier and is converted by monoamine oxidase B (MAO-B) to MPP^+ within astrocytes. From the astrocytes with the help of cation transporter 3 (OCT-3), MPP^+ is released into extracellular spaces. From extracellular spaces, MPP^+ is taken up by the dopamine (DA) transporter. Since midbrain neurons contain the highest concentration of dopamine transporters/cell once in the cell, MPP^+ can move through several cellular compartments: it can enter into mitochondria where it interferes with complex I of the electron transport chain and inhibits it. Blockade of this complex I leads to a reduction in cellular ATP. Inhibition of mitochondrial complex I not only interferes with adenosine triphosphate (ATP) synthesis, but also results in enhanced production of superoxide anion radical (Fig. 2).

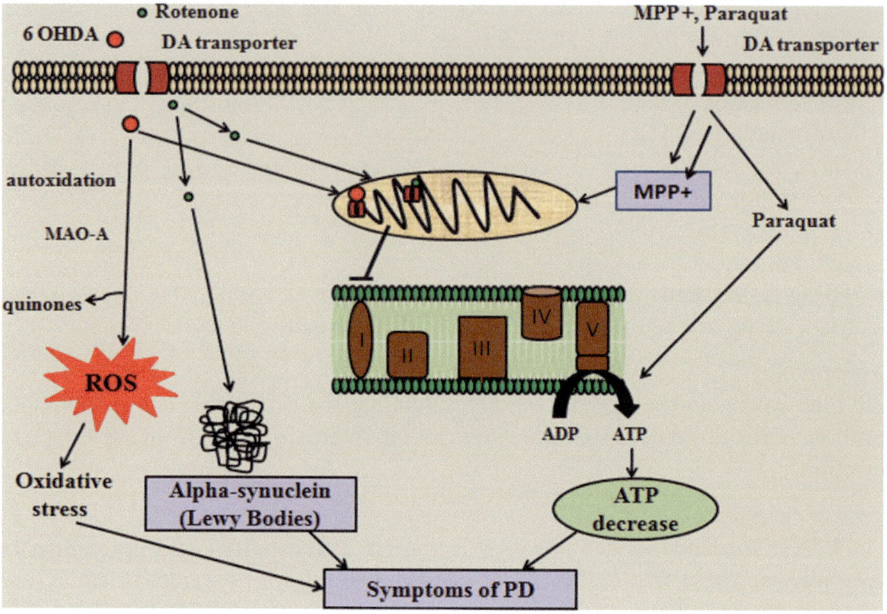

Fig. 2 Schematic representation of various neurotoxins causing PD

Different doses & routes of administration of MPTP

MPTP can be given by a number of different routes, including subcutaneous, intravenous, intraperitoneal, and stereotaxic injection into the brain.

S. No	Dose	Route	Species	References
1.	MPTP (30 mg/kg per day) for 10 consecutive days	i.p.	Male C57BL/6 mice	Nicholas (2007)
2.	Chronic MPTP/Probenecid model-10 doses of MPTP (25 mg/kg s.c) and probenecid 250 mg/kg for over 5 weeks at 3.5 days interval	s.c.	Male C57Bl/6 mice	Meredith et al. (2008)
3.	Single-dose administration of MPTP (1 μm/2 μl)	Bilaterally SNPc	Male Wistar rats	Wang et al. (2010)
4.	MPTP (40 mg/kg) four injections of 10 mg/kg i.p. in divided doses at an interval of 1 h	i.p.	Male Laca mice	Gupta et al. (2011)
5.	MPTP (100 μg/μl in 3 doses 1st, 7th and 14th day)	Intranigral	Male Wistar rats	Sharma et al. (2015)

Procedure:

- Brain surgery is performed by using a stereotaxic apparatus.
- The rats are anaesthetized using ketamine (80 mg/kg, i.p) and xylazine (5 mg/kg, i.p) and the cannulas are implanted bilaterally 2 mm above the SNpc using following coordinate's anteroposterior (AP) −5.0 mm from the bregma, mediolateral (ML) ±2.1 mm from the midline, and dorsoventral (DV) −7.7 mm from the skull and AP: −5.0 mm, ML: ±2.1 mm, DV: 8.0 mm from the bregma, midline, and skull surface.
- MPTP (100 μg/μl) is administered bilaterally on 1st, 4th, and 7th day.
- Immediately after surgery, the rats are injected with gentamycin (5 mg/kg i.p) to avoid infection and are left in observational boxes until they had recovered from anesthesia.
- Then they are housed individually in polypropylene cages (30 cm × 19 cm 12 cm) for a week.

Species

The most commonly used regimens of MPTP in monkeys are the multiple intraperitoneal or intramuscular injections and the intracarotid infusion. It is easy to perform and produce a bilateral Parkinsonian syndrome. Mice are much less sensitive to MPTP than monkeys; thus, to produce significant dopaminergic damage in this animal species much higher amount of doses are required presenting a much more difficult situation. Second, in contradiction to the monkeys, mice treated with MPTP do not develop PD. The rats are relatively insensitive to MPTP, but despite of drawback, the MPTP is continuously used in rats.

Advantages

1. The MPTP model exhibits many features of PD, including dopaminergic neurodegeneration, motor deficits, alpha-synuclein inclusion formation, and neuroinflammation. This model is thus an appealing choice for developing biomarkers for early detection of PD.
2. MPTP has been used as a toxin in large range of species.
3. MPTP has been administered by all common routes to the rodents.

Drawbacks

1. The effect of MPTP on the mouse DA system could be obtained depending upon the dose, dosing interval, animal strain etc.

Clinical relevance

Metabolite of MPTP is inhibitor of mitochondrial complex I. Dysfunction of complex increases the levels of free radicals. A reduced activity of the mitochondrial complexes I and IV have been observed in patients with Parkinson's disease in the SNPc. In vitro experiments on mitochondria isolated from whole brain demonstrate that complex I activity must be inhibited by approximately 70% to significantly impair ATP production. But PD postmortem tissues demonstrate only 40% inhibition of mitochondrial complex I activity.

Rotenone-Induced Parkinsonism

Rotenone is widely used as a pesticide and is derived from the roots of tropical plants in the Pisum *sativum* family that are native of Malaysia, South America, and East Africa. In 1895, A French botanist Emmanuel Geoffroy isolated the active chemical constituent from these plants and named it "nicouline." Later on, a Japanese chemist, Nagai isolated a pure crystalline compound which he called rotenone from Derris *elliptica*, after the Japanese name of the plant "roten."

Rotenone exposure replicates all the hallmarks of PD including complex inhibition, behavioral alterations, inflammation, alpha-synuclein aggregation, Lewy body formation, and oxidative stress. In 1985, Rotenone was first used as animal model for PD when its direct injection (5 mm) into the left median forebrain bundle (MFB) causes degeneration of dopaminergic nigrostriatal neurons. Rotenone is highly lipophilic in nature, and it easily crosses the blood–brain barrier, unlike many other toxic agents bypasses the dopamine transporter (DAT) for cellular entry. After entry into brain, it accumulates in subcellular organelles including the mitochondria, and specifically inhibit complex I, thereby disrupting mitochondrial respiration, increases "reactive oxygen species" (ROS) production and oxidative stress. Due to these actions, rotenone is used as classic mitochondrial poison in various in vitro and in vivo models. Rotenone acts systemically as a potent inhibitor of mitochondrial complex I (NADH ubiquinone oxido-reductase), whose activity levels are usually reduced in SNPc, frontal cortex, blood platelets, and skeletal muscle of Parkinson's patients. Recently, it has been found that systemic long-term administration of rotenone produces selective degeneration of dopaminergic neurons and induces PD like locomotor symptoms in rats. In the case of DA neurons, the cargo is dopamine; oxidation of dopamine produces large quantities of ROS and may provoke cell death (Fig. 2).

Different doses & routes of administration of Rotenone

Rotenone has been administered by various routes in animals. Oral administration of rotenone appears to cause less neurotoxicity.

S. no.	Dose	Route	Species	References
1.	Single-dose administration of Rotenone dissolved in absolute alcohol at (0.4 μg/μl) Coordinates A: 4.9, L: 1.6, D: −2.3)	MFB (left median forebrain bundle)	Female Sprague–Dawley rats	Heikkila et al. (1985)
2.	Rotenone (3.0 mg/kg/day) rotenone for up to 5 weeks.	s.c.	Male Lewis rats	Sherer et al. (2003)
	Rotenone (2.5 mg/kg) once daily for 60 days.	i.p.	Male Wistar rats	

(continued)

(continued)

S. no.	Dose	Route	Species	References
3.	Rotenone (2 mg/kg) daily for 5 weeks	s.c.	Male Sprague–Dawley	Sharma et al. (2013)
4.	Rotenone (2–3 mg/kg) for 28–36 days	i.v.	Lewis rats	Betarbet al. (2000)
5.	Rotenone (1.5 and 2.5 mg/kg) daily for 60 days	i.p.	Male Sprague–Dawley	Alam and Schmidt (2002)
6.	Rotenone (1.5 mg/kg) three times per week for 2 weeks	s.c.	Male Sprague–Dawley rats	Meabdel-Salam et al. (2014)
7.	Rotenone (12 mg/kg) for a period of 12 days	Per oral (p.o)	Male Wistar rats	Ittiyavirah and Ruby (2014)
8.	Rotenone (3, 6, 12 μg) for 7 days Coordinates for SN are AP: 5.3; L: 2.0; DV: 7.8	Intranigral	Sprague–Dawley rats	Swarnkar et al. (2013)

PD is induced by rotenone by different doses as given in above table. It produces a loss of striatal DA terminals results from the progressive degeneration of dopaminergic neurons of SNpc neurons. Notably, dying DA neurons contain cytoplasmic inclusions, which like Lewy bodies, are immune positive for alpha-synuclein and ubiquitin.

Advantages of Rotenone model of PD

1. Rotenone represents an easy model to induce PD because it can be administered peripherally and locally in the rodents. The neurodegeneration is slow which makes the rotenone model appropriate to study neuroprotective agents for PD.
2. Rotenone-induced PD is the only model in which formation of α-synuclein occurs.
3. Rotenone and other pesticides are powerful inhibitors of mitochondrial respiration. The model mimics general complex I deficiency which is considered to play a role in PD.

Drawbacks

1. In comparison to MPTP or 6-OHDA, the lesions produced by rotenone are more variable and the extent of the lesion varies considerably.
2. Rotenone does not induce Parkinsonism in all animals.
3. Individual rodents develop the symptoms of PD varies considerably in this model.

Clinical relevance
Rotenone is the specific inhibitor of mitochondrial complex I. A reduced activity of the mitochondrial complexes I and IV has been observed in patients with Parkinson's disease in the SNPc. In early 1924, researchers found that autopsy results showed that the regions with the greatest iron deposition were the substantia nigra (SN) and the globus pallidus (Xiong et al. 2012) Iron also accumulates in the SN in the rotenone-induced rat model of PD.

Paraquat-Induced Parkinsonism

The herbicide (weed killer) paraquat (N, N'-dimethyl-4-4'-bipyridinium) also induces a toxic model of PD. Paraquat shows structural similarity to MPP^+ and is present in the environment. Exposure to paraquat cause an increased risk for PD toxicity appears to be mediated by the formation of superoxide radicals. The levels of alpha-synuclein are significantly elevated in both SNPc and frontal cortex.

Paraquat is administered systemically in (10 mg/kg i.p.) once a week for three weeks. Paraquat leads to degeneration of SNPc dopaminergic neuron following by α-synuclein containing inclusions, as well as there is advancement in α-synuclein immunostaining in frontal cortex.

Clinical relevance
Paraquat causes mitochondrial dysfunction or oxidative stress is associated with PD and produced clinical features of PD in humans.

Maneb-Induced PD

Maneb is a manganese containing fungicide and is extensively used in agriculture. Parkinson's like symptoms have been reported to develop after chronic exposure to manganese containing fungicides like maneb. Surprisingly, in combination, the neurotoxic effects of maneb or paraquat on the nigrostriatal DA system in mice are synergistically potentiated. The specific toxicity of these environmental neurotoxins on nigral dopaminergic neurons remains unexplored. The marked use of paraquat and maneb in agricultural applications means that people living in these areas are more prone to nigral neurodegeneration.

Different doses & routes of administration of Maneb

S. no.	Dose	Route	Species	References
1.	Single-dose administration of maneb (100 mg/2 ml/kg of body weight)	i.p.	Female Sprague–Dawley rats	Konno et al. (2001)
2.	Paraquat (5 mg/kg) or Paraquat (5 mg/kg) + maneb (15 mg/kg) twice a week for 9 weeks (a total of 18 injections)	i.p.	Non-transgenic mice	Thiruchelvam et al. (2005)

Clinical relevance
Maneb increases the levels of ROS and produces the symptoms of PD. Oxidative stress has been involved in the pathogenesis of PD.

2.1.3 Metal-Induced Parkinsonism

Manganese

The manganese (Mn) was firstly recognized as an element in 1774 by Carl Wilhelm Scheele the Swedish chemist, and it is isolated by Johan Gottlieb Gahn Coevally. During the last three decades, a high-risk occupational factor for the development of neurological defects associated with exposure to Mn gradually come into focus in workers engaged in welding and workers engaged in the ferromanganese-alloy industry and the manufacturing of dry-cell batteries.

Oxidative stress has been involved as a contributing factor by which manganese may be cytotoxic. The oxidation of DA by Mn is a potential mechanism by which Mn-induced oxidative stress can be induced after the prolonged exposure to rodents and primates, Mn can accumulate in DA-rich brain regions of rodents and primates (for example, basal ganglia). Another possible mechanism is that, through dysfunction in mitochondria, interferes with proper respiration, thereby causing the excessive production of ROS. Mn efflux has been suggested to account for the net accumulation of Mn in mitochondria. Mn alone inhibits the antioxidant system and disturbs mitochondrial respiratory chain.

Different doses & routes of administration of Manganese

S. no.	Dose	Route	Species	References
1.	$MnCl_2$ (6 mg/kg) once daily for 4 weeks	i.p.	Male Sprague–Dawley rats	Zheng et al. (1999)
2.	$MnCl_2$ (0.04 M) and $Mn(Oac)_3$ (0.02 M) twice a week for 5 months		Male CD-1 mice	Ordonez-Librado et al. (2011)
	$MnCl_2$ (0.04 M) and $Mn(Oac)_3$ (0.02 M) three times a week for 6 months	Inhalation		
3.	Mn (0, 25, or 50 mg Mn/kg/d) from PND 1-21	Orally	Male Long–Evans rats	Beaudin et al. (2013)

Vanadium-Induced Neurotoxicity

1.	V_2O_5 (182 μg/50 μL) for 3 times a week for 1 month.	Intranasally	Male C57BL/6 mice	Ngwa et al. (2014)

Subchronic Diesel Exhaust Exposure

1.	Diesel exhaust (991.8, 311.2, 100.3, 34.9, and 0 μg PM/m3) daily for 6 months	Subchronic Diesel exhaust exposure	Male Fischer 344 rats	Levesque et al. (2011)

Advantage

1. Mn-induced PD is more reliable model of Parkinsonism because it is progressive and bilateral.

Drawbacks

1. Despite the similarities in extra pyramidal symptoms between Mn neurotoxicity and PD,

   Mn-induced lesions are approximately different from those lesions observed in PD.

Clinical relevance
The oxidation of DA by Mn is a potential mechanism by which Mn-induced oxidative stress can be induced in rodents. Oxidative stress has been involved as a contributing factor for PD. The first report of neurological effects associated with exposure to manganese (Mn) in the scientific literature in 1837 by John Couper when he described motor symptoms salivation, weakness in muscles, limb tremor, whispering speech, and posture abnormalities in five men working in a Mn ore crushing plant in France. He called this compilation of symptoms "manganese crusher's disease."

2.1.4 Neuro-inflammatory Models

LPS-Induced Parkinsonism

LPS is large molecule consisting of lipids and sugars joined by chemical bonds. LPS are the major outer surface membrane components present in almost all Gram-negative bacteria and act as exceptionally strong stimulators of immune system almost in all the organisms. In 1998, Bing and Castano reported that intranigral LPS injection induces inflammation on the nigrostriatal system and produces symptoms of PD.

LPS is well known to produce neurodegeneration, and this is closely linked to neuroinflammation. Numerous evidences indicate inflammation plays a major role in the pathogenesis of PD. Intranigral administration of LPS results in neuroinflammation in the nigrostriatal pathway, long-lasting elevations in tumor necrosis factor alpha (TNF-α), increased microglia, increased levels of pro-oxidants, and reduced amounts of antioxidant glutathione (GSH). Increased activity of pro-inflammatory cytokines such as TNF-α, IL-1β, and IL-6, and inducible nitric oxide synthase (iNOS) has been found in cerebrospinal fluid of patients with PD. Systemic LPS injection also activates apoptotic cell death in SNpc.

Different doses & routes of administration of LPS

S. no.	Dose	Route	Species	References
1.	Single-dose administration of LPS (10 mg/kg) or LPS (5 mg/kg) repeatedly for 5 days	i.p.	Male Wistar–Han rats	Noworyta-Sokolowska et al. (2013)
2.	Single-dose administration of LPS (5 and 10 μg/4 μl)	Intrastriatal SNPc	Wistar–Han rats	Noworyta-Sokolowska et al. (2013)
3.	LPS (10 μl) into nasal cavity for 5 months	Unilateral Intranasal	Female C57BL/6 mice	He et al. (2013)
4.	Single-dose administration of LPS (2.5 μg/μl or 5 μg/μl)	Right or left striatum	Male Sprague–Dawley rats	Choi et al. (2009)
5.	Single-dose administration of LPS (10 μg/4 μl, bilaterally)	Intrapallidal	Fisher F344 rats	Zhang et al. (2005)

Drawbacks

1. In LPS-induced dopaminergic neurodegeneration, gender differences seem to be an important factor in the sensitivity to the LPS-induced PD. In systemic LPS administration, C57BL/6 female mice are more resistant than male mice.

Clinical relevance
The first evidence for a role of inflammation in PD came from McGeer and colleagues who observed activated microglia and T cells in the *postmortem* SNpc of a PD patient. LPS causes neuroinflammation in the nigrostriatal pathway, including lifelong elevations in TNF-α, increased levels of oxidants, and reduced amounts of antioxidant enzymes. Inflammation plays a major role in the pathogenesis of PD.

2.2 Drug-Induced Parkinsonism

Drug-induced Parkinsonism (DIP) is the second most common cause of Parkinsonism after idiopathic PD and has been assume the most prevalent form of secondary Parkinsonism in the Western world. Common clinically used drug-induced PD is (Fig. 3):

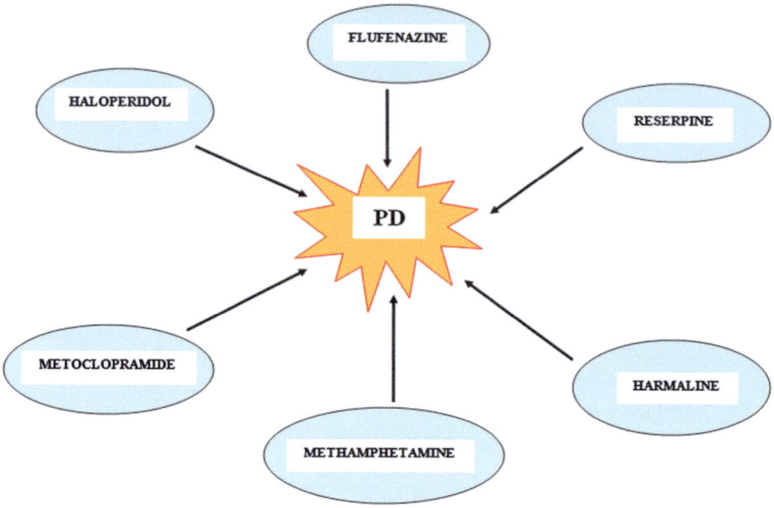

Fig. 3 Drugs causing PD

2.2.1 Antipsychotic Drugs-Induced Parkinsonism

Haloperidol

In one of the largest series, an overall incidence of 15% for neuroleptics-induced Parkinson's syndrome (NIPS) in over 3700 patients has been reported. The symptoms of muscle rigidity and catalepsy induced within 60 min of haloperidol (0.5–5 mg/kg, i.p.) injection. Haloperidol catalepsy is produced by blockade of postsynaptic striated DA receptors, leading to the functional lack of DA at postsynaptic receptor sites. Neurotoxic effect is caused by an alteration in the iron transport to the CNS and subsequent iron deposition in the basal ganglia. It was also suggested that the favorable side effect profile of the atypical antipsychotics might be related to the dual blockade of D_2 and 5-HT2_A.

Different doses & routes of administration of Haloperidol

S. no.	Dose	Route	Species	Reference
1.	Administration of Haloperidol (1 mg/kg, for 1 week or 21 days)	i.p.	Male Wistar albino rats Male Swiss albino rats–Laca mice of either sex	Naidu et al. (2000)

Flufenazine

Fluphenazine is a typical antipsychotic drug used for the treatment of chronic psychoses such as schizophrenia. Flufenazine is a phenothiazine derivative with D_1 and D_2 receptor blockade activity.

Different doses & routes of administration of Flufenazine

S. no.	Dose	Route	Species	References
1.	Single-dose administration of Flufenazine (0.3–3.0 μg/μl) Stereotaxic coordinates −2.4 mm from bregma; ±2.3 mm from midline; −2.0 mm from dura; Incisor bar +0.5 mm	Bilaterally Intrastriatal (i.s)	Male random-bred hooded rats	Elliott et al. (1990)
2.	Single-dose administration of Flufenazine (2 mg/kg)	s.c.	Male random-bred hooded rats	Elliott et al. (1990)

Clinical relevance

Haloperidol and fluphenazine causes blockade of postsynaptic striated DA receptors, leading to the functional lack of DA at postsynaptic receptor sites and level of dopamine is already decreasing in PD.

2.2.2 Indole Alkaloids-Induced Parkinsonism

Reserpine

Reserpine is a dopamine depletor that blocks vesicular storage of monoamines. In 1950s, Carlsson first demonstrated that reserpine can be used to develop animal model for PD using rabbits. Reserpine is first recognized pharmacological model of PD. He showed that treatment of rats with reserpine results in deficiency of dopamine and other catecholamines in brain, thereby resulting in akinesia, which could be successfully reversed by L-dopa. The akinetic state arising due to dopamine depletion in the caudate and putamen and helped Carlsson to speculate that PD results from striatal dopamine depletion. Reserpine acts by inhibiting the vesicular monoamine transporter, VMAT 2. This leads to loss of storage capacity and subsequent depletion of brain (and peripheral) monoamines including noradrenaline, dopamine, and 5-HT.

Different doses & routes of administration of Reserpine

S. no.	Dose	Route	Species	Reference
1.	Administration of Reserpine (2, 2.5 & 5 mg/kg)	i.p.	Mice or rats	Kumar and Kulkarni (2006)

Harmaline

Harmaline is a fluorescent psychoactive indole alkaloid from the group of harmala alkaloids and beta-carbolines.

Different doses & routes of administration of Harmaline

S. no.	Dose	Route	Species	Reference
1.	Single-dose administration of harmaline (20 mg/kg)	i.p. or s.c	Male Wistar rats or mice	Iseri et al. (2011)

2.2.3 CNS Stimulants

Methamphetamine

Methamphetamine is one of a family of drugs called amphetamines, which acts as CNS stimulants. Methamphetamine increases the auto-oxidation of dopamine, because methamphetamine displaces the dopamine in the vesicles; large amounts of dopamine are released in synaptic cleft and cytosol where it then undergoes auto-oxidation. Hydrogen peroxide is produced as a by-product in the synthesis and metabolism of dopamine, leading to increased signaling of oxidative stress.

Different doses & routes of administration of Methamphetamine

S. no.	Dose	Route	Species	References
1.	Methamphetamine (15 mg/kg) once in every 8 h for a total of 5 injections	s.c.	Male Holtzman rats	Robinson et al. (1990)
2.	D-Methamphetamine (10 mg/kg) administered at 2 h intervals for a total of 4 injections	i.p.	Male Sprague–Dawley rats	Moszczynska et al. (2011)

Clinical relevance

Methamphetamine increases the auto-oxidation of dopamine. Hydrogen peroxide is produced as a by-product in the synthesis and metabolism of dopamine, leading to increased signaling of oxidative stress. Oxidative stress is the contributing factor in the development of PD.

2.2.4 Anti-emetics-Induced Parkinsonism

Metoclopramide

Anti-emetics are usually overlooked as a causative agent of DIP. Metoclopramide, the prototypical antiemetic, accounts for nearly a third of all drug-induced movement disorders. It is commonly prescribed in the elderly. Metoclopramide-induced

Parkinsonism is typically encountered within the first 3 months of Metoclopramide therapy; the parkinsonian findings resolve in most patients within 2 months after drug therapy is discontinued.

2.3 Others Models of PD

2.3.1 Drosophila as a Model of PD

Seventy-seven percentage of human disease genes are conserved in the complete sequence of the Drosophila genome. The clusters of dopaminergic neurons are present in adult brain of Drosophila and these neurons degenerate when rotenone (a complex I inhibitor) are fed to flies that also triggers dopaminergic neuronal degeneration in mammals. Among the genes that mediate familial PD, only a-synuclein does not have a homolog in Drosophila.

Advantages

1. Drosophila melanogaster has emerged as an especially effective tool to study the genes of PD. Drosophila has been used extensively for investigating biological processes, such as cell death as well as cell proliferation, growth, and migration.

Clinical relevance

These features make flies an excellent model system in which to study the function of disease genes including those involved in neurodegenerative diseases.

2.3.2 Zebrafish as a New Model for PD

Zebrafish are among few newly emerged animal models for neurodegenerative diseases. They come under the vertebrates and therefore more closely relate to humans than other animal models such as Drosophila or *C. elegans*. Their dopaminergic nervous system is well studied and characterized. Zebrafish embryos are transparent, and this transparency allows visualization and study of various macroscopic and microscopic changes that occurs during embryogenesis and organogenesis. In addition, transgenic zebrafish allows gene expression studies during early development and create easy access to new animal models.

Clinical relevance

The zebrafish brain does not contain a mesencephalic region comparable to the substantia nigra, treatment with MPTP (a PD-inducing drug in humans, and to a lesser extent mice) showed a direct effect on diencephalic dopaminergic neurons, illustrating that zebrafish can develop a phenotype comparable to PD.

2.3.3 Encephalitis Virus-Induced PD

A number of viruses like Japanese Encephalitis Virus (JEV) have ability to penetrate BBB and selectively infect the substantia nigra and induce PD. JEV infection is one of the most common cause of arthropod-borne human encephalitis worldwide. JEV infection is associated with postencephalitis Parkinsonism syndrome in man. In vitro studies have suggested that JEV inhibit the proliferation of T-lymphocytes and evades host immune system. It remains unclear why the neurons of the substantia nigra remain susceptible to JEV infection as opposed to other regions of the brain.

Procedure:

- Thirteen days after birth, Fischer rats were infected with Japanese encephalitis virus (JEV) and sacrificed 12 weeks later.
- Brain histological analysis revealed similar changes as found in PD. TH-positive neurons was significantly decreased in the substantia nigra. Specifically, there was neuronal loss mainly to the zona compacta of the substantia nigra.
- Specifically when animals infected at 1–3 days of age developed widespread infection throughout the brain, whereas in animals infected at age 14-day old, JEV antigen was essentially restricted to the basal ganglia and substantia nigra.

Clinical relevance
JEV antigen was essentially restricted to the basal ganglia and substantia nigra. The limitation of infection to these areas suggested that infection of rats between days 12 and 14 after birth may result in pathological and clinical processes which might resemble PD.

3 Conclusion

Rodents and non-human primates are important tools for better understanding of PD and also helpful in understanding of the pathophysiology of PD. These newer animal models for PD have been less commonly used for drug development and also represent an easy model to induce Parkinsonism. In our review, we have also explored some toxins and drugs that can be used as targets for future research to replace classical toxic PD models. No model is perfect, but in various models rodents can demonstrate many pathophysiological features of PD.

Ethical Statement

All institutional guidelines, national guidelines, and state and local laws and regulations with professional standards for the care and use of laboratory animals should be followed. Studies involving animals must state that the institutional animal ethical committee has approved the protocol. For authors using experimental animals, a statement should be made that the animals' care is in accordance with institutional guidelines, and animals used have been treated humanely and with regard for the alleviation of suffering. Researchers should treat animals as sentient and must consider their proper care and use and the avoidance or minimization of discomfort, distress, or pain as imperatives. Animal experiments should be designed only after due consideration of animal health. It should be ensured that all researchers who are using animals have received instruction in research methods and in the care, maintenance, and handling of the species being used. All the surgical procedures should be performed under appropriate anesthesia and follow only those procedures which avoid infection and minimize pain during and after surgery.

References

Abdel-Salam OM, Omara EA, Youness ER, Khadrawy YA, Mohammed NA, Sleem AA (2014) Rotenone-induced nigrostriatal toxicity is reduced by methylene blue. J Neurorestoratol 2:65–80

Alam M, Schmidt WJ (2002) Rotenone destroys dopaminergic neurons and induces parkinsonian symptoms in rats. Behav Brain Res 136(1):317–324

Beaudin SA, Nisam S, Smith DR (2013) Early life versus lifelong oral manganese exposure differently impairs skilled forelimb performance in adult rats. Neurotoxicol Teratol 38:36–45

Betarbet R, Sherer TB, MacKenzie G, Garcia-Osuna M, Panov AV, Greenamyre JT (2000) Chronic systemic pesticide exposure reproduces features of Parkinson's disease. Nat Neurosci 3(12):1301–1306

Choi DY, Liu M, Hunter RL, Cass WA, Pandya JD, Sullivan PG, Shin EJ, Kim HC, Gash DM, Bing G (2009) Striatal neuroinflammation promotes Parkinsonism in rats. PLoS One 4(5): e5482

Elliott PJ, Close SP, Walsh DM, Hayes AG, Marriott AS (1990) Neuroleptic-induced catalepsy as a model of Parkinson's disease II. Effect of glutamate antagonists. J Neural Transm Parkinson's Dis Dement Sect 2(2):91–100

Gupta A, Kumar A, Kulkarni SK (2011) Targeting oxidative stress, mitochondrial dysfunction and neuroinflammatory signaling by selective cyclooxygenase (COX)-2 inhibitors mitigates MPTP-induced neurotoxicity in mice. Prog Neuro-Psychopharmacol Biol Psychiatry 35(4): 974–981

He Q, Yu W, Wu J, Chen C, Lou Z, Zhang Q, Zhao J, Wang J, Xiao B (2013) Intranasal LPS-mediated Parkinson's model challenges the pathogenesis of nasal cavity and environmental toxins. PLoS One 8(11):e78418

Heikkila RE, Nicklas WJ, Vyas I, Duvoisin RC (1985) Dopaminergic toxicity of rotenone and the 1-methyl-4-phenylpyridinium ion after their stereotaxic administration to rats: implication for the mechanism of 1-methyl-4-phenyl-1, 2, 3, 6-tetrahydropyridine toxicity. Neurosci Lett 62(3):389–394

Hritcu L (2008) Hematological disorders in 6-hydroxydopamine-induced rat model of Parkinson's disease. Turk J Hematol 25:140–144

Iseri PK, Karson A, Gullu KM, Akman O, Kokturk S, Yardýmoglu M, Erturk S, Ates N (2011) The effect of memantine in harmaline-induced tremor and neurodegeneration. Neuropharmacology 61(4):715–723

Ittiyavirah SP, Ruby R (2014) Effect of hydro-alcoholic root extract of Plumbago zeylanica l alone and its combination with aqueous leaf extract of Camellia sinensis on haloperidol induced parkinsonism in wistar rats. Ann Neurosci 21(2):47

Kahale V, Mhaiskar A, Shelat P, Pooja RU, Gaikwad NJ, Mundhada DR (2014) To determine the Effect of Berberine on 6-OHDA induced memory impairment in Parkinson's disease in rodents. Pharma Innov J 3(7):101–108

Konno N, Tsunoda M, Nakano K (2001) The effect ofin vitro andin vivo ethylenbis dithiocarbamate fungicides on NMDA receptors in rat brain membranes. Environ Health Prev Med 6(1):54–59

Kumar A, Kulkarni SK (2006) Effect of BR-16A (Mentat), a polyherbal formulation on drug-induced catalepsy in mice. Indian J Exp Biol 44(1):45–48

Levesque S, Surace MJ, McDonald J, Block ML (2011) Air pollution & the brain: subchronic diesel exhaust exposure causes neuroinflammation and elevates early markers of neurodegenerative disease. J Neuroinflammation 8(1):1

Meredith GE, Sonsalla PK, Chesselet MF (2008) Animal models of Parkinson's disease progression. Acta Neuropathol 115(4):385–398

Moszczynska A, Yamamoto BK (2011) Methamphetamine oxidatively damages parkin and decreases the activity of 26S proteasome in vivo. J Neurochem 116(6):1005–1017

Naidu PS, Singh A, Kulkarni SK (2003) Quercetin, a bioflavonoid, attenuates haloperidol-induced orofacial dyskinesia. Neuropharmacology 44:1100–1106

Ngwa HA, Kanthasamy A, Jin H, Anantharam V, Kanthasamy AG (2014) Vanadium exposure induces olfactory dysfunction in an animal model of metal neurotoxicity. Neurotoxicology 43:73–81

Nicholas AP (2007) Levodopa-induced hyperactivity in mice treated with 1-methyl-4-phenyl-1, 2, 3, 6-tetrahydropyridine. Mov Disord 22(1):99–104

Noworyta-Sokolowska K, Gorska A, Golembiowska K (2013) LPS-induced oxidative stress and inflammatory reaction in the rat striatum pharmacological reports, vol 65, pp 863–869

Ordoñez-Librado JL, Anaya-Martínez V, Gutierrez-Valdez AL, Colín-Barenque L, Montiel-Flores E, Avila-Costa MR (2010) Manganese inhalation as a Parkinson disease model. Parkinson's Disease, 2011

Robinson TE, Yew J, Paulson PE, Camp DM (1990) The long-term effects of neurotoxic doses of methamphetamine on the extracellular concentration of dopamine measured with microdialysis in striatum. Neurosci Lett 110(1):193–198

Sharma S, Deshmukh R (2015) Vinpocetine attenuates MPTP-induced motor deficit and biochemical abnormalities in Wistar rats. Neuroscience 286:393–403

Sharma N, Nehru B (2013) Beneficial effect of vitamin E in rotenone induced model of PD: behavioural, neurochemical and biochemical study. Exp Neurobiol 22(3):214–223

Sherer TB, Betarbet R, Testa CM, Seo BB, Richardson JR, Kim JH, Miller GW, Yagi T, Matsuno-Yagi A, Greenamyre JT (2003) Mechanism of toxicity in rotenone models of Parkinson's disease. J Neurosci 23(34):10756–10764

Swarnkar S, Goswami P, Kamat PK, Patro IK, Singh S, Nath C (2013) Rotenone-induced neurotoxicity in rat brain areas: a study on neuronal and neuronal supportive cells. Neuroscience 230:172–183

Thiruchelvam M, Prokopenko O, Cory-Slechta DA, Richfield EK, Buckley B, Mirochnitchenko O (2005) Overexpression of superoxide dismutase or glutathione peroxidase protects against the paraquat+ manebinduced Parkinson disease phenotype. J Biol Chem 280:22530–22539

Ungerstedt U (1968) 6-Hydroxy-dopamine induced degeneration of central monoamine neurons. Eur J Pharmacol 5(1):107–110

Wang AL, Liou YM, Pawlak CR, Ho YJ (2010) Involvement of NMDA receptors in both MPTP-induced neuroinflammation and deficits in episodic-like memory in Wistar rats. Behav Brain Res 208(1):38–46

Xiong P, Chen X, Guo C, Zhang N, Ma B (2012) Baicalin and deferoxamine alleviate iron accumulation in different brain regions of Parkinson's disease rats. Neural Regeneration Res 7(27):2092

Zhang J, Stanton DM, Nguyen XV, Liu M, Zhang Z, Gash D, Bing G (2005) Intrapallidal lipopolysaccharide injection increases iron and ferritin levels in glia of the rat substantia nigra and induces locomotor deficits. Neuroscience 135(3):829–838

Zheng W, Zhao Q, Slavkovich V, Aschner M, Graziano JH (1999) Alteration of iron homeostasis following chronic exposure to manganese in rats. Brain Res 833(1):125–132

Animal Models of Huntington's Disease

Navneet Kaur, Sumit Jamwal, Harsimran Kaur Gill and Puneet Kumar Bansal

1 Introduction

Huntington's disease (HD) is an autosomal dominant, hyperkinetic movement disorder, and progressive neurodegenerative disorder characterized by choreiform abnormal movements, cognitive deficits, psychiatric manifestations, emotional disturbances, and dementia associated with progressive striatal atrophy. The disorder was first described in 1872 by Dr. George Huntington in his article titled "On Chorea." Dr. Huntington entitled the disease "Hereditary Chorea," whereas observations of some cases which do not present with chorea later led to the acceptance of the "Huntington's Disease" label. The HD gene (IT15) is sited on the short arm of chromosome 4, and the normal number of CAG (glutamine) repeats is expanded (generally >40). Genetic mutation in HD gene results in an expanded polyglutamine stretch in the NH_2 terminus of huntingtin protein (HTT).

N. Kaur · S. Jamwal · P.K. Bansal (✉)
Department of Pharmacology, ISF College of Pharmacy, Moga 142001, Punjab, India
e-mail: punnubansal79@gmail.com

N. Kaur
e-mail: nav_neet@hotmail.co.uk

S. Jamwal
e-mail: jamwalisf@gmail.com

H. Kaur Gill
Insectary Lab, Cornell University, 169, Helios Circle, Ithaca, NY 14853, USA
e-mail: harsimrangill.pau@gmail.com

P.K. Bansal
Department of Pharmaceutical Sciences and Technology, Maharaja Ranjit Singh Punjab Technical University (State Govt. University), Bathinda, Punjab, India

P.K. Bansal and R. Deshmukh (eds.), *Animal Models of Neurological Disorders*, https://doi.org/10.1007/978-981-10-5981-0_4

The major pathophysiological hallmarks of neurodegenerative process in HD are degeneration of GABAergic medium spiny striatal neurons (MSN), cortical neurons, and striatal apathy. GABAergic and enkephalin neurons of the basal ganglia are most susceptible to neurodegeneration in HD. The mechanism of the selective striatal damage in HD is remained elusive. However, few theories have been proposed as pathogenic mechanisms of HD such as mutant huntingtin protein-induced mitochondrial dysfunction, excitotoxicity, oxidative stress, apoptosis, neurotransmitters imbalance, and neuroinflammation. Mitochondrial abnormalities play a central role in HD pathology, and disruption of mitochondrial enzyme complex activity is well reported in HD patients and animal models. Reduced activity of the complex II enzyme succinate dehydrogenase (SDH) has been reported in postmortem HD brain tissue. Changes in the activity of mitochondrial enzymes further support the mitochondrial dysfunction hypothesis in HD.

Excitotoxic events are associated with hyperactivity of glutamate neurotransmission and followed by persistent elevations in mitochondrial calcium levels. The hyperactivation of NMDA receptors results in oxido-nitrosative stress, opening of the mitochondrial permeability transition pore and neuroinflammation. The theory of excitotoxicity is supported by fact that intrastriatal injection of excitotoxins such as quinolinic acid (QA) and kainic acid produces neurodegeneration of GABAergic MSN similar to as seen in HD. Also, postmortem studies have reported high levels of activated microglia and macrophages in the degenerating neurons in HD. Increased level of pro-inflammatory cytokines such as IL-6, IL-1β, and TNF-α have been shown in striatum and plasma of HD patients as well as in animal models of HD. Despite number of theories proposed, the exact pathogenic mechanisms are still unclear; however, the scientific advances led to the better understanding of the disease. Animal models emerge as an alternative to know the molecular pathogenesis of HD because they imitate the clinical and neurobiological symptoms of disease. The most significant rationale for modeling human disorders in a nonhuman organism is to provide new insights into development of existing therapies and novel therapeutic strategies for HD along with the evaluation of the efficacy and safety of potential new drugs. Rodents and nonhuman primates are the most commonly used animal models but nonmammalian HD models also exist like *Caenorhabditis elegans* and *Drosophila melanogaster*.

2 Classification of Animal Models

Animal models of HD are basically categorized into three major classes: mitochondrial toxin models, excitotoxic models, and genetic models as shown in Fig. 1.

Mitochondrial toxin models and excitotoxic models are discussed as a group under the title of toxin models. Mitochondrial toxins include 3-nitropropionic acid and malonate. Administration of these toxins to both rats and primates leads to striatal lesions which resemble clinical HD. These toxins act by inhibiting mitochondrial complex II and further lead to attenuation in ATP levels. Excitotoxic

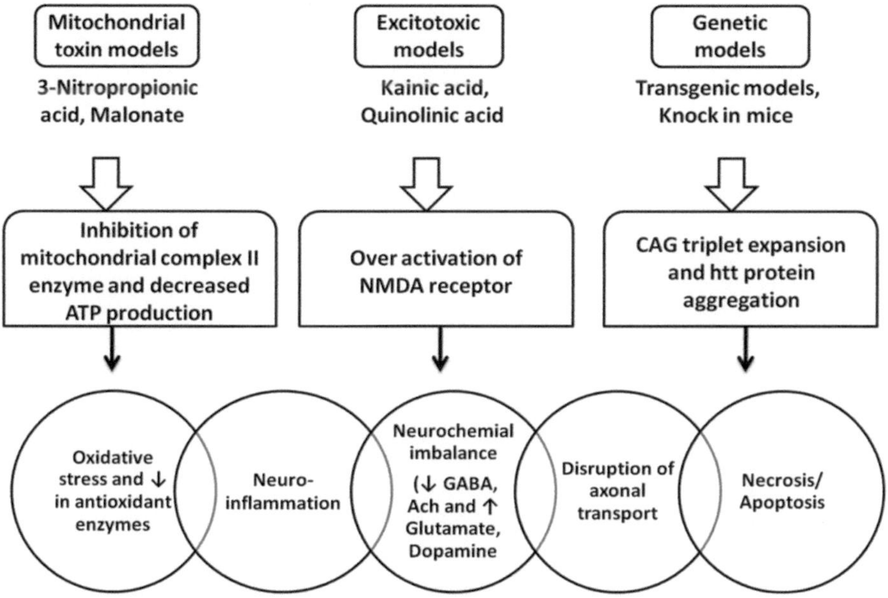

Fig. 1 Overview of animal models of Huntington's disease

models include kainic acid and quinolinic acid. These amino acids produce excitotoxicity by binding to non-NMDA and NMDA receptors, respectively, on striatal neurons.

Although the details of genetic model systems of HD are not discussed here, in brief rodent genetic models can be alienated into two categories: transgenic models and knock-in models. Transgenic models include fragment models like R6/2, R6/1, transgenic mice, and full length mouse models such as YAC72, BACHD, and knock-in models include Hdh/Q72–80 and CAG140. Choice of experimental models should be made accordingly because it depends on the validity and characteristics of models, and the type of experiments planned.

2.1 Toxin Models

2.1.1 Energy Metabolism Deficit Models

Numerous clinical and preclinical studies provide strong evidence for reduced glucose consumption in the brain, especially in the basal ganglia. Magnetic resonance spectroscopy (MRS) imaging has shown increased lactate levels in the striatum and occipital cortex of HD patients, suggesting a compensatory glycolytic response to impaired mitochondrial function. In order to mimic metabolic alterations observed in HD at experimental level, different animal models have been designed like 3-NP and malonic acid.

3-Nitropropionic Acid Model

3-nitropropionic acid (3-NP) is a mycotoxin, synthesized by fungi (*Aspergillus flavus; Astragalus, Arthrinium*) and plants (*Indigoferaendecapylla*). The first identification regarding toxicity of 3-NP was revealed in Western regions of North America, when animals intoxicated with leguminous plants such as *Indigofera* or *Astragalus* presented a range of motor abnormalities consisting of general weakness, impaired coordination of the hind limbs further developed into paralysis. The same effects of 3-NP again came into picture in the brain when children from China ingested *Arthrinium*-infested sugarcane. 3-NP was produced by the metabolism of *Arthrinium* fungus, which perpetually caused neuronal death in the caudate and putamen nuclei of striatum and escorted severe dystonia in these children. The pioneering study in this field was performed by Chinese researchers Hamilton and Gould. M. Flint Beal was the first to theorize that chronic treatment in rats with 3-NP could imitate the partial and steady mitochondrial defects seen in clinical HD, and escort to selective striatal degeneration with histological and neurochemical distinctiveness redolent of HD.

3-NP causes irreversible inhibition of the mitochondrial complex II enzyme (succinate dehydrogenase). Complex II is a member of the Krebs tricarboxylic acid cycle (oxidizing succinate to fumarate) and a doorway for electrons into the respiratory chain at the level of ubiquinol. Chronic administration of 3-NP causes prolonged energy impairments by mitochondrial dysfunction and mimics nearly all pathophysiological features of HD, including preferential degeneration of medium spiny GABAergic neurons of straitum. In humans, ingestion of 3-NP persuades cognitive impairment and motor abnormalities, including dystonia, involuntary jerky movements, torsion spasms, and facial grimaces. In vitro, studies have revealed that 3-NP diminishes ATP at cellular levels and leads to neuronal damage by an excitotoxic mechanism as shown in Fig. 2. In rats, 3-NP-induced lesions in the basal ganglia linked to enhanced lactate levels resulted in augmented NMDA-receptor binding. Thus, 3-NP toxicity leads to reduction of ATP and bioenergetic failure, calcium overload, excitotoxic events, and neuronal death. In turn, excitotoxic cell death has been allied with reactive oxygen species (ROS) formation and oxidative stress. 3-NP-induced oxidative stress also stimulates aging process, as demonstrated by increased DNA fragmentation and reduced expressions of the DNA repair enzyme apurinic/apyrimidinic endonuclease in older mice.

Diverse range of animal species and strains can be used to build up this model with parallel profiles of neurotoxicity to those seen in HD brains. Rats are more sensitive than mice to 3-NP treatment. Systemic administration of 3-NP to rats, mice, and nonhuman primates shows good results because 3-NP readily crosses blood-brain barrier. Both the subcutaneous osmotic pump and direct subcutaneous injection are effective for the 3-NP systemic administration to the rat, but as the injection is adjusted daily according to the animal's weight, it conveys a more precise dose.

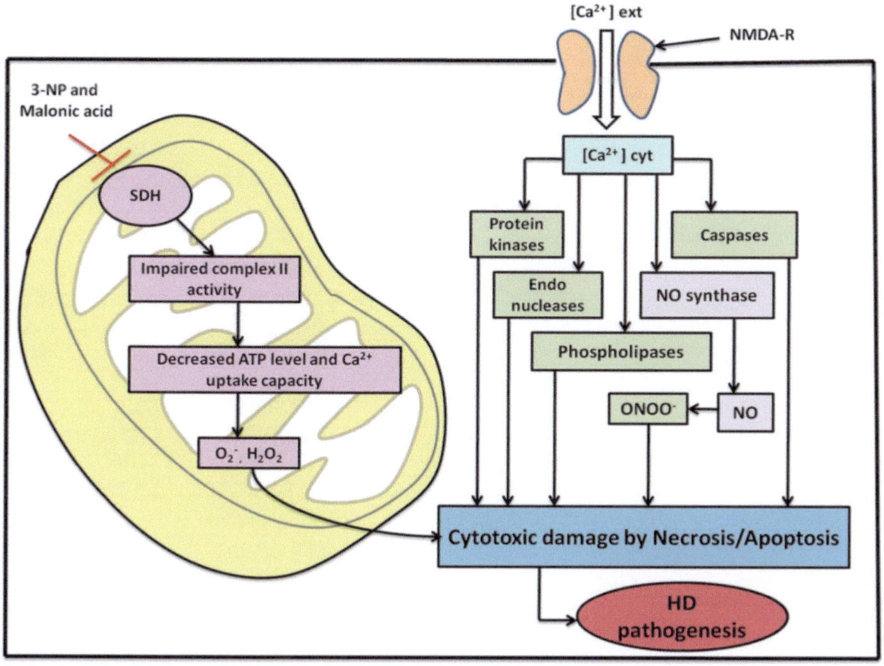

Fig. 2 Mechanism of action of 3-nitropropionic acid and malonic acid

Procedure: 3-NP can be administered by s.c or i.p route directly. It has been administered at various doses. Some of them given are as in Table 1.

Malonic Acid Model

Malonic acid is another selective inhibitor of complex II enzyme (succinate dehydrogenase) that causes motor impairment and neuronal pathology similar to HD after intrastriatal administration in rodents. Unlike 3-NP, malonic acid does not cross the blood-brain barrier (BBB). Malonic acid induces mitochondrial potential collapse, mitochondrial swelling, and cytochrome *c* (Cyt *c*) release and diminishes glutathione (GSH), and nicotinamide adenine dinucleotide coenzyme (NADPH) stores in brain-isolated mitochondria. By inhibition of succinate dehydrogenase and depletion of striatal ATP as shown in Fig. 2, malonic acid has been shown to cause dose-dependent neurotoxicity both "in vivo" and "in vitro," resulting in neuronal depolarization and secondary excitotoxicity. It has been suggested that malonic acid toxicity involves neurons dying not only by secondary excitotoxicity but also by postponed caspase activation and apoptosis. Thus, the exact mechanism by which malonic acid induces toxicity remains unclear. Further, mitochondrial permeability transition pore opening and the subsequent release of proapoptotic factors and Cyt c might account for malonic acid-induced toxicity. Malonic acid induces an augmentation in the conversion of salicylate to 2,3-dihydroxybenzoic acid and

Table 1 Different doses of 3-NP to induce HD

S. No.	Treatment schedule	Animals used	References
1.	120 mg/kg/day for two days, i.p	Male Swiss Webster mice	Gould and Gustine (1982)
2.	30 mg/kg/day (1–2 days) and 10 mg/kg/day (3–4 days), s.c	Male Sprague-Dawley rats	Hamilton and Gould (1987)
3.	7.5 mg/kg twice daily for 10 days, i.p	Male Sprague-Dawley rats	Guyot et al. (1997)
4.	2 × 7.5 mg/kg during first two days; 2 × 3.75 mg/kg during seven days; then 2 × 5 mg/kg during nine days (Total = 18 days), i.p	Male Sprague-Dawley rats	Guyot et al. (1997)
5.	20 mg/kg once a day for two or three days, s.c and 20 mg/kg, for four days, i.p	Male Wistar rats	Nishino et al. (1997)
6.	10 mg/kg/day, once every four days for 28 days, i. p	Male Sprague-Dawley rats	Borlongan et al. (1995)
	10 mg/kg for five days, i.p	Male Wistar rats	Szabo et al. (2005)
	10 mg/kg once a day for 14 days, i.p	Male Wistar rats	Kumar et al. (2009)
7.	Bilateral intrastriatal injection of 3-NP (750 nmol)/(375 nmol)	Male Sprague-Dawley rats	Shear et al. (1998)
8.	60 mg/kg twice a day for five days	Male CD1 mice	Kim et al. (2000)
9.	63 mg/kg/day for five days, s.c	Male Lewis rats	Park et al. (2008)
10.	60 mg/kg twice a day on first day and 80 mg/kg twice a day on second day, i.p	Male C57BL/6 mice	Jang et al. (2014)

2,5-dihydroxybenzoic acid, a catalog of free radical generation, which is worsened in mice lacking the free radical scavenger glutathione peroxidase. A mouse lacking a neuronal isoform of the NOS gene, with damaged nitric oxide generation, shows decline in the sizes of malonic acid-induced striatal lesions.

Procedure:

- Anesthetize the animal using suitable anesthesia and then expose the skull by making incision on the scalp.
- Malonic acid is administered by intrastriatal injection through hole made in the skull by a small hand drill at respective stereotaxic coordinates. Intrastriatal injections of malonic acid are given at various doses and some of them are given as in Table 2.

Table 2 Different doses of malonic acid to induce HD

S. No.	Treatment schedule	Animals used	References
1.	1.33 M/1 μl, intrastriatal injection into left striatum (coordinates are bregma: anterior 0; lateral: 2.8 mm; ventral: 3.5 mm from the brain surface)	Male Sprague-Dawley rats	Toulmond et al. (2004)
2.	2 M/1 μl, intrastriatal injection into left striatum (coordinates are +0.8 mm anterior, +2.9 mm lateral from the bregma, −4.5 mm ventral from the dura mater)	Male Sprague-Dawley rats	Sagredo et al. (2009), Valdeolivas et al. (2012, 2013)
3.	6 μmol/4 μl, intrastriatal injection into right striatum (coordinates are anterior +1.7 mm; lateral ± 2.7 mm; ventral −4.8 mm from bregma and dura)	Male Wistar rats	Kalonia et al. (2010a, b), Kumar et al. (2013)

2.1.2 Excitotoxin Models

One of the most primitive experimental models of HD was established by administering brain delivery of excitatory agonists to rats and mice. It has been reported that systemic administration of glutamate linking neurotoxicity is connected to the overactivation of excitatory amino acid receptors. Excitotoxicity is pathologic neurodegeneration mediated by excessive activation of non-NMDA and NMDA GluRs and also voltage-dependent ion channels. It leads to increase in intracellular concentration of Ca^{2+}. Ca^{2+}-activated enzymes such as proteases, phospholipases, and endonucleases add to the degradation of various cell components and neuronal death. The excessive interaction of ligand with GluR subtypes leads to pathophysiological modifications in intracellular ion concentrations, pH, protein phosphorylation, mitochondrial function, energy metabolism, and movement.

Kainic Acid Model

Specific glutamate analogues have been shown to produce axon-sparing selective lesions in the brain. As striatal cell loss is the principal neuropathological hallmark in HD, one of the first rodent models used the kainic acid as excitotoxin to selectively destroy striatal medium spiny neurons. Coyle and Schwarcz (1976) first described the neurotoxic penalty of direct infusion of the potent glutamate analog kainic acid (KA) in the rat neostriatum. The injection of KA produces selective damage to neurons with cell bodies intrinsic to the striatum, whereas spared the extrinsic axons passing through or terminating in the region. In the beginning, KA was isolated from the seaweed called "Kaininsou" or "Makuri." Kainic acid (KA) is an agonist for ionotropic glutamate receptor subtype, and KA administration has been shown to be responsible for enhanced production of reactive oxygen species, mitochondrial dysfunction as well as apoptosis in neurons of many regions of the brain as shown in (Fig. 4). Administration of KA is known to induce a sequence of distorted behavioral events characterized by epileptiform seizures, which are followed by neurodegeneration in specific brain regions including the hippocampus, cortex, thalamus, and amygdala.

Procedure:

- Anesthetize the animal using suitable anesthesia and expose the surface of the skull by making incision on the scalp.
- Kainic acid is administered by intrastriatal injection through hole made in the skull by a small hand drill at respective stereotaxic coordinates. Intrastriatal injections of kainic acid are given at various doses and some of them are given as in Table 3.

S. No.	Treatment schedule	References
1.	2.2 nmol into striatum	Arvin et al. (1998)
2.	9.3 nmol into striatum	Zaczek et al. (1980)
3.	2.5 μg of kainic acid into striatum	Coyle and Schwartz (1976)

Table 3 Different doses of quinolinic acid used to induce HD

S. No.	Treatment schedule	Animals used	References
1.	210 nmol/0.7 μl, bilateral intrastriatal injection (coordinates are anterior +1.7 mm from bregma, lateral +2.7 mm from sagittal suture, ventral −6.2 mm from dura)	Male Wistar rats	Pintor et al. (2006)
2.	120 nmol/μl per side, bilateral intrastriatal injection (coordinates are 0.5 mm anterior to bregma, ±2.6 mm lateral to bregma and 4.5 mm ventral to the dura)	Male Wistar rats	Pérez-De La Cruz et al. (2009)
3.	300 nmol/4 μl, unilateral intrastriatal injection in right striata (coordinates are anterior +1.7 mm; lateral ±2.7 mm; ventral −4.8 mm from bregma and dura)	Male Wistar rats	Kalonia et al. (2011)
4.	240 nmol/1 μl, unilateral intrastriatal injection in right striata (coordinates are anterior-posterior 0.5, lateral 2.6, relative to bregma and ventral 4.5, relative to the dura)	Male Wistar rats	Antunes Wilhelm et al. (2013)
5.	200 nmol/2 μl, bilateral intrastriatal injection (coordinates are anterior +1.7 mm; lateral ±2.7 mm; ventral −4.8 mm from bregma and dura)	Male Sprague-Dawley rats	Mishra and Kumar (2014)

Quinolinic Acid Model

Initial toxin models that relied on excitotoxic mechanisms made use of kainic acid (KA). Although these experiments were milestone studies, investigators learned that KA produced remote lesions and, destroyed fibers of passage at higher concentrations, and so they tested substitute excitotoxins such as ibotenic acid and quinolinic acid. Quinolinic acid (QA) is an endogenous intermediate in the kynurenine pathway of tryptophan metabolism as explained in Fig. 3. Tryptophan easily crosses the blood-brain barrier with the help of amino acid transporters and is taken up by astrocytes, microglia, and macrophages in the brain and converted into kynurenine, which is further converted into QA with the help of 3-hydroxyanthranilic acid oxygenase (3HAO) enzyme. Usual levels of QA do not produce any toxic effect on brain cell, but only minute augment in QA levels causes toxicity. In the postmortem brains of HD sufferers, increased activity of enzyme 3HAO in brains has been reported, with the greatest increase in striatal brain region.

The major mechanism of QA toxicity is principally related to overstimulation of NMDA receptors and Ca^{2+} overload, followed by mitochondrial dysfunction, cyt c release, ATP depletion, and oxidative damage as shown in (Fig. 4). QA has differential affect on striatal neurons as compared to kainic acid. Both GABAergic and substance P-containing neurons are damaged by QA, with comparative sparing of NADPH-diaphorase and cholinergic neurons, which are found to be spared in HD. QA is incapable of crossing the BBB and is, therefore, experimentally administered directly into the striatum via intrastriatal injection.

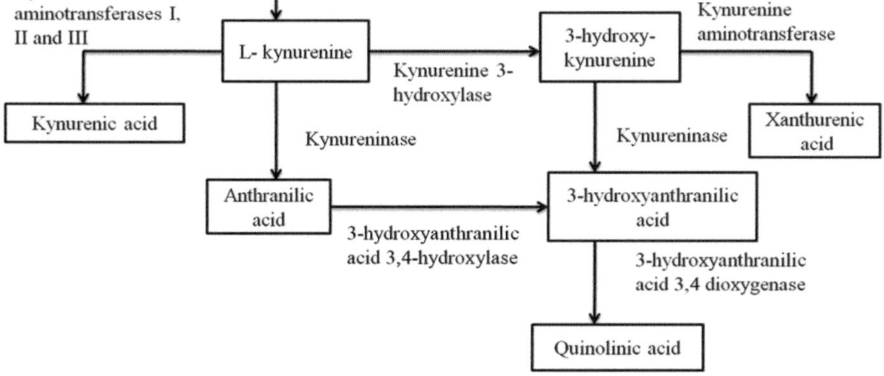

Fig. 3 Kynurenine pathway

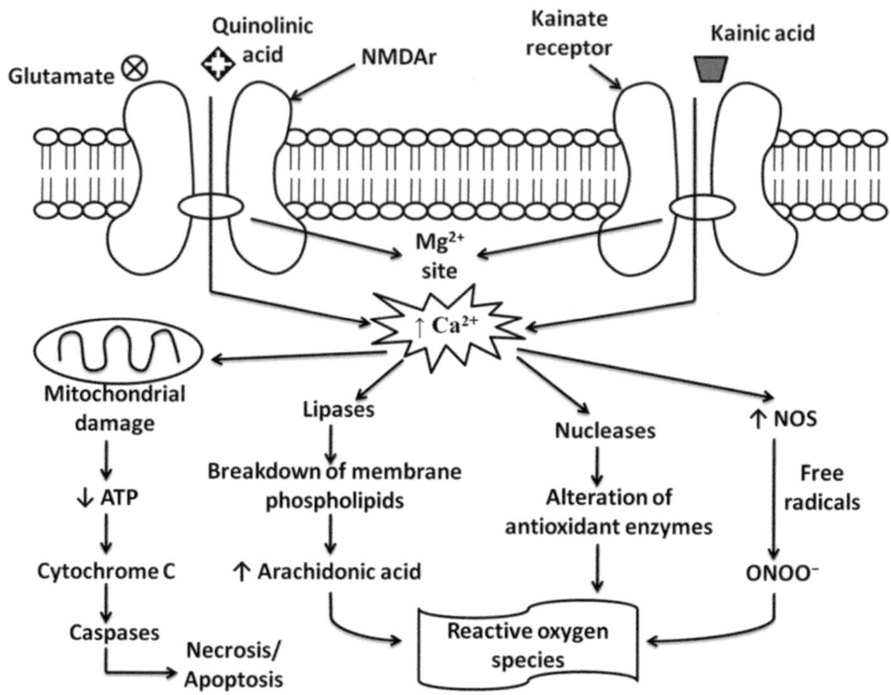

Fig. 4 Mechanism of action of quinolinic acid/kainic acid

Excitotoxin lesions in the monkey by using quinolinic acid present an experimental primate model that closely resembles the neurochemical, neuropathological, and clinical features of HD. Thus, when injected into the brain, QA mimics neurodegenerative measures in rodents, similar to those observed in the brains of patients with Huntington's disease. Interestingly, intrastriatal administration of quinolinic acid in rodents has been shown to raise HTT immunoreactivity, leading to the suggestion that HTT may be induced as a cytoprotective agent after activation of the kynurenine pathway, again emphasizing the close links between this pathway and HD pathogenesis.

Procedure:

- Anesthetize the animal using suitable anesthesia and then expose the surface of the skull by making incision on the scalp to perform stereotaxic surgery.
- Dissolve QA in normal saline and administer unilaterally or bilaterally using Hamilton micro-syringe.
- Deliver QA by intrastriatal injection through hole made in the skull by a small hand drill at respective stereotaxic coordinates.
- Quinolinic acid has been administered to produce HD models at different doses. Some of them are discussed in Table 3.

Advantages:

- These toxins lead to enormous cell death specifically in striatal brain region, which makes them valuable models for studying neurotoxicity phenomena.
- These types of models produce symptoms analogous to HD pathology.
- These models are beneficial in understanding various mechanisms involved in pathogenesis of HD (ROS formation, protease activation, etc.).
- These models can be valuable for investigation of neuroprotective and neurorestorative therapy for clinical cases of HD.
- It is more easy and feasible to induce HD by using toxins. So these models can be established easily in the research laboratories.
- As compared to other models, induction of toxin models of HD is more economical.

Disadvantages:

- In case of toxin models, production or misfolding of mHTT does not occur, and therefore cytoplasmic and neuronal inclusions are not reported, which explain no obvious connection between the genetic root of HD and mechanism of action of these toxins.
- The beginning of cellular death is progressive in clinical HD and the onset age is inversely proportional to the number of CAG triplets, and this situation is not simulated in these toxic models because immediate cell death occurs in this case which is not dependent on mHTT.
- Besides cognitive and behavioral characteristics of HD phenotype, toxin models do not produce other aspects of disease like suicidal tendencies and obsessive compulsive disorder, etc.
- The model induced by toxin is restricted to the toxicant bioavailability; the effect gets ceased once the toxin get metabolized and eliminated and permits the tissues to respond to the insult.

Clinical relevance:

Toxin models include mitochondrial toxins and excitotoxicants. 3-NP and malonate are mitochondrial toxins which act by inhibiting mitochondrial complex II enzyme and lead to ATP depletion and free radical generation. Moreover, 3-NP causes degeneration of GABAergic MSNs selectively in striatum. On the other hand, excitotoxic models like QA and KA primarily act through the mechanism of excitotoxicity. Clinical reports of HD clearly evidence complex II inhibition, selective vulnerability of GABAergic MSNs of striatal region and excitotoxicity as a major feature of HD pathogenesis. These models are very closely related to the actual clinical condition of HD. Owing to these benefits, toxin models can be used in assessment of various novel therapies which are thought to act by targeting mitochondria, excitotoxicity, and GABAergic neurodegeneration.

2.2 Nonhuman Primate Models

Nonhuman primates have a central nervous system that is similar to humans in terms of neural circuitry and this, along with similarities with human physiological and behavioral characteristics, makes them more important and true models of neurological and psychiatric diseases as compared to other animal. Nonhuman primates act as a bridge between rodents and humans. Neurotoxins have been administered in several nonhuman primates at various doses to induce Huntington's disease like symptoms and are discussed in Table 4.

Advantages:

- Terminal and essential invasive experiments can be performed to monitor neurodegeneration, and intranuclear inclusions of huntingtin, as well as other characteristics of the disease at the neural level.
- A considerable number of independent assessments can be done in long-term study of a single primate, which results in consistent statistical answers from comparatively small numbers of animals.
- Nonhuman primates have a vertebrate brain and a neural circuit organization that resembles with humans.

Disadvantages:

- Nonhuman primate models are more costly, and their housing and food costs are comparatively more.
- These models lift more ethical alarms than rodent models. The debate arises from the fact that nonhuman primates have close phylogenic relationship with humans and share many anatomical, behavioral, and physiological characteristics with humans, which implies that they can suffer in the same way as humans.

Table 4 Different doses of toxins to induce HD in nonhuman primates

S. No.	Treatment schedule	Animals used	References
1.	Ten injections of 360 nmol of QA placed in adjacent parasagittal lines 2 mm apart. Five injections placed in the left caudate nucleus and five within the left putamen, extending rostrocaudally from the head of the caudate nucleus posteriorly to the level of the anterior commissure	*Macaca mulatta* rhesus monkey	Ferrante et al. (1993)
2.	3-NP at a dose of 16 mg/kg per day, in two i.m. injections of 8 mg/kg each for 3–6 weeks	Macaques and *Papio anubis* baboons	Brouillet et al. (1995)
3.	3-NP at a dose of 14 mg/kg per day, in two i.m. injections of 7 mg/kg each for 20 weeks	Macaques and *Papio anubis* baboons	Palfi et al. (1996)
4.	3-NP at a dose of 12 mg/kg/day and then weekly increased by 2 mg/kg, i.m. injections for 8–10 weeks	*Papio anubis* baboons	Dautry et al. (2000)

Clinical relevance:

Nonhuman primates are predecessors of humans, so they are very analogous to humans. The main motive to use nonhuman primates for inducing HD is the close similarities between them and human brains in terms of overall anatomy, chemical communication and cellular structure, functional and cognitive abilities, neural circuitry. Moreover, research studies have enlightened genetic makeup of nonhuman primates, and humans also show a very well-built resemblance. All the behavioral, biochemical, and neurochemical changes are more prominent in nonhuman primate models as compared to other animal models. These models can serve an important role in testing of neuroprotective therapies because nonhuman primates can respond to test drugs in a very similar manner as humans do. On the other hand, ethical issues are continuously barring the use of nonhuman primates in research due to a very close resemblance between humans and nonhuman primates which evidence possibilities of an unacceptable level of suffering.

3 Conclusion

Huntington's disease (HD) is a neurodegenerative disorder, characterized by progressive motor dysfunction, emotional disturbances, and dementia due to degeneration of GABAergic MSNs in striatum. No therapy is available to delay the progression of Huntington's disease, and only symptomatic relief is provided by various drugs. Tetrabenazine, a dopamine depletory drug, is the only US-FDA approved drug for treating Huntington's chorea. Need for novel effective therapy is insistent. Different animal models can be used to explore the pathogenic mechanism and to test novel drugs in this rare disease. Every model targets some particular pathological features of the disease like mitochondrial inhibition, selective neurodegeneration, and excitotoxicity. A specific model can be picked to induce the disease on the basis of expected mechanism of action of test drug.

Ethical Statement
All institutional guidelines, national guidelines, state and local laws, and regulations with professional standards for the care and use of laboratory animals should be followed. Studies involving animals must state that the Institutional Animal Ethical Committee has approved the protocol. For authors using experimental animals, a statement should be made that the animals' care is in accordance with institutional guidelines and animals used have been treated humanely and with regard to the alleviation of suffering. Researchers should treat animals as sentient and must consider their proper care and use and the avoidance or minimization of discomfort, distress, or pain as imperatives. Animal experiments should be designed only after due consideration of animal health. It should be ensured that all researchers who are using animals have received instruction in research methods and in the care,

maintenance, and handling of the species being used. All the surgical procedures should be performed under appropriate anesthesia and follow only those procedures which avoid infection and minimize pain during and after surgery.

References

Borlongan C, Koutouzis T, Randall T et al (1995) Systemic 3-nitropropionic acid: behavioral deficits and striatal damage in adult rats. Brain Res Bull 36:549–556

Brouillet E, Hantraye P, Ferrante R et al (1995) Chronic mitochondrial energy impairment produces selective striatal degeneration and abnormal choreiform movements in primates. Proc Natl Acad Sci 92:7105–7109

Coyle JT, Schwarcz R (1976) Lesion of striatal neurons with kainic acid provides a model for Huntington's chorea

Dautry C, Vaufrey F, Brouillet E et al (2000) Early N-acetylaspartate depletion is a marker of neuronal dysfunction in rats and primates chronically treated with the mitochondrial toxin 3-nitropropionic acid. J Cereb Blood Flow Metab 20:789–799

Ferrante R, Kowall N, Cipolloni P et al (1993) Excitotoxin lesions in primates as a model for Huntington's disease: histopathologic and neurochemical characterization. Exp Neurol 119:46–71

Gould D, Gustine D (1982) Basal ganglia degeneration, myelin alterations, and enzyme inhibition induced in mice by the plant toxin 3-nitropropanoic acid. Neuropathol Appl Neurobiol 8:377–393

Guyot M, Hantraye P, Dolan R et al (1997) Quantifiable bradykinesia, gait abnormalities and Huntington's disease-like striatal lesions in rats chronically treated with 3-nitropropionic acid. Neuroscience 79:45–56

Hamilton B, Gould D (1987) Nature and distribution of brain lesions in rats intoxicated with 3-nitropropionic acid: a type of hypoxic (energy deficient) brain damage. Acta Neuropathol 72:286–297

Jang M, Lee M, Cho I (2014) Ethyl pyruvate ameliorates 3-nitropropionic acid-induced striatal toxicity through anti-neuronal cell death and anti-inflammatory mechanisms. Brain Behav Immun 38:151–165

Kalonia H, Kumar P, Kumar A et al (2010a) Protective effect of montelukast against quinolinic acid/malonic acid induced neurotoxicity: possible behavioral, biochemical, mitochondrial and tumor necrosis factor-α level alterations in rats. Neuroscience 171:284–299

Kalonia H, Kumar P, Kumar A et al (2010b) Protective effect of rofecoxib and nimesulide against intra-striatal quinolinic acid-induced behavioral, oxidative stress and mitochondrial dysfunctions in rats. Neurotoxicology 31:195–203

Kalonia H, Kumar P, Kumar A (2011) Attenuation of proinflammatory cytokines and apoptotic process by verapamil and diltiazem against quinolinic acid induced Huntington like alterations in rats. Brain Res 1372:115–126

Kumar A, Sharma N, Mishra J et al (2013) Synergistical neuroprotection of rofecoxib and statins against malonic acid induced Huntington's disease like symptoms and related cognitive dysfunction in rats. Eur J Pharmacol 709:1–12

Kumar P, Kumar A (2009) Effect of lycopene and epigallocatechin-3-gallate against 3-nitropropionic acid induced cognitive dysfunction and glutathione depletion in rat: a novel nitric oxide mechanism. Food Chem Toxicol 47:2522–2530

Mishra J, Kumar A (2014) Improvement of mitochondrial NAD+/FAD+-linked state-3 respiration by caffeine attenuates quinolinic acid induced motor impairment in rats: Implications in Huntington's disease. Pharmacol Rep 66:1148–1155

Nishino H, Kumazaki M, Fukuda A et al (1997) Acute 3-nitropropionic acid intoxication induces striatal astrocytic cell death and dysfunction of the blood-brain barrier: involvement of dopamine toxicity. Neurosci Res 27:343–355

Palfi S, Ferrante RJ, Brouillet E, Beal MF, Dolan R, Guyot MC, Peschanski M, Hantraye P (1996) Chronic 3-nitropropionic acid treatment in baboons replicates the cognitive and motor deficits of Huntington's disease. J Neurosci 16(9):3019–3025

Park J, Lee S, Chu K et al (2008) Galantamine reduces striatal degeneration in 3-nitropropionic acid model of Huntington's disease. Neurosci Lett 448:143–147

Pérez-De La Cruz V, Elinos-Calderón D, Robledo-Arratia Y et al (2009) Targeting oxidative/nitrergic stress ameliorates motor impairment, and attenuates synaptic mitochondrial dysfunction and lipid peroxidation in two models of Huntington's disease. Behav Brain Res 199:210–217

Pintor A, Tebano M, Martire A et al (2006) The cannabinoid receptor agonist WIN 55,212-2 attenuates the effects induced by quinolinic acid in the rat striatum. Neuropharmacology 51:1004–1012

Sagredo O, González S, Aroyo I et al (2009) Cannabinoid CB2 receptor agonists protect the striatum against malonate toxicity: relevance for Huntington's disease. Glia 57:1154–1167

Szabo A, Papp A, Nagymajtényi L (2005) Effects of 3-nitropropionic acid in rats: general toxicity and functional neurotoxicity. Ar Hig Rada Toksikol 56:297–302

Toulmond S, Tang K, Bureau Y et al (2004) Neuroprotective effects of M826, a reversible caspase-3 inhibitor, in the rat malonate model of Huntington's disease. Br J Pharmacol 141:689–697

Valdeolivas S, Pazos M, Bisogno T et al (2013) The inhibition of 2-arachidonoyl-glycerol (2-AG) biosynthesis, rather than enhancing striatal damage, protects striatal neurons from malonate-induced death: a potential role of cyclooxygenase-2-dependent metabolism of 2-AG. Cell Death Dis 4:862

Valdeolivas S, Satta V, Pertwee R et al (2012) Sativex-like combination of phytocannabinoids is neuroprotective in malonate-lesioned rats, an inflammatory model of Huntington's disease: role of CB1 and CB2 receptors. ACS Chem Neurosci 3:400–406

Wilhelm E, Jesse C, Bortolatto C et al (2013) Correlations between behavioural and oxidative parameters in a rat quinolinic acid model of Huntington's disease: protective effect of melatonin. Eur J Pharmacol 701:65–72

Animal Models of Epilepsy

Prashant Koshal, Sumit Jamwal, Kiran Kumar Akula
and Puneet Kumar Bansal

1 Introduction

Epilepsy is the second most common neurological disorder with a predisposition to the occurrence of frequent seizures, affecting around 1.4–1.5 per 1000 in Asian countries and about 65 million people worldwide, and 3 million people in USA. Epileptic seizures are characterized by abnormal and excessive electrical discharge in a population of neuron due to imbalance between excitation and inhibition. It is proposed that changes in the inhibition have underpinned the development of epilepsy and predisposition to seizures in brain. γ-Amino butyric acid (GABA) a principle inhibitory neurotransmitter in central nervous system (CNS), formed within presynaptic GABAergic axon terminals and act on $GABA_A$ and $GABA_B$ receptors, hyperpolarize the neuronal cell by increasing chloride conductance and opening of potassium channels with decreasing Ca^{2+} entry, respectively. Several point mutations altering the synaptic and extrasynaptic functioning of $GABA_A$

P. Koshal · S. Jamwal · P.K. Bansal (✉)
Department of Pharmacology, ISF College of Pharmacy, Moga 142001, Punjab, India
e-mail: punnubansal79@gmail.com

P. Koshal
e-mail: prashantkoshal@gmail.com

S. Jamwal
e-mail: jamwalisf@gmail.com

K.K. Akula
Department of Biology and Biochemisty, University of Houston,
4800 Calhoun Rd, Houston TX 77004, USA
e-mail: kakula@central.uh.edu

P.K. Bansal
Department of Pharmaceutical Sciences and Technology, Maharaja Ranjit Singh Punjab
Technical University (State Govt. University), Bathinda, Punjab, India

P.K. Bansal and R. Deshmukh (eds.), *Animal Models of Neurological Disorders*, https://doi.org/10.1007/978-981-10-5981-0_5

receptor are implicated in various types of epilepsies. Most often seizures occur at the neonatal stage and reflect the intense and long-term outcomes such as epilepsy, cognitive impairment, and motor disorder. Phenobarbitals and benzodiazepines are most effective in the treatment of epilepsy in adults but failed to produce positive results in neonates. Several co-morbidities such as depression and cognitive deficit in the epileptic patients majorly affect the overall lifestyle and healthcare cost. Antiepileptic drugs, which are clinically used to treat the epileptic seizures, can also produce certain psychiatric co-morbidities. Moreover, clinically used antiepileptic drugs also reported to produce secondary complications.

2 Need of Animal Models

In present decennary, several antiepileptic drugs (AEDs) are available but the development of new antiepileptic drugs with the more appropriate tolerability and efficacy is still a chief concern. An ideal animal model is always required to determine the safety and efficacy of newer drug, prior to be tested in human subjects, and to determine relationship between epilepsy and its related secondary complications. Animal models play important role to understand the pathophysiology and pattern of disease progression. Animal models can be used to establish safety and efficacy of novel AEDs over clinically used AEDs. However, a single animal model cannot be efficient for the all above purposes. After the establishment of anticonvulsant potential of new AEDs in a simple model like maximal electroconvulsive seizures (MES) or pentylenetetrazol (PTZ) test, different animal models like kindling model of temporal lobe can be used to investigate the anticonvulsant spectrum of novel AEDs.

2.1 Characteristics of an Ideal Animal Model of Epilepsy

An experimental animal model is designed in such a way that it should possess the following features as described in Table 1.

Table 1 Characteristics of seizure in animal model

Sr. No	Characteristics of seizure model
1.	Seizures should be spontaneous and recurrent
2.	Clinically relevant to human epileptic seizure
3.	Electroencephalography (EEG) pattern should be similar to clinical epileptic observations
4.	Intensity of seizures should be sufficient for acute and chronic dosing of drugs in experimental studies
5.	It should posses cognitive and neurobehavioural alteration
6.	Latent period should occur between brain insult and seizure

2.2 Score

(1) Seizure severity—With respect to the strength, motor seizure is classified on a five-point scale, at the beginning little or no motor activity with after discharge (AD) exploratory behaviour. Jaw opening and head nodding is the first sign of skeletal seizure, which occurs 2–4 days before the occurrence of full motor seizure, next element of motor seizure is forelimb contraction and falling, and the full motor seizure rearing and falling.

Stage 1—Mouth and facial movements
Stage 2—Head nodding
Stage 3—Forelimb clonus
Stage 4—Rearing
Stage 5—Rearing and falling

3 Classification of Seizure

Partial seizure (Focal onset)
Simple partial (Consciousness is not impaired)

- With motor signs
- With somatosensory and special sensory symptoms
- With autonomic symptoms
- With psychic symptoms

Complex partial (With impairment of consciousness)

- Starts as simple partial and lead to impairment of consciousness
- With impairment of consciousness at onset

Partial seizure to the secondary generalization

- Simple partial to the secondary generalization
- Complex partial to the secondary generalization
- Simple partial involving complex partial

Generalized seizure
Absence seizure

- Impairment of consciousness only
- With mild clonic component
- With mild atonic component
- With automatism

Myoclonic seizure
Clonic seizure
Tonic seizure
Tonic-clonic seizure
Atonic seizure

4 Pathophysiology of Seizures

An absence seizure is non-convulsive and causes brief loss of consciousness, which arises and terminates suddenly. It is most privileged type of seizure in childhood age. The EEG study of patient with absence epilepsy showed slow spike wave discharge (2–4 Hz) bilaterally. Thalamus is the major relay centre for all sensory and motor perception, and it is also involved in spread of absence seizure. Reticular thalamic nucleus (TRN) covers rostral, lateral, and ventral parts of thalamus, and spreads spike wave discharge (SWD). Reticular thalamic nucleus made up of GABAergic interneuron and thalamocortical nucleus (TC), which is primarily excitatory in nature, contains T-type Ca^{2+} channel and involved in rebound burst firing of neuron (Figs. 1, 2).

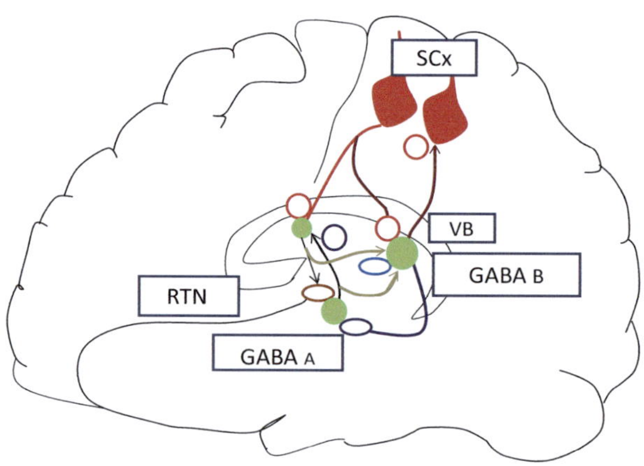

Fig. 1 Neuronal pathways involved in pathophysiology of absence seizures. Glutamatergic projection appears from somatosensory cortex (SCx), and it forms synapse with the ventrobasal posterior thalamus (VB), further forming excitatory synapse with SCx. Thalamic reticular neurons are primarily GABAergic and make coordination with these excitatory loops, by causing hyperpolarization. During the 'absence seizure', neuron of RTN became more hyperpolarized by activation of T-type Ca^{2+} channels and opening of these channels due to depolarization on VB region. RTN neuron hyperpolarizes the VB neurons by $GABA_A$ and $GABA_B$ and initiates the opening of T-type Ca^{2+} channels present on this cells; thus, rebound burst firing occurs from thalamus to SCx

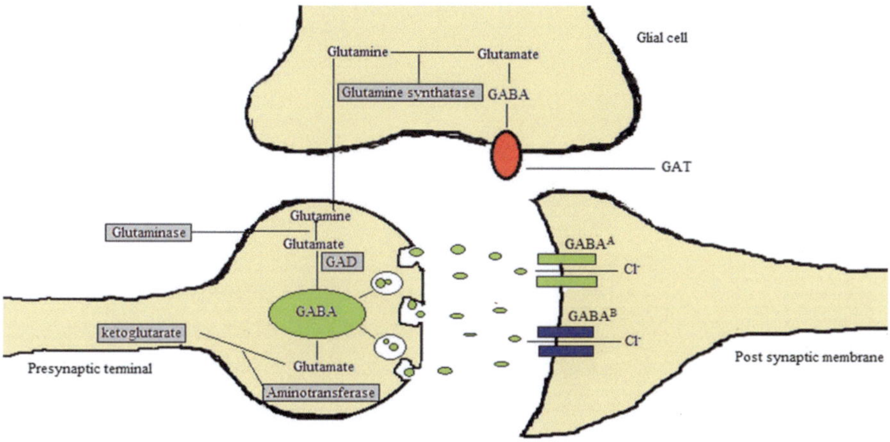

Fig. 2 GABAergic pathway in brain

5 Classification of Animal Models of Epilepsy (Fig. 3)

5.1 Chemical-Induced Convulsions

5.1.1 Pentylenetetrazol-Induced Convulsions

Principle—The convulsing action of PTZ is due to the blocking of $GABA_A$ receptor and inhibition of the chloride channels opening. Usually, it binds to the picrotoxin binding site on $GABA_A$ receptor. Single dose of PTZ in rats significantly decreases the mRNA level of $GABA_A$ receptors and its surface availability. Single pentylenetetrazol dose in albino mice causes myoclonic, clonic than generalized tonic-clonic convulsions (Fig. 4).

Acute treatment: PTZ-induced convulsion is validated model of generalized absence, myoclonic seizure, and clonic convulsion, due to the cortical stimulation. Different dose of PTZ in different species is despite below.

Sr. No	Species	Dose (single)	Reference
1	Male Swiss mice	95 mg/kg s.c	Kaminski et al. (2001)
2	Wistar rat	60 mg/kg i.p	Malhotra and Gupta (1997)
3	Male SD rat	45 mg/kg i.p	Walsh et al. (1999)
4	Male NMRI mice	60 mg/kg i.p	Ahmadini et al. (2003)
5	Swiss mice	50 mg/kg i.p	Medina et al. (2001)

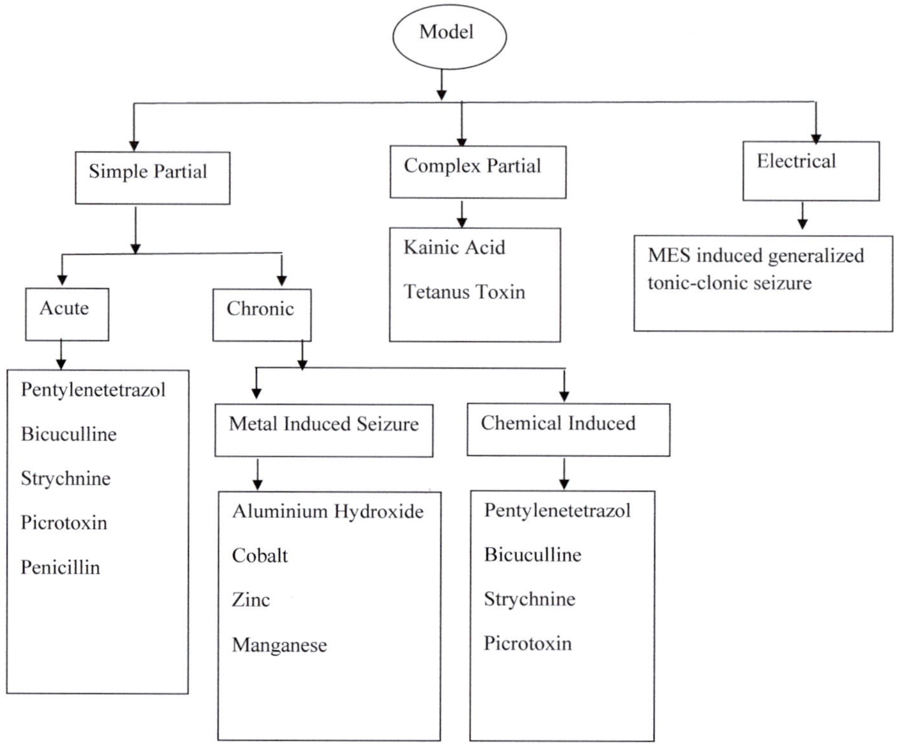

Fig. 3 Classification of animal models of epilepsy

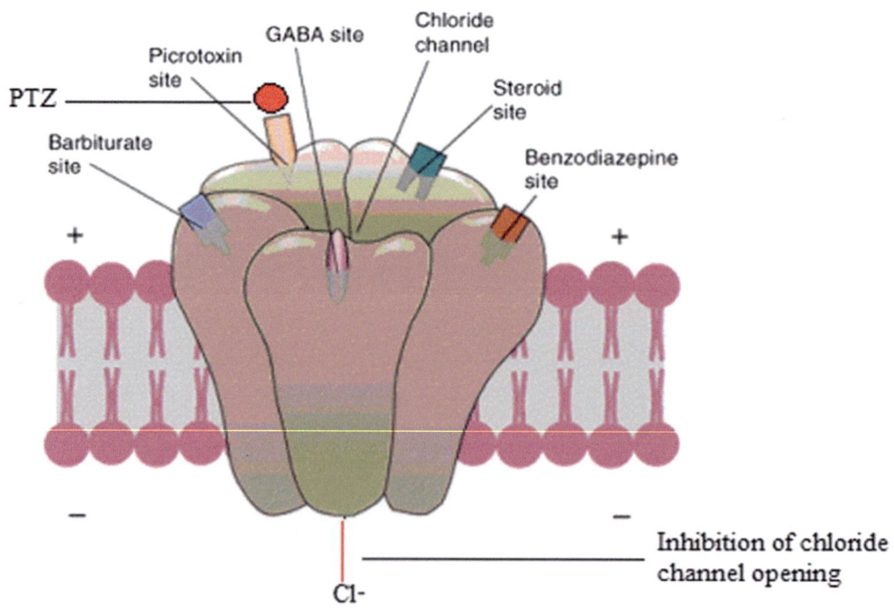

Fig. 4 Pentylenetetrazol (PTZ)-induced convulsions

Chronic treatment: Goddard in 1967 described that kindling is a validated model of epilepsy which causes generalized seizure. Repeated administration of sub-convulsive dose of chemoconvulsant like PTZ causes tonic-clonic seizure, and at this stage, animal is thought to be fully kindled.

Sr. No	Species	PTZ dose	No. of injection	References
1	Male Swiss albino mice	35 mg/kg i.p	11 (on alternate days)	Ilhan et al. (2005)
2	Male Wistar rat	30 mg/kg i.p	13 (on alternate days)	Atack et al. (2000)
3	Male Wistar rat	30 mg/kg i.p	21 (on alternate days)	Atapour et al. (2000)
4	Male Wistar rat	45 mg/kg	13 (on alternate days)	Phole et al. (1997)

Advantage—PTZ administration produces myoclonic jerk followed by the generalized tonic-clonic seizures. In the process of kindling, there is increase in seizure severity with reduction of seizure threshold as well as neurodegeneration in limbic region is found similar to human mesotemporal epilepsy. This pattern has also been observed in the electrical amygdala kindling.

Disadvantage—Phenytoin, carbamazepine, oxcarbazepine are not effective against pentylenetetrazol-induced seizures while ethosuximide, benzodiazepines, vigabatrin, tiagabine are effective in this model.

5.1.2 Bicuculline-Induced Convulsions

Principle—Bicuculline acts on $GABA_A$ receptor in a competitive manner and inhibits the chloride channel opening. Bicuculline causes spike wave discharge (SWD) after focal and systemic application and tonic-clonic seizure activity.

Acute treatment:

Sr. No	Species	Weight	Dose (single)	Reference
1	Male Laca albino mice	22–30 gm	4 mg/kg i.p.	Dhir et al. (2006)
2	Male SD rat	150–200 gm	0.375 mg/kg i.v.	Bowdler et al. (1980)

Chronic treatment:

Bicuculline Methionide Kindling Chemitrode, stainless steel guide cannulae (0.60 mm in o.d and 0.32 mm in i.d), which contain bipolar stainless steel electrode on its external wall and Mandrell (0.29 mm in diameter) were implanted in left Basolateral Amygdala. Stainless steel Mandrell (0.29 mm in diameter) then removed and replaced by the stainless steel injection cannulae (0.29 mm in o.d).

After 2 weeks of implantation, implanted animals were injected Bicuculline methionide injection (0.2 n mole dissolved in 0.8 μl sterile water containing NaCl with pH 6.2) once every fourth day. Injection was delivered over 30 s by using Hamilton syringe. Epileptic seizure is produced 10 min after injection. Additional 0.2 n mole Bicuculline methionide was injected if first injection failed to produce response.

Assessment of Seizure

Stage 0—No symptoms
Stage 1—Mouth and facial movement
Stage 2—Head nodding
Stage 3—Forelimb clonus
Stage 4—Rearing
Stage 5—Rearing and falling

Advantage—Bicuculline methionide induces chronic and abnormal functional changes instead of structural lesion. Bicuculline methionide-induced seizure lasts for at least 6 months and similar to electrical amygdala kindling.

Disadvantage—Many clinically available anticonvulsant drugs like diazepam are not effective towards the Bicuculline-induced seizure, but carbamazepine and phenytoin are effective.

5.1.3 Strychnine-Induced Convulsions

Principle—Glycine is the inhibitory neurotransmitter, which controls the motor rhythm generation in the brain stem and spinal cord. Glycine acts by strychnine-sensitive glycine receptor (GlyRs), which is highly found in medulla oblongata, spinal cord, and brain stem (Betz and Laube 2006). Strychnine is competitive antagonist of glycine receptor, and by antagonizing the inhibitory effect of glycine on brain stem and spinal cord, it causes convulsions.

Acute treatment:

Sr. No	Species	Dose (single)	References
1	Male mice	2 mg/kg s.c	Bogdanova et al. (1997)
2	Male albino mice	4 mg/kg i.m	Olatokunboh et al. (2009)

Chronic treatment—Strychnine is given to the Wistar albino rat at the dose of 15 mg/kg i.p, or 30 mg/kg i.p on alternate days for 15 days. Seizure are evaluated as

Score 0 = no seizure
Score 1 = jerks
Score 2 = straub tail
Score 3 = clonic convulsions

Advantage—Similar pattern of generalized tonic-clonic convulsions can be produced by the strychnine.

Disadvantage—Strychnine-induced seizure pattern is different from the seizures produced by primary GABA antagonist.

5.1.4 Picrotoxin-Induced Convulsions

Principle—Picrotoxin is a $GABA_A$ receptor antagonist. It binds to the β_2/β_3 subunit of $GABA_A$ receptor and blocks the opening of chlorine channel and causes intense tonic-clonic seizure.

Procedure (Acute treatment)

Sr. No	Species	Dose (single)	Reference
1	Laka albino mice	4 or 8 mg/kg i.p	Akula et al.,2007
2	Male Wistar rat	5 mg/kg i.p	Paul and Subramanian (2002)
3	Wistar rat	3 mg/kg i.p	Paul and Krishnamoorthy (1998)

Chronic treatment—Male Wistar rats (200–320 gm) were treated with sub-convulsive dose of picrotoxin (1.5 mg/kg i.p) 20 injections per day for 2–3 days. Stages 3–4 (hindlimb clonus with rearing as well as generalized tonic clonic with rearing and falling) were observed.

Advantage—Many drugs are used to produce partial seizure by focal application, but picrotoxin produces tonic-clonic seizure by the systemic application. It also causes burst firing of dopaminergic neuron.

Disadvantage—Valproate and ethosuximide have very low action towards the pilocarpine-induced convulsions compared to the diazepam, carbamazepine, and phenytoin. Small neurotoxic dose of valproate is required to attenuate the convulsions.

5.1.5 Kainic Acid-Induced Convulsions

Principle—Kainic acid (KA) is analogue of glutamate, which is primarily excitatory neurotransmitter and agonist of AMPA/kainate class of glutamate receptor. Kainic acid-induced limbic seizures are validated models, which on systemic or intracerebral administration produces limbic seizure. Hippocampus is more vulnerable for kainic acid-induced injury. Study revealed that administration of kainic acid decreased the size of hippocampus and increased the size of ventricles. Kainic acid produces complex partial seizures, leading to the secondary generalization in a dose-dependent manner. The first symptom of kainic acid administration is automatism, which is known as ‘wet dog shakes’, and at the sufficient dose, this automatism is followed by motor seizure for several hours. Thus, KA injections represent the model of convulsive status epilepticus.

Treatment:

Sr. No	Species	Dose (single)	Reference
1.	Wistar rat	10 mg/kg i.p	Gupta et al. (2006)
2.	Albino Wistar rat	6 mg/kg i.p	Konrad et al. (1999)

Advantage—Less mortality rate can be achieved by direct application of kainic acid in brain compared to its systemic application. It is best animal model for mesial temporal lobe epilepsy (MTLE).

Disadvantage—Kainic acid sensitivity varies from different mouse strains, whereas C57BL/6, C57BL/10, and F1 C57BL/6 strains are resistant to systemic administration.

5.1.6 Tetanus Toxin-Induced Convulsions

Principle—Synaptobrevin is a protein and required for the neurotransmitter release. Tetanus toxin causes breakdown of this protein and inhibits the release of inhibitory neurotransmitter, which leads to chronic epileptic stage. Little mossy fibre sprouting also occurs in hippocampus with tetanus toxin administration. Study revealed that single micro-injection of tetanus toxin produces the model of complex partial seizures clinically similar to the human complex partial seizures, and these seizures occur for at least 3 min spontaneously and occasionally for several weeks or months.

Dorsal Hippocampal Administration of Tetanus Toxin (TT)

Animal	Anaesthesia	Craniotomy	Co-ordinates/dose	Reference
Male SD rats	1.5–3.0% inhaled isoflurane with 0.05 mg/kg buprenorphine	Over right hippocampus (3.5 mm posterior and 2.8 mm lateral to bregma)	25 ng of TT in 0.5 μl phosphate-buffered saline with 0.2% bovine serum albumin is administered at 3.3 mm posterior to bregma, 3.2 mm lateral, and 3.1 mm ventral	Rolston et al. (2010)

Advantage—Cortical application of tetanus toxin in cats produces Epilepsia Partialis Continua (EPC) syndrome similar to humans. It causes long-lasting changes in the synaptic excitation with no/minimal neuronal damage. Seizure induced by tetanus toxin continues for at least 1–3 weeks and thus offers long-term treatment studies.

Disadvantage—Tetanus toxin administration in brain is used to produce the model of MTLE. However, seizures induced by the tetanus toxin is week or short-term. Therefore, this model has low acceptance compared to the kindling and post-status epilepticus model.

5.1.7 Penicillin-Induced Convulsions

Principle—Penicillin produces experimental model of partial seizure by selectively antagonizing the inhibitory effect of $GABA_A$ receptor. In consideration to clinical data, systemic administration of penicillin at high dose produces myoclonic and tonic-clonic seizures in human.

Treatment:

Animal	Anaesthesia	Co-ordinates	Dose
Male Wistar rat	Ketamine hydrochloride 100 mg/kg i.p	Cannulae were implanted to right lateral cerebral ventricle (0.6 mm posterior to bregma, 2.0 mm lateral to midline, 4.2 mm below the surface of scull)	Penicillin G potassium 300 IU, in 3 μl.

Behavioural Observations

Score 0—No response
Score 1—Twitching of face and ear, fictive scratching
Score 2—Myoclonic twitching and tremor
Score 3—Bilateral forelimb clonus
Score 4—Tonic-clonic seizure
Score 5—Tonic-clonic seizure without retention of postural control (Bostamci and Bagirici, 2007)

Advantage—It is appropriate model to investigate the spread of seizure activity. Penicillin-induced epileptic seizure starts locally, but latter it spreads and converts to generalized epilepsy, thus resembling the model of grand mal epilepsy.

(H) Metal-induced convulsions

(i) Aluminium hydroxide

Principle—This is the validated model of temporal lobe epilepsy. Aluminium hydroxide leads to partial seizures (rhythmic jerking) followed by secondary generalized tonic-clonic seizure. Similar pattern can be seen in cat and monkey as in humans, such as gliosis, cell loss, and neovascularization at the implantation site of aluminium.

Procedure

(1) Animal—Adult male cats

Anaesthesia—Suitable anaesthesia with adjuvants

Aluminium hydroxide administration—Cannulae made up of 22.0 g spinal quincke needles were implanted to the anterior sigmoid gyrus (anterior +27 mm, lateral 3.5 mm, and ventral 1.0 mm below the dural surface) and in the basolateral amygdala (anterior +1.20 mm, lateral 10.0 mm, ventral −7.0 mm). 0.1 ml sterile aluminium hydroxide was injected.

EEG recording—Insulated bipolar electrode measuring 1.0 mm at the tip was placed at the right and left posterior sigmoid gyrus (anterior +23.0, lateral 5.0), right and left anterior suprasylvian gyrus (anterior +16.0, lateral 10.0), right caudate nucleus (anterior +15.0, lateral 6.0, ventral +6.07), left and right rostral thalamus (anterior +13.0, lateral 2.0, ventral −3.9), and right mesencephalic reticular formation (anterior +2.0, lateral 2.0, ventral −2.5) (Feeney et al. 1998).

(2) Animal—Male rhesus monkey

Anaesthesia—Pentobarbital 35 mg/kg i.p., 4 mg/kg i.m. used to reduce oedema, atropine sulphate 0.05 mg/kg s.c. used to reduce the salivation.

Aluminium hydroxide administration—Cannulae are implanted on the temporal lobe in the following co-ordinates:

(1) Anterior hippocampus—Anterior 9.0 mm, lateral 10.1 mm, horizontal −4.0 mm

(2) Middle hippocampus—Anterior 6.0 mm, lateral 10.1 mm, horizontal −9.0 mm

(3) Posterior hippocampus—Anterior 3.0 mm, lateral 10.7 mm, horizontal 0.2 mm

Aluminium hydroxide injection was administered in volume between 0.1 and 0.3 ml.

EEG recording—Stainless steel bipolar depth electrode was used to record the EEG. It is placed over anterior and posterior hippocampus bilaterally adjacent to the aluminium injection site. EEG recording was done by using 6.4-mm-diameter, no. 4 self-tapping stainless steel screw.

Advantage—Important neuronal loss of dendritic cell in the epileptic locus can be achieved by this model. Neuronal loss decreased GABAergic neuron as well as decreased the positive terminal of GAD (glutamate decarboxylase) in the epileptic focus.

Disadvantage—This model requires six weeks to three months to get developed.

(ii) Cobalt

Principle—It is a chronic model of epilepsy. Implantation of cobalt in the cortex of animal produces primary and secondary epileptogenic locus that continues for several weeks. Study showed that tonic-clonic jerks produced by cobalt implantation might be due to increase in the glycogen concentration in the sensory-motor cortex. Study also revealed that injury made by cobalt in cortex decreased the GABAergic immunoreactive cells.

Procedure—Cobalt metal disc (diameter 3.2 mm and thickness 0.003 mm) was implanted on the sensory-motor cortex of male Wistar rat (200–300 gm) under suitable anaesthesia.

Electrocorticograph (ECoG) recording—Four stainless steel screws can be placed over sensory-motor cortex.

Advantage—Bonvall et al. developed this model. Cobalt produces chronic epileptic seizure by inhibiting the synthesis of GABA and GAD. Around 20 days of cobalt treatment, there is neuronal loss in CA1 region of hippocampus.

Disadvantage—Cobalt-induced epileptic seizures are not long-lasting, which affect the long-term epileptological study.

6 Model of Status Epilepticus

6.1 Lithium-Pilocarpine Model of Status Epilepticus

Principle—Status epilepticus (SE) may be defined as the neurological state in which repeated generalized convulsions occur lasting for 30 min which further causes neuronal injury in brain, and it is well-established model of SE related to temporal lobe. Pilocarpine-induced SE occurs within 50–60 min after injection and stage of SE is continued to the 8–12 h then it is aggravated into the recurrent seizure till the 14–25 days. This phase continued with the latent phase that leads to neurological damage followed by spontaneous and recurrent in animal for long time. Similarly in lithium-pilocarpine model, neuronal damage occurs in hippocampus, piriform cortex, entorhinal cortex, amygdala, thalamus, and septum. Among all of the structures, piriformcortex is more sensitive to the damage during SE and involved in focus of epileptic activity causing secondary generalization.

Treatment:

Animal	Dose of lithium and pilocarpine
Sprague Dawley rats	Lithium is injected at a dose of 3 mEq/kg i.p and after 24 h pilocarpine at a dose of 30 mg/kg i.p are administered (Han et al., 2009)
Albino rats	Lithium chloride 3 nmol/kg i.p is administered 24 h before pilocarpine 40 mg/kg i.p

Advantage—In this model, lithium is given to increase the effect of pilocarpine. This also reduces the dose of pilocarpine required to produce SE and induces more stable convulsions in rats.

Disadvantage—Benzodiazepines are effective against the lithium-pilocarpine-induced seizures if given earlier because seizure becomes resistant with time.

6.2 Perforant Path Stimulation Model of Status Epilepticus

Principle—Perforant path stimulation induces damage to the hippocampal neuron particularly in the CA1, CA2 region of hippocampus and dentate hilus in rodent, which is similar to clinical SE conditions. Electrical stimulation also leads to mossy fibre sprouting in hillar region of dentate gyrus.

Procedure

Animal—Male Sprague Dawley rat (250–320 gm)

Anaesthesia—Suitable

Electrode implantation—Twisted 125-μm Teflon-coated stainless steel wire is implanted 4.4 mm lateral, 8.0 mm caudal from bregma to stimulate the perforant path.

Stimulation—2–3 mA, 50 μs alternating monopolar pulses at 20 Hz for 2 h were used to stimulate perforant path. Behaviour is recorded by Racine classification.

Advantage—Necrosis can be induced by the perforant path stimulation model in the region of CA1, CA3, and hillar neurons. Moreover, hippocampal status epilepticus can be induced without any intervention of excitotoxic damage, which is caused by kainic acid or pilocarpine model.

Disadvantage—Perforant path stimulation causes limited damage to the dorsal hippocampus. This model shows difference in brain lesion, according to the site of stimulation and the intensity of stimulation.

7 Maximal Electroshock (MES)-Induced Convulsions

Principle—Animals can be used to study the different types of epileptic seizures such as grand mal and petitmal. Grand mal type of epilepsy can be induced in mice or rat by the maximal electroshock, in which corneal electrodes are used to induce cortical stimulation, and the seizure consequences such as tonic flexion, tonic extensor, clonic convulsion, stupor, recovery, or death can be studied in the animals.

Procedure—Tetracaine in a concentration of 0.5% in saline is applied on the eyes before the experiment and then 50 mA, 60 Hz, 0.2 s current in mice and 150 mA, 60 Hz, 0.2 s current in rat were applied through corneal electrode. The entire animals were subjected to the 6 Hz/MES test. The current intensity of 32 or 44 mA at 6 Hz for 3 s was applied by corneal electrode to check the effect of drug.

Advantage—MES induces single generalized tonic-clonic convulsion in animals which can be seen in drug-resistant epileptic patient.

Disadvantage—This model is not effective for the anticonvulsant drugs which increase the seizure threshold but not have enough power to increase the threshold above 50 mA for mice and 150 mA for rats.

Acquired Epilepsy (Chronic model) Pathogenesis of acquired epilepsy (AE) is thought to be due to brain insult, which causes abnormal remodelling of neuron or neuronal network. At the time of brain insult, there is elevation of [Ca^{2+}] overload. If the injury or insult is very intense, it causes neuronal death. Development of AE occurs in different stages like injury to the brain followed by epiliptogenesis and spontaneous recurrent seizures. The ratio of acquired epileptic patients is about 50%, and rest of these are idiopathic.

8 Animal Models of Temporal Lobe Epilepsy

The animal model of TLE is widely used to understand the underlying basic mechanism of epiliptogenesis, and it resembles the similar characteristics as human epilepsy like mossy fibre sprouting and neuronal loss. It is also beneficial to know the basic structural and neuronal changes, causes interictal behavioural disturbance, and is often associated with TLE. Epileptic seizure originated from temporal lobe is always manifested with aura (an unusual sensation) like epigastric discomfort, hallucination, sensory and motor perceptions. Psychiatric problems such as depression, anxiety are the most common interictal emotional disturbance in TLE patients compared with schizophrenic patients (Klynchuk et al. 2000).

8.1 Amygdala Kindling

Principle—Amygdala is made up of 10 nuclei, among all the nuclei; basolateral amygdala (BLA) plays a crucial role for the development and spreading of epileptic seizure. Excessive glutamatergic activity is the key component of hyperexcitability and epiliptogenesis. In the chemical and electrical kindling, overactivation of glutamatergic NMDA and AMPA receptor due to excessive glutamatergic activity leads to intense calcium influx from intracellular store, and via voltage-gated calcium channels. This leads to persistent depolarization, epiliptogenesis, and spike wave discharge (SWD). Somatostatin-immunoreactive interneuron (GABAergic interneurons of BLA) is more venerable for damage during kindling and seizure activity. Reduced GABAergic inhibition due to interneuronal loss may lead to imbalance between excitation and inhibition, resulting in epiliptogenesis in amygdala (Aroniadou-Anderjaska et al. 2008).

Right amygdala kindling

Species	Anaesthesia	Co-ordinates	EEG recording	Frequency
Male Wistar rat	Pentobarbital 50 mg/kg i.p	Bipolar electrode is implanted, anterior posterior 1.5 mm, lateral 4.4 mm, ventral 8.5 mm from bregma	From site of electrode implantation	1 s train of 50 Hz, 1 ms biphasic square wave pulse, with 500 μA amplitude is delivered until 10 fully kindled Stage 5 is not obtained

Seizure assessment—According to the Racine scale

Score 0—No seizure response
Score 1—Immobility, eye closure, ear twitching, snuffing, and facial clonus
Score 2—Head nodding
Score 3—Forelimb clonus
Score 3.5—Bilateral forelimb clonus without rearing
Score 4—Bilateral forelimb clonus with rearing
Score 4.5—Falling on a side, loss of rightning reflex accompanied by generalized clonic seizure
Score 5—Rearing and falling (Borowicz et al. 2002).

Advantage—Amygdala kindling may replicate the aspects of clinical cardiovascular complications associated with epilepsy such as hypertension and bradycardia, and it produces the similar epileptic zone which is also seen in temporal lobe epilepsy.

8.2 Hippocampal Kindling

Principle—Study reveals that kindling causes irreversible structural changes in the morphology of neuron in various brain regions, such as sprouting of mossy fibres in the granular cell of dentate gyrus in hippocampus. Sprouting may be defined as the neuronal remodelling in the dentate gyrus, in which mossy fibres grow abnormally. Entorhinal cortex (EC) gives projections to the granular cell of dentate gyrus (DG). The mossy fibres, which originate from the dentate gyrus, forward these projections to CA3 region (made up of small pyramidal cell), and Schaffer collaterals complete this circuit at CA1 region (made up of broad pyramidal cell). There is permanent loss of hillar mossy cells of DG region which provide excitatory input to the inhibitory (GABAergic) interneuron and maintain the excitation and inhibition.

Procedure Ventral hippocampal kindling

Species	Male SD rat (260–300 gm)
Anaesthesia	Chloral hydrate (400 mg/kg)
Electrode	Twisted teflon-coated stainless steel wires (diameter 0.13 mm)
EEG recording	From the tip which is not insulated
Co-ordinates	−5.4 anterior posterior from bregma, −5.2 mm lateral, −6.5 mm ventral
Frequency	To determine after discharge threshold, administer 50 μA, 60 Hz, 2 s, monophasic square waves at 1 ms/pulse. If no after discharge is found, then stimulus can be increased by 20 μA per stimulus until Stage 5 is not reached. Then, procedure is continued by 12 kindling per day at 30-min interval for 4 consecutive days with the intensity of 450 μA, 60 Hz for 10 s at 1 ms square wave pulses
Reference	Ding et al. (2010)

Advantage—Hippocampal kindling resembles the features with the patients who suffer from temporal lobe epilepsy, and it prevents memory loss associated with resective surgery.

8.3 Corneal Kindling

Principle—Corneal kindling is the easier model among all the animal models of electrical kindling. It can be used to predict the behaviour and cognitive impairment caused by the drugs like NMDA receptor antagonist and investigational new antiepileptic drug (AED). But mortality rate is relatively high compared to the classical kindling models.

Treatment:

(a) Mice—Stimulated by 3 mA, 60 Hz, 2 s once daily

(b) Rate—Stimulated by 8 mA, 60 Hz, 4 s twice daily

Stimulation is applied until Stage five on the Racine scale is not found (Kupferberg 2001)

Species	Electrode	Frequency
Male and female NMRI mice	Corneally placed saline-soaked cooper electrode	3 mA, for 3 s, 50 Hz twice daily with 6-h interval for 12 days

Seizure assessment

Score 1—Mild facial clonus and eye blinking
Score 2—Sever facial clonus, head nodding, chewing
Score 3—Unilateral or alternating forelimb clonus
Score 4—Bilateral forelimb clonus and rearing
Score 5—Bilateral forelimb clonus with rearing and falling
Score 6—Tonic hindlimb extension

Advantage—A large number of animal can be kindled within short duration. The test compound required in less amount. That is cost-effective (Kupferberg 2001).

8.4 Piriform Cortex Kindling

Principle—Piriform cortex has direct connection with the olfactory bulb, and it is highly sensitive to generate limbic motor seizure. The piriform cortex is made up of anterior and posterior cortex, and generation of seizure might be due to reduced GABAergic inhibition in the anterior piriform cortex compared to posterior. Piriform cortex is made up of three layers. Primary layer is made up of GABAergic neurons, afferent tract that originated from the olfactory bulb, and afferent fibres that originated from other neurons of piriform cortex. Secondary and tertiary layers are comprised of pyramidal cells and cell bodies, respectively.

Treatment:

Species	Anaesthesia	Electrode	Co-ordinates	Frequency	Reference
Male Wistar rat	Mixture of ketamine (100 mg/kg i. p) and diazepam (8 mg/kg i.p)	Twisted pair of 0.2-mm-diameter Teflon-coated stainless steel wires, whose tip was exposed by 0.2 mm	0.8 mm posterior from bregma, 4.9 mm left, 8.8 mm ventral	To check the after discharge threshold, 1 s train of 50 Hz square wave, for 1 ms, separated by 1-min interval. Current intensity is 7 μA and increases gradually by 20% per step to 500 μA or until a behavioural seizure occurred.	Gallego et al. (2010)

Advantage—It resembles the simple model compared to the neocortex which is six-layered and requires less stimulation compared to the hippocampal and amygdaloidal stimulation.

Disadvantage—Kindling animal model of temporal lobe epilepsy is labour intensive and requires much stimulation to elicit spontaneous seizures. Electrode implantation and chemical vaccination to the brain are little complex procedure. The lesions caused by kindling in the brain are not quite similar to the clinical observations of mesial temporal lobe epilepsy.

Ethical Statement All institutional guidelines, national guidelines, state and local laws, and regulations with professional standards for the care and use of laboratory animals should be followed. Studies involving animals must state that the institutional animal ethical committee has approved the protocol. For authors using experimental animals, a statement should be made that the animals' care is in accordance with institutional guidelines and animals used have been treated humanely and with regard for the alleviation of suffering. Researchers should treat animals as sentient and must consider their proper care and use and the avoidance or minimization of discomfort, distress, or pain as imperatives. Animal experiments should be designed only after due consideration of animal health. It should be ensured that all researchers who are using animals have received instruction in research methods and in the care, maintenance, and handling of the species being used. All the surgical procedures should be performed under appropriate anaesthesia and follow only those procedures which avoid infection and minimize pain during and after surgery.

References

Ahmadiani A, Mandgary A et al (2003) Anticonvulsant effect of flutamide on seizures induced by pentylenetetrazole: involvement of benzodiazepine receptors. Epilepsia 44:629–635

Aroniadou-Anderjaska V, Fritsch B et al (2008) Pathology and pathophysiology of the amygdala in epileptogenesis and epilepsy. Epilepsy res 78:102–116

Atack JR et al (2000) Kindling induced by pentylenetetrazole in rats is not directly associated with changes in the expression of NMDA or benzodiazepine receptors. Pharmacol Biochem Behav 65:743–750

Atapour N, Kalantaripour TP (2000) Chemical kindling and seizure susceptibility in morphine dependent rats. Eur Neuropsychopharmacol 10:483–487

Betz H, Laube B et al (2006) Glycine receptors: recent insights into their structural organization and functional diversity. J Neurochem 97:1600–1610

Bogdanov NN et al (1997) Pentylenetetrazol and strychnine convulsions in brain weight selected mice. Seizure 6:135–138

Borowicz KK et al (2002) Niguldipine impairs the protective activity of carbamazepine and phenobarbital in amygdala-kindled seizures in rats. Eur Neuropsychopharmcol 12:225–233

Bostanci MÖ, Bagirici F et al (2007) Anticonvulsive effects of quinine on penicillin-induced epileptiform activity: An <i> in vivo </i> study. Seizure 16:166–172

Bowdler JM et al (1980) Brain and serum calcium concentrations following electroconvulsive shock or bicuculline-induced convulsions in rats. Brit J Pharmacol 71:321–325

Dhir A, Naidu PS et al (2006) Effect of cyclooxygenase-2 (COX-2) inhibitors in various animal models (bicuculline, picrotoxin, maximal electroshock-induced convulsions) of epilepsy with possible mechanism of action. Indian Exp Bio 44:286

Ding Y, Wang S et al (2010) Fructose-1, 6-diphosphate inhibits seizure acquisition in fast hippocampal kindling. Neuroscilett 477:33–36

Feeney DM et al (1998) Hyposexuality produced by temporal lobe epilepsy in the cat. Epilepsia39:140–149
Gallego JM, Sancho et al (2010) Injection of embryonic median ganglionic eminence cells or fibroblasts within the amygdala in rats kindled from the piriform cortex. Seizure 19:461–466
Gupta YK et al (2006) Protective effect of vineatrol against kainic acid induced seizures, oxidative stress and on the expression of heat shock proteins in rats. Eur Neuropsychopharmacol 16:85–91
Han SR et al (2009) Differential expression of activating transcription factor-2 and c-Jun in the immature and adult rat hippocampus following lithium-pilocarpine induced status epilepticus. Yonsei Med J 50:200–205
Ilhan A, GUREL et al (2005) Antiepileptogenic and antioxidant effects of <i> Nigella sativa </i> oil against pentylenetetrazol-induced kindling in mice. Neuropharmacology 49:456–464
Kalynchuk L (2000) Long-term amygdala kindling in rats as a model for the study of interictal emotionality in temporal lobe epilepsy. Neurosci Biobehav R 24:691–704
Kamiński RM et al (2001) Amlodipine enhances the activity of antiepileptic drugs against pentylenetetrazole-induced seizures. Pharmacol Biochem Behav 68:661–668
Kupferberg H (2001) Animal models used in the screening of antiepileptic drugs. Epilepsia 42:7–12
Malhotra J, Gupta YK et al (1997) Effect of adenosine receptor modulation on pentylenetetrazole-induced seizures in rats. Brit J Pharmacol 120:282–288

Animal Models of Tardive Dyskinesia

Swati Datta, Sumit Jamwal, Kiran Kumar Akula
and Puneet Kumar Bansal

1 Introduction

Today we are in an era of utmost modernization; technology used in medical science says hats off to human mind. We have left no pebbles unturned in discovering human "BRAIN." Researchers have gone deep into lobes then whether its physiology of brain or vast anatomical features. Irrespective of all the laurels attained in it, there is no suitable treatment for some of the CNS disorders for which patients are paying its cost with their lives. One such example is of schizophrenia from which tardive dyskinesia (TD) occurs as a side effect. Throwing some light on the preclinical work done on TD, a review is presented to put together the toxic agents causing TD. Hoping it could prove fruitful in a process of attenuation or abolition of TD. Schizophrenia was conceived by Eugen Bleuler in 1950. It has its onset during

S. Datta · S. Jamwal · P.K. Bansal (✉)
Department of Pharmacology, ISF College of Pharmacy, Moga 142001, Punjab, India
e-mail: punnubansal79@gmail.com

S. Datta
e-mail: dutta.swati90@gmail.com

S. Jamwal
e-mail: jamwalisf@gmail.com

K.K. Akula
Department of Biology and Biochemistry, University of Houston,
4800 Calhoun Rd, Houston, TX 77004, USA
e-mail: kakula@central.uh.edu

P.K. Bansal
Department of Pharmaceutical Sciences and Technology, Maharaja Ranjit Singh Punjab
Technical University (State Govt. University), Bathinda, Punjab, India

P.K. Bansal and R. Deshmukh (eds.), *Animal Models of Neurological Disorders*, https://doi.org/10.1007/978-981-10-5981-0_6

puberty and lasts throughout life. Neuroleptics are universally prescribed psychotropic drugs for schizophrenia from which TD emerged as side effect.

Tardive is a French word which means "late onset." TD occurs late after the onset of neuroleptic treatment, also termed as medication-induced movement disorder. It is irreversible offender which persists even after the discontinuation of neuroleptics. TD or orofacial dyskinesia or bucco-lingual masticatory syndrome is a knotty hyperkinetic syndrome playing a role of a complex motor side effect of chronic treatment for prevalent schizophrenia worldwide. TD renders one to socially handicapped, due to irregular abnormal rhythmic movements, tongue protrusions, vacuous chewing movements, licking, smacking, puckering, grimacing, panting. It is characterized by choreiform, athetosis, vacuous chewing movements, tongue protrusions, facial jerking. TD is manifested by dopamine supersensitivity, GABAergic neuronal hypofunction, excitotoxicity, oxidative stress.

Following phenothiazines (chlorpromazine) in 1950s, enormous neuroleptics were produced for next 30 years with extrapyramidal side effects, with a root mechanism of blocking post-synaptic dopamine receptors. Tardive dyskinesia emerged as a major limitation of chronic neuroleptic treatment. TD came into forefront 5 years after the instigation of chlorpromazine. Recent clinical data states that TD has been reported with the involvement of speech.

Dopamine supersensitivity has been reported leading TD, particularly dopamine D2 receptor blockade. Dopamine is released in the basal ganglia circuit, but sufficient receptors are not available to act on it, and here receptors become supersensitive for even a bit of dopamine available leading to dopamine supersensitivity. This hypothesis was put forward by Klawans, but it has not gained much evidence. As dopamine sensitivity occurs after a few weeks of exposure to neuroleptics, but TD occurs after long-term exposure. All the more clinical data adds to it by enumeration of postmortem studies of TD patient showing significant increase in dopamine receptors, but only a portion of them exhibits dyskinesia. Secondly, glutamate excitotoxicity plays a part for TD by excessive stimulation of NMDA receptors through blockage of D_2 receptor on glutamatergic terminals in striatum, leads to persistent enhanced release of glutamate that kills striatal neurons.

Now what cannot be left behind is free radical generation, leading to "malicious oxidative stress" without which a pathogenesis of disease could be completed. Dopamine supersensitivity is correlated with enhanced dopamine metabolism which is natural to take place as sufficient receptors are not available to compensate dopamine, so it will be metabolized by the action of MAO-B leading to reactive oxygen species. GABAergic hypofunction in the striatal neurons is well supported through rodent and clinical data forming basis of TD. Loss of these neurons was confined to ventrolateral striatum (area concerned with innervations of oral musculature). Various assumptions have been laid down by the researchers, like increased smoking by individuals, genetic predisposition, and contrasting results are available about age factor, sex ratio for TD without the availability of sound evidence.

Much of the work has been done on the protective drugs that could be used against TD for which positive as well as conflicting results are available but an adequate one is not yet available. To explore the occurrence, prevalence, and numerous factors, we need animal models to understand the pathological, physiological, clinical manifestations, different pathways followed by different antipsychotics. Moreover, animal models preferably primate models persuade more closely to human condition in terms of therapeutic concern. The prevalence of TD in the society and extending horizons of diseased condition in which antipsychotics are prescribed makes it necessary to understand the pathological role of anatomical factors involved along with the new therapeutic approaches to target it, and the best possible way to achieve it before administrating it to human lives is the use of animal models which is ineluctable.

Word neuroleptic has its roots in Greek, according to which "neuro" corresponds to "neuron" and "leptic" corresponds to "seizure." Neuroleptics or antipsychotic or major tranquilizers are the drugs given to treat psychotic disorder schizophrenia (biological illness). It decreases the intensity of hallucinations and delusions. They have been used in westerly medicines since 1950s. Neuroleptics are targeted in the brain having basal ganglia their main site of action. Basal ganglion is the region involved for motor coordination with the involvement of various inhibitory and excitatory neurotransmitters. Any disturbance in it or any abnormality occurring in the complexity of network of nuclei and neuronal structures present in it which are still not fully elucidated leads to abnormal motor coordination associated with oral dyskinesias. Neuroleptics block membrane's viability for numerous chemical entities by interfering with their signaling. This leads to the generation of plotters and culprits of the pathological condition of patient. Extent of occurrence of extrapyramidal side effects due to antipsychotics is directly related to possession over D_2 receptors. The reason for this is that the antipsychotics have marked targeted action in striatum which is rich with D_2 receptor. It has been proved with proof specs of abnormalities in striatum with the brain imaging technique of TD patients. Neuroleptics also block the complex I of electron transport chain by contributing to the formation of free radicals and generating oxidative stress.

Putting attention on subcategories of neuroleptic discrimination between typical or classic and atypical should be known. Major difference lies on their ability to cause TD at therapeutic doses. In that respect, typical ones occupy the throne recognized with bitter tablets producing "medication-induced movement disorder." They are first to be discovered, first to be used, and first to be regarded as sinners. But nowadays preference is given to atypical for achieving therapeutic treatment along with the minimization of hyperkinetic syndrome.

Typical neuroleptics increase the level of prolactin, but there are no such findings with atypical. Typical neuroleptics cause TD with high affinity for D_2 receptor blocking profile and occur frequently at even low therapeutic doses, whereas atypical neuroleptics do share receptor affinity with D_1, D_2, and serotonergic receptors. Typical neuroleptics increase basal ganglia volume along with changes in cortical areas, whereas atypical neuroleptics produce changes in thalamus. There was a little relief with the atypical ones, but they emerged with a new hindrances

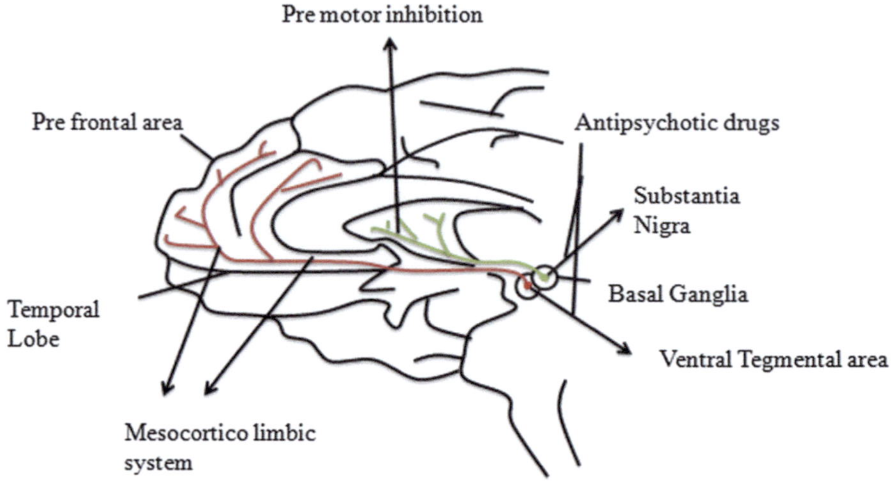

Fig. 1 Antipsychotics action on mesocortical, mesolimbic, and nigrostriatal pathway

which we call in scientific paradigms as adverse effects such as agranulocytosis, seizures, weight gain, sedation, whereas reports with typical ones are missing. After recognizing the drawbacks of typical, researchers felt need for the development of neuroleptics with a modification and then came atypical with an aim to abolish the occurrence of TD. But to our fate researchers were not fully successful in their mission; hence, there was some improvement that atypical ones are efficacious without TD at low doses. Reports of occurrence of TD with them at low doses are less, but at high doses they do share the same pharmacological profile (Fig. 1).

2 Classification of Animal Models of TD (Fig. 2)

2.1 Neuroleptic-Induced TD

Neuroleptics provide an adequate data and represent an excellent model to give an explanatory justification for the occurrence of TD using neuroleptics. Neuroleptics being lipophilic in nature have the ability to change the permeability of various receptors. Clinical data report states about the relevance of neuroleptic treatment in association with TD patients on showing striatal abnormalities during brain imaging. Animals treated with neuroleptics show frequent signs of orofacial dyskinesia (VCM's, TP, and FT). This leads to the alteration in the range of neuropeptide, neurotransmitter, antioxidants, and receptor in basal ganglia of animals treated with neuroleptics showing orofacial movements. Adequate methods have been adopted to quantify them. Different reports are available about the time taken by

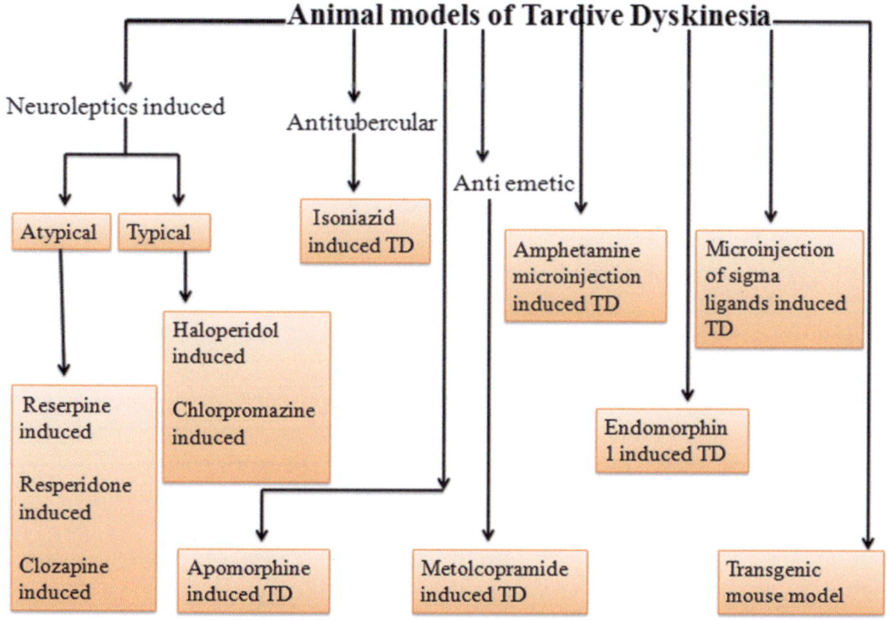

Fig. 2 Classification of Animal Models of TD

neuroleptics showing symptoms of TD from steady occurrence in several months to early onset. No rigid data is available due to variability in various factors age, gender, genetics, hereditary, medical history.

2.1.1 Typical Neuroleptics

Classical neuroleptics typically block D_2 receptors. This pays the path for dopamine hypothesis. Blocking of D_2 receptors is linked with cognition impairment corresponding to extrapyramidal side effects of neuroleptics following a cascade of dopamine supersensitivity, decreased striatal GABAergic and cholinergic neurons. Repetitive and enhanced stimulation of D_2 receptors located on glutamatergic neurons produces excitotoxic levels. They are associated with six times enhanced ROS generation following nigrostriatal dopamine system which has proved to be analogues with the generation of extrapyramidal side effects. They are associated with prominent changes in synaptic areas of basal ganglia. Electron microscopy of basal ganglia region showed altered osmiophilia and decrease in size of mitochondria, cytoplasm of neurons along with their processes.

Chlorpromazine-Induced TD

Principle: Emergence of chlorpromazine brought revolution in psychiatry with a drastic improvement in schizophrenic patients. Soon, its neurobiology is elucidated which showed its association with D_2 receptor blocking agents producing

symptoms of TD. Chlorpromazine is a classic typical dopaminergic D_2 receptor blocker leading to dopamine supersensitivity, enhanced dopamine metabolism. Clinically effective dose of chlorpromazine is 10–25 mg, p.o., 3 times a day. Blocking of D_2 receptor results in dopaminergic supersensitivity, glutamate excitotoxicity due to increased activation of NMDA receptor and reactive oxygen species. This action is responsible for antipsychotic activity along with other neuroleptics which showed extrapyramidal side effects.
Procedure: In preclinical studies, chlorpromazine 5 mg/kg, i.p., is given to animals for 21 days for the induction of TD (VCM's, facial jerking, and tongue protrusions) (Naidu et al. 2002).

Haloperidol-Induced TD

Principle: Haloperidol has been one of the extensively and widely used classical antipsychotics in psychiatry, obstetrics, and anesthesiology. It is one of the preferable drugs for the preclinical studies. Much of the work has been done on it for the elucidation of its complete pharmacological profile. Haloperidol is a dopamine antagonist. It disrupts the activity of neurotransmitter by blocking dopaminergic D_2 receptor which further leads to the cascade of dopamine supersensitivity, enhanced dopamine metabolism, free radical generation (Fig. 3).
Procedure: Haloperidol 1 mg/kg, i.p., is given to animals for 21 days for the induction of tardive dyskinesia. Haloperidol is not teratogenic as evidenced by animal data, but it could lead to embryo toxicity and also found in breast milk.

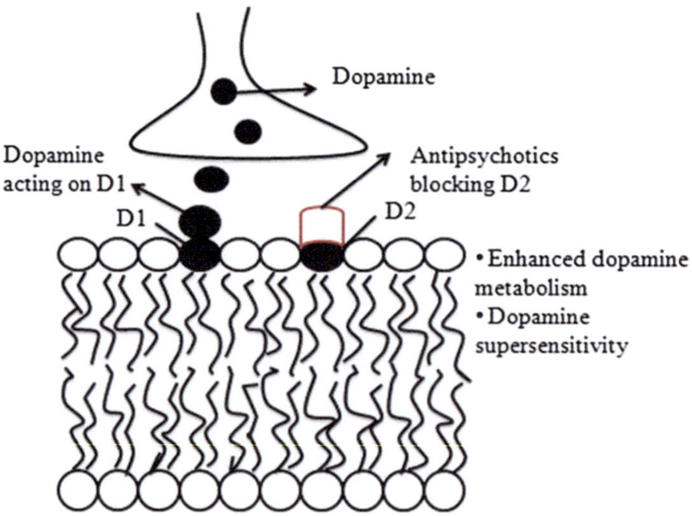

Fig. 3 Hypothetical diagrammatic representation of: (i) antipsychotic action on dopaminergic receptors. (ii) Amphetamine-induced TD

2.1.2 Atypical Antipsychotics

Atypical antipsychotics come under the category of "second-generation" antipsychotics. They are accompanied with minimal risk of causing structural disruption and triggering neurotransmitter alterations contributing to cognition impairment. They follow mesolimbic pathway and block 5-HT_2 and interpret dopaminergic transmission by loosely binding to D_2 receptor. According to the clinical data, available occurrence of TD is more prevalent in females as compared to males for which a possible reason could be the influence of estrogen in females for developing TD. Estrogen may have an additive effect with atypical antipsychotics on dopamine blockade. Declining estrogen in postmenopausal women also precipitate TD by reliving the blockade of dopaminergic receptors. Results of atypical are assessed by scores obtained through "a trend toward greater improvement in Quality of Life Scale and symptom scores." Atypical antipsychotics were developed with the aim to diminish or mimic the side effects of typical one (TD), but researchers were not fully successful in their objectives; hence, we can conclude it by saying through the data yet available that atypical antipsychotics could be preferred over typical antipsychotics.

Reserpine-Induced Tardive Dyskinesia

Principle: Regarding the critical mater of evaluating animal model of TD, reserpine, a monoamine depleting agent recognized for depleting dopamine, serotonin, and norepinephrine nonselectively proved to be better in fulfilling the needs. Reserpine-induced orofacial dyskinesia occurs late during the course of administration with prolonged persistence after the termination of administration. Though reserpine is not categorized as a neuroleptics, yet it is used as an antipsychotic having association with TD, showing similar features to TD. Reserpine-induced orofacial movements are reduced with D2 antagonist and are aggravated with dopamine agonist.
Procedure: Reserpine at a dose of 1 mg/kg, s.c., for 3 days is used for the induction of orofacial dyskinesia (Kulkarni et al. 2001).

Risperidone-Induced Tardive Dyskinesia

Switching over to risperidone one can have a sigh of relief to some extent as the data reports that the occurrence of TD is not associated with low doses of risperidone whereas high doses do not spare minds portion with the plethora of TD. Risperidone is a baby compound obtained by the derivatization of benzisoxazol. Belonging to the family of second-generation antipsychotics, possesses affinity and ability of blocking D_2 as well as 5HT2A receptors.
Procedure: Reserpine at a dose of 6 mg/kg, p.o., for 6 months is used for the induction of orofacial dyskinesia. One advantage as already mentioned above is procuring control model of VCM's at dose of 1.5 mg/kg, p.o., for 6 months.

Clozapine-Induced Tardive Dyskinesia

Principle: Clozapine being a atypical molecule is a drug of choice for schizophrenia. But as we have learnt, atypical antipsychotics do have the propensity

to cause TD. Clozapine has been defined as advantageous and novel by due to its affinity for serotonergic $5HT_{2A}$ and D_4 receptor antagonism along with weak D_2 receptor blockage. Serotonin receptor blockade has shown protective effect in TD which suggests that genetic variations in serotonin functions of schizophrenic patients may alter the risk for TD who are chronically exposed to dopamine receptor antagonist drugs.
Procedure: Dyskinetic movements occur in animals at the dose of 2 mg/kg, i.p., for 21 days (Naidu et al. 2002).

2.2 Antitubercular Drugs-Induced Tardive Dyskinesia

Part played by neuroleptic drugs for the occurrence of TD is not finding a solution yet, when another hindrance came for medical health professionals when an antitubercular drug isoniazid started showing the signs and symptoms of tardive dyskinesia.

2.2.1 Isoniazid-Induced Tardive Dyskinesia

Principle: Isoniazid has a potential of inhibiting GAD (a master component for the synthesis of GABA). Then, automatically it will be rewarded by the name of GABA depletor. As well-established dearth of GABA in the nigral regions contributes to the pathophysiology of TD. Decline in the activity of GAD is in association with the occurrence of characteristic symptoms of TD (Kulkarniet al. 2001).
Procedure: Doses at which dyskinetic symptoms occur are 1, 2, 5, 10 μmol/rat, i.c. v. Highest permissible effect occurs at 5 μmol. Peak of isoniazid is attained after 30 min of administration which remains for 60 min (Kulkarniet et al. 2001).

2.3 Amphetamine Microinjection-Induced TD

Principle: In general, amphetamine is a CNS stimulant. It increases heart rate and blood pressure. It treats attention deficit disorder and narcolepsy. But a research done by Rubovits et al. showed the characteristic ability of amphetamine to induce stereotype behavior in rodents by acting on striatal dopamine level (Fig. 3) (Rubovitset al. 1972).
Procedure: Dose of 20 μg/0.5 μl of amphetamine into confined subregion of striatum produces dyskinetic movements (Rubovits et al. 1972).

2.4 Endomorphin-1-Induced Tardive Dyskinesia

Principle: Endomorphins are endogenous opioid peptides exhibiting agonistic properties for mu receptors. Basal ganglia, a part of the brain, is involved in the locomotor activities and other functions too through various subregions and neurotransmitters involved which are having their influence through direct and indirect pathway. Globus pallidus (GP) is also one of them containing a heavy number of mu receptor. Endomorphin when administered in the striatum acts as a shooter of dopamine efflux with the involvement of substantia nigra, contributing to the hypothesis of dopamine supersensitivity (Fichna et al. 2007).

Procedure: Endomorphin-1 is administered at dose of 18 pmol bilaterally into GP therefore exhibiting the characteristic features of orofacial dyskinesia.

3 Conclusion

Oxidative stress is surely related to neuroleptics and their use in schizophrenic patient with both typical and atypical ones for which various antioxidant as adjuvants are tried in animal models, but satisfactorily they are not able to diminish the occurrence of TD rather they can only minimize it and that too in specific conditions. TD is not a disease, it is such a horrible side effect of a disease (an outcome of a bitter pill)." Much investigation is yet to be done on it; it is a tiny piece of the literature to throw some light on TD with an aim to develop efficient and possible therapy for TD.

Ethical Statement

All institutional guidelines, national guidelines, state and local laws and regulations with professional standards for the care and use of laboratory animals should be followed. Studies involving animals must state that the institutional animal ethical committee has approved the protocol. For authors using experimental animals, a statement should be made that the animals' care is in accordance with institutional guidelines and animals used have been treated humanely and with regard to the alleviation of suffering. Researchers should treat animals as sentient and must consider their proper care and use and the avoidance or minimization of discomfort, distress, or pain as imperatives. Animal experiments should be designed only after due consideration of animal health. It should be ensured that all researchers who are using animals have received instruction in research methods and in the care, maintenance, and handling of the species being used. All the surgical procedures should be performed under appropriate anesthesia and follow only those procedures which avoid infection and minimize pain during and after surgery.

References

Fichna J, Janecka Costentin J et al (2007) The endomorphin system and its evolving neurophysiological role. Pharmacol Rew 59:88–133

Kulkarni SK, Naidu PS (2001) Drugs of today tardive dyskinesia. An Update Pharmacology Division University Institute of Pharmaceutical Sciences, Punjab University Chandigarh, vol 37, pp 97–119

Naidu PS, Kulkarni SK (2002) Carvedilol attenuates neuroleptic induced orofacial dyskinesia: possible antioxidant mechanism. Br J Pharmacol 136:193–200. UIPS Panjab University Chandigarh

Rubovits R, Klawans HK (1972) Implications of amphetamine induced stereotyped behavior as a model for tardive dyskinesias. Arch Gen Psychiatry 27:502–507

Animal Models of Traumatic Brain Injury

Tavleen Kaur, Sumit Jamwal and Puneet Kumar Bansal

1 Introduction

Traumatic brain injury (TBI) is defined as the damage to the brain as a result of mechanical forces like crush, violent blow, or jolt to the head from blunt or penetrating object into the skull like a bullet or a sharp object. TBI may or may not alter the consciousness of person, but it is one of the leading factors responsible for impairment of cognitive ability or physical functioning. It is well depicted in clinical reports that around 10 million of deaths and hospitalizations annually are directly attributable to TBI. Head injuries are mainly of two types, i.e., primary head injury is an injury sustained by the brain at the time of impact, e.g., brain laceration, brain contusion whereas secondary head injury may be delayed neuronal damage or cell loss over a period of hours, days, weeks, or months. This injury involves biochemical and molecular changes in the distant tissues lead to secondary injuries (such as hypoxia, hypotension, seizures, or repeated TBI).

TBI is one of the leading causes of death and disability among the children's and young adults. Researchers discerned that TBI is a frequent injury occurs in the victims of sports, and during motor vehicle clashes. This also produces short- and

T. Kaur · S. Jamwal · P.K. Bansal (✉)
Department of Pharmacology, ISF College of Pharmacy, Moga 142001, Punjab, India
e-mail: punnubansal79@gmail.com

T. Kaur
e-mail: prtavleen@gmail.com

S. Jamwal
e-mail: jamwalisf@gmail.com

P.K. Bansal
Department of Pharmaceutical Sciences and Technology, Maharaja Ranjit Singh Punjab Technical University (State Govt. University), Bathinda, Punjab, India

P.K. Bansal and R. Deshmukh (eds.), *Animal Models of Neurological Disorders*, https://doi.org/10.1007/978-981-10-5981-0_7

long-term physical, cognitive, behavioral, and emotional impairments. Even though after the availability of extensive literature on the brain trauma, there is no reliable curative drug available for the treatment of patients suffering from TBI till date. TBI patients frequently suffer from long-term personality changes, cognitive deficits, and motor performances (post-concussion syndrome), making it need of hour for novel therapeutic interventions. Worldwide, motor vehicle accidents and military combats are the prevalence factors responsible for TBI in developing nations. TBI contributes a third (30.5%) of all the injury-related deaths in the USA. According to Centers for Disease Control and Prevention 2014 reports, 1.7 million Americans sustain a TBI/year out of which 27.5 thousand hospitalizations, 80 thousand disabilities, and 52 thousand deaths, creating a significant socioeconomic and emotional burden on the families and society.

The mechanical insult to the brain may lead to decreased cerebral blood flow (CBF) and thus produces ischemia-like condition. This ischemia is thought to be the first step in pathology of TBI (Fig. 1). Similarly, in anaerobic glycolysis, lactate accumulates and there is reduction in oxygen metabolism, increased membrane permeability leading to edema and decreased glucose uptake in the affected area of the brain. Altered metabolic function in the cells may initiate glutamate-induced excitotoxicity and neuronal cell death. Also, altered calcium homeostasis increases the reactive oxygen species, and generation of inflammatory mediators may also lead to the cell death.

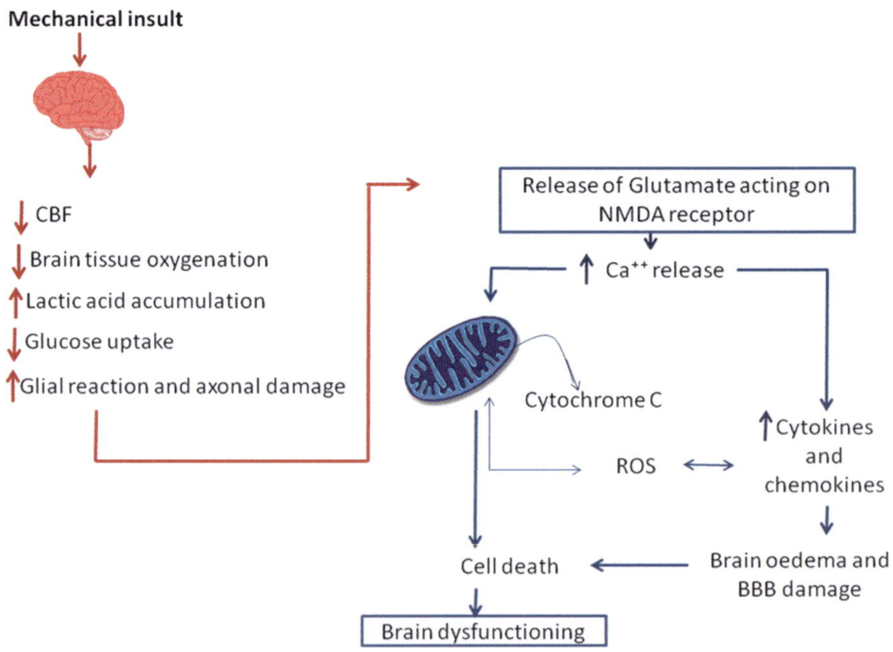

Fig. 1 Molecular mechanism involved in TBI

2 Types of Traumatic Brain Injury (TBI)

Depending upon the degree of severity, TBI can range from mild to severe with an extended period of unconsciousness. TBI occurs when an outside force impacts the head causing the brain to move, a direct blow to the head or a rapid acceleration and deceleration of the head mainly caused by motor vehicle accidents, sporting or leisure, workplace injuries, assaults, blasts, etc.

Traumatic brain injuries (TBIs) are classified into many types depending upon the tools used for analysis such as CT scans and magnetic resonance imaging (MRI). These techniques easily help to differentiate the multiple types of brain injury and variety of host factors and other confounders that might influence the yield of clinical trials (Fig. 2).

Based on severity of traumatic brain injury			
Injury	GCS	PTA	LOC
Mild TBI	13–15	<1 day	0–30 min
Moderate TBI	9–12	>1 to <7 days	>30 min to <24 h
Severe TBI	3–8	>7 days	>24 h
Vegetative state	<3	–	Coma
Persistent vegetative state	<3	–	Coma longer than one month
Brain death	–	–	–

where *GCS* Glasgow Coma Scale; *PTA* post-traumatic amnesia; *LOC* loss of consciousness

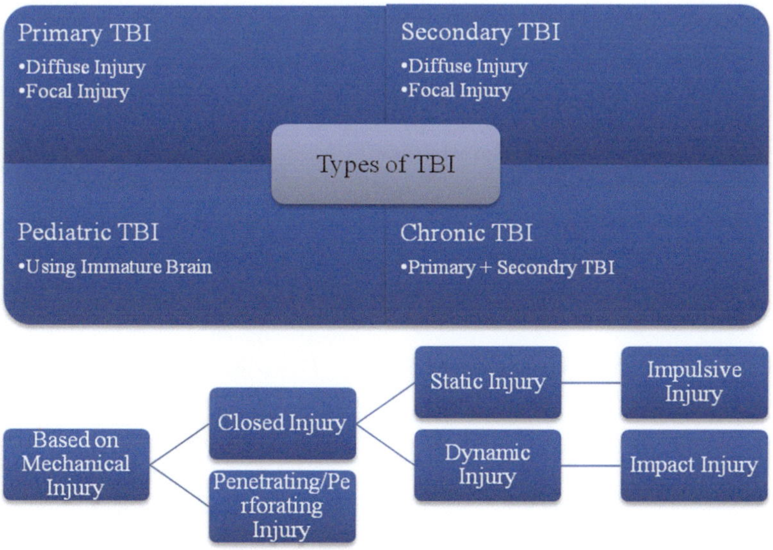

Fig. 2 Classification of TBI

The human head injury is very heterogeneous, and it is very hard to conduct controlled studies on human beings according to ethical guidelines. So the animal models are the powerful tools help to determine the typical patterns of dysfunction, perceive neurobiological mechanisms, and explore potential therapeutic interventions. Still at present, no FDA-approved pharmacological treatment available for the detrimental consequences (e.g., cognitive, emotional, behavioral impairment) is occurred due to TBI. However, the neuroprotective drugs, which were earlier identified to be effective in animal models of TBI, had not shown significant effectiveness in phase II or phase III of clinical trials. This failure in the preclinical studies highlights a compelling need to revisit the current status of animal models of TBI and therapeutic strategies.

However, many mild TBIs have functional effects that last for a considerable amount of time and the underlying factors remain to be established. In most models, the mechanical input is controlled and results in injury that is reproducible, quantifiable, and clinically relevant. No single animal model is reliable for providing the full spectrum of human TBI, so that is why different models with different pathologies were proposed to know the underline mechanism.

3 Classification of TBI Models

The classification of TBI based on GCS for trial inclusion and targeted therapies is important, but mechanistic classification has great utility in modeling injuries and developing preventive measures. The most commonly used TBI models are described below:

Injury/mechanism	Animal model
Acceleration/deceleration TBI	• Feeney's weight-drop model • Shohami's weight-drop model • Marmarou's weight-drop model • Maryland's model
Impact acceleration (direct brain deformation models)	• Fluid percussion injury model • Controlled cortical impact model
Penetrating TBI models (skull perforation by missile, impact energy)	• Penetrating ballistic brain injury model • Pellet accelerated penetrating trauma model
Blast TBI models (high velocity, impact brain injury)	• Open-field blast • Blast tubes for explosive • Shock tubes with compressed air or gas • Fragment penetration model
Miscellaneous	• Cell culture system • Cranium only blast injury apparatus

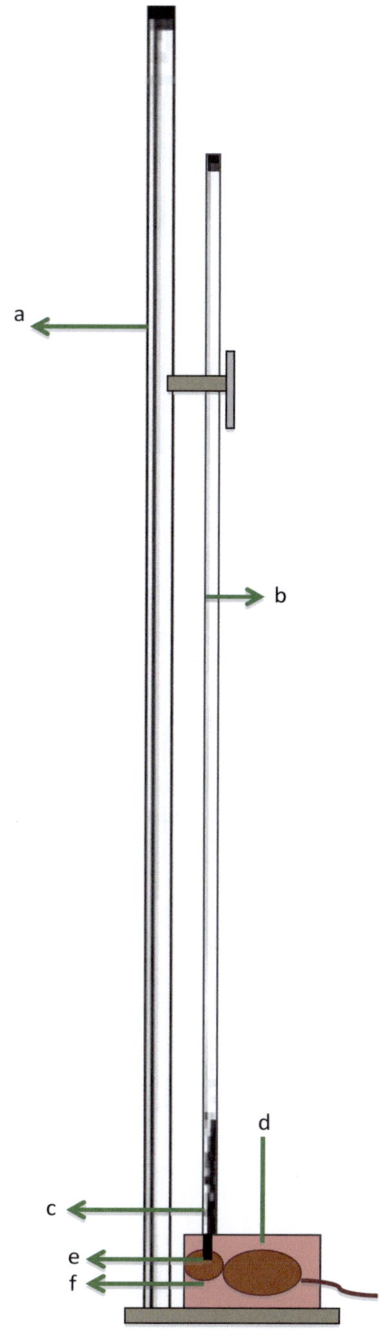

Fig. 3 Design of Weight-drop model assembly where: **a** Clamp stand; **b** Guide pipe 1–2 m; **c** Cylindrical slotted weights 450 gm; **d** Anesthetized rat lying on foam bed; **e** Metallic disk surgically adhered between bregma and lambda; **f** Foam bed of (10 cm thick)

3.1 Models for Acceleration/Deceleration TBI

Weight-drop TBI model: Weight-drop TBI model closely resembles with the real-life injuries and symptoms as observed in mild traumatic brain injury patients; therefore, weight-drop model is considered as original TBI model and is commonly used to deliver a traumatic brain injury to rodents like rats/mice. TBI patients often experience cognitive, behavioral, and emotional disturbances which are closely mimicked by this model (Fig. 3).

Procedure:

- In weight-drop model, firstly animal is anesthetized, and then, skull is exposed for the free-falling weight to generate a direct focal cortical compression.
- The apparatus consists of a guide pipe of length 1–2 m, either made up of Plexiglas or metal through which weights (composed of either metal or acrylic) are passed.
- For the induction of a mild injury, the weight should be less than 50 g, while for moderate, it should be 51–99 g weight; severe trauma corresponds to a 100–149 g weight, and for induction of ultra-severe injury, weights greater than 150 g should be used.
- Usually, a metal plate is fixed to the cranium or by craniotomy directly into the brain to reduce the risk of skull fracture.
- The animals are placed on non-flexible platforms for inducing focal brain injury in order to reduce loss of energy whereas flexible platforms like foam or platforms with elastic springs are used for inducing a diffuse brain injury so that impact is distributed over the skull.
- The induced brain injury such as hemorrhages, neuronal cell death, astrogliosis, diffuse axonal injury, or cytotoxic brain edema is purely dependent upon the severity of the impact. Weight-drop models of TBI vary with regard to the method of induction of injury and location of impact.

3.1.1 Feeney's Weight-Drop Model

It is consisted of a rat model in which craniotomy is induced prior to TBI resulting in a cortical contusion with hemorrhage and damage to the blood–brain barrier. This leads to activation of microglia and astrocytes with activation of the complement system and invasion of neutrophils and macrophages. Cortical spreading depression and delayed microcirculatory disturbances have also been reported in this model. The severity of injury determines the pattern of post-traumatic cell death (Feeney et al. 1981).

3.1.2 Shohami's Weight-Drop model

This rodent model is mostly used for closed head injury (CHI), consisting of a weight-drop impact delivered to one side of an unprotected skull by the head on the hard surface. The height and mass of the falling weights determine the severity of injury in this model. Therefore, heavier weights or increased falling height produces an ipsilateral cortical brain contusion and blood–brain barrier disruption followed by brain edema and finally cell death. Recently, neurobehavioral deficits, activation of microglia and astrocytes, neurodegeneration, and morphological changes detected by MRI have been demonstrated in this mouse CHI model indicating this model resembles the clinical condition in human CHI.

3.1.3 Marmarou's Weight-Drop Model

The impact acceleration model of diffuse traumatic brain injury (DTBI), commonly referred to as the "Marmarou" weight-drop model. This model depicts the clinical aspects of DTBI a type of TBI in humans, which is mainly caused by falls or motor vehicle accidents. The device consists of sectioned weights set that fall freely from a designated height through a Plexiglas tube.

Procedure:

- In this, surgical procedure is performed with anesthesia given to the rats and are subjected to a midline incision to expose their skull.
- A metallic disk is adhered surgically between lambda and bregma so as to prevent skull fracture.
- The rats are then placed on a foam bed and subjected to the impact by dropping the sectioned weight onto the metallic disk.
- The impact that is induced by a falling of a 450-g weight from a 2-m height causes a mortality rate of 44% with 12.5% incidence of skull fracture.
- The mortality in this model is primarily caused by respiratory depression; therefore, mechanical ventilation after the impact greatly reduces the mortality rate after severe injury (Foda and Marmarou 1994).

3.1.4 Maryland's Model

In this model, TBI can be produced by applying the impact force to the anterior part of the cranium, causing anterior–posterior plus sagittal rotational acceleration of the brain inside the intact cranium. The characteristics of animals are absence of mortality, absence of cortical contusions, skull fractures, prolonged apnea, but the chances of hemorrhages and diffuse axonal injury are more. Also, the neurobehavioral dysfunction is manifested as reduced spontaneous exploration persists for more than 1 week. However, there is a need of more studies for exploring the pathological causes (Xiong et al. 2013).

Advantages:

- This model is useful to investigate mild to severe diffuse brain injuries.
- This model provides ease to perform and is able to produce graded diffuse axonal injury similar to human TBI.
- Weight-drop model produces variable brain injury and is used to assess focal to diffuse brain injuries as seen clinically.
- This model does not produce any structural damage to the mouse's brain as confirmed by MRI.

- It has been widely used in preclinical settings to assess the effect of pharmacological intervention to treat diffuse brain injury.
- This model produces irreversible learning and memory deficits, escorted by a depression-like behavior in mice as evidenced even 90 days post-injury.

Disadvantages:

- Weight-drop models show high variability in producing brain injury.
- Unintended skull fractures and imprecision with regard to the impact site is commonly seen in this model.
- These models have been criticized in the past due to their reduced level of impact control and measurement.

3.2 Impact Acceleration Models (Direct Brain Deformation Models)

The direct brain deformation models (through craniotomy) and penetrating head injury (through skull perforation by a missile) are caused by the impact energy.

3.2.1 Fluid Percussion Injury Model

Principle: The model of closed head injury with fluid pressure is an important model of cerebral concussion. It was first developed by Lindgren and Rinder (1969) in order to produce an "experimental brain concussion." The model was originally developed in rats and has been modified for use in mice to create similar injuries related to focal injury models. Thompson et al. (2005) concluded that it is currently the most widely employed animal model of TBI which appropriately elicits the human TBI. As a result of fluid percussion, this model corresponds to mortality rate 20–25% in animals after the acute post-traumatic period (15 min). Generally, the common features are respiratory failure and pulmonary edema which replicates (Fluid Percussion Injury) FPI clinically like TBI without skull fracture.

Procedure:

- Firstly, anesthetize the animals and then apply fluid pulse to the intact dura mater through a craniotomy made centrally around the midline between bregma and lambda or laterally over the parietal bone between bregma and lambda in a stereotaxic frame for inducing diffuse brain injury.
- A reservoir cylinder filled with sterile solution of either saline or distilled water is attached to the cap cemented on the craniotomy of the animal's skull.
- Now with the help of pendulum, generate a rapid ($\sim$20 ms) fluid pulse which causes an insult, through the craniotomy onto the intact dura following the inner curvature of the skull and creates an elastic decompression of the brain.
- The mechanical forces disrupt cell membrane, blood vessels, and neuronal processes.

- The severity of the injury depends on the strength of the pressure pulse, which can be adjusted by selecting the angles of the pendulum. The brain injury caused by this model replicates human TBI without skull fracture mixed with diffuse injury characteristics (Thompson et al. 2005).

Traumatic pathology involves:

- Cortical contusion, hemorrhage, and brain edema (cytotoxic or vasogenic) typically either bilateral for central FP injury or ipsilateral for lateral fluid percussion injury.
- Changes in the blood pressure, elevated craniocerebral pressure, decreased cerebral perfusion pressure, reduced cerebral blood flow directly promote respiratory arrest, and increased cerebral vascular resistance have been shown to produce fluid percussion.
- Neuronal cell death, necrosis, and apoptosis are found to be major hallmarks which improve the reproducibility of fluid percussion model.

Advantages:

- FPI has been regarded as the best model to represent intracerebral hematoma.
- The model provides graded level of injury severity by adjusting the force of the fluid pressure pulse.
- The model increases the relevance as the recovery period after surgery returns the animals to a condition that more closely resembles the human condition.
- FPI significantly alters the cognitive dysfunction irrespective of the location of injury; hence, it can be a useful model for post-traumatic dementia.
- It is widely accepted in neurotrauma research for both mechanistic studies and for drug screening because it can reproduce intracranial hemorrhage, brain swelling, and progressive gray matter damage that are all pathophysiological hallmarks of human TBI.
- This model induces the injury which directly replicates the clinical contusion without skull fracture.

Disadvantages:

- Fluid percussion at high velocity levels produces an injury primarily associated with the brain stem which may also result in histopathological changes that are not typically characteristic of severe human TBI.
- Fluid percussion model is complicated in terms of clinical aspects because the pressure from injection of the fluid pulse is not directly associated with mechanical impact to the brain.
- This model includes the fluid pulse that disperses diffusely within the epidural space, making the tissue displacement difficult to compute, along with the lack of a cortical contusion.

- It increases the severity or morbidity mainly due to disproportional involvement of the brainstem and development of neurogenic pulmonary edema.
- The fluid flow characteristics (i.e., direction, displacement, and velocity) are dependent on brain geometry and the species so it is difficult to achieve the accurate biomechanical analyses after injury.
- High mortality rate due to apnea is evident.

3.2.2 Controlled Cortical Impact Injury Model (CCI)

Principle: This model was the first to be developed by Lighthall (1988) with the use of ferrets. The CCI model is widely used in TBI research because of its simplicity and high reproducibility. CCI is an invasive impact method that causes a measurable brain displacement using a solid percussion device applied through a cranial opening.

Procedure:

- It consisted of a rigid impactor to generate the mechanical energy to the intact dura with the head of the animal usually for delivering the good impact.
- The compressed air mechanically drove the actuated piston which is rigidly mounted on a crossbar in either an angled or vertical position to control the velocity and depth of cortical impact causing deformation in the parenchyma with an intact dura.
- The impactor is attached to the lower end of the metallic piston, and pressurized air acts as the source of the mechanical energy.
- A typical penetration depth for this device is 2.6–2.8 mm with a velocity of 4.0 m s^{-1} and a dwell time of 50–150 ms consistently produces an injury of moderate severity. However, it is useful for producing controlled focal or cortical deformations in the rodents that mimic histological, physiological, neurochemical, and functional aspects of human traumatic brain injury.

Advantages:

- The model is reproducible and induces varying severity of brain injury.
- This model provides high level of correlation between the degree of cortical distortion and histological damage.
- This model provides ease to control various mechanical factors such as time, velocity, and intensity of impact.
- This model minimizes the risk of a rebound injury.
- This model does not produce stem deformation and has very less mortality rate.
- This model replicates similar neurobehavioral and cognitive deficits as seen clinically.

Disadvantages:

- This model fails to produce symptoms of pure diffuse brain injury.

3.3 Penetrating TBI Models

Missile injuries, such as gunshot wounds, are a common cause of military TBI. The injury occurs when an object with high velocity penetrates the skin, skull, and meninges directly injured to the brain tissue. These injuries are of two types penetrating and perforating depending on how the missile traverses the head. When the object enters and dwells within the cranial cavity, it is called penetrating injury, but when the object passes through the cranial cavity and leaves an exit wound, it is called perforating injuries. The severity of the injury depends upon missile shape and mass and also the direction and velocity. At present, no high-speed penetrating rodent injury model has been available except a ballistic brain injury model using a balloon inflation technique. At current, there are only two rodent penetrating TBI models.

3.3.1 Penetrating Ballistic Brain Injury Model (PBBI)

The PBBI model in rats was developed by Williams et al. (2005), in which bullet-like round inflatable penetrating probe is used to produce a large temporary cavity in the brain. The model reflects the features both the permanent injury and large temporary injury. For the permanent injury, the tract is created by selecting the specific path of the bullet itself, but the large temporary cavity or injury is generated by energy dissipation from the bullet. The procedure utilizes the probe with balloon, which is to be inserted in the brain at the desired location and with rapid inflation commonly mimics the temporary cavity.

Advantage:

- The model has been characterized in a large number of studies and can presumably generate important knowledge about cavity formation during fragment penetration.

Disadvantages:

- The major obstacle with this model is high mortality rate in high-speed penetrating TBI.

3.3.2 Pellet Accelerated Penetrating Trauma Model

Principle: This is a recently developed penetrating TBI model (Plantman et al. 2012), in which the object penetrates in the brain parenchyma at approximately 90 m/s after being hit by a pellet of an anesthetized rat. The pellet which hits the probe is accelerated by a modified air rifle. The speed and the depth of penetration are adjustable and highly reproducible (Plantman et al. 2012). Neurodegeneration occurs in the injured cortex 24 h after injury. After a period of 14 days, the injury develops into a large cavity. It also produces extracellular perivascular edema.

Advantage:

- The biomechanics of this TBI model enables survival of the animals following a high-speed penetration of brain tissue.
- The model produces motor disturbances with severe memory impairment.

3.4 Models for Blast (High Velocity, Impact Brain Injury)

Traumatic brain injuries (TBI) related to high velocity is more common among the military personnel worldwide. However, the mechanism how the blast waves affect the brain is not well understood. Experimental animal models provide significant correlation by mimicking mild TBI due to blast and also provide an insight pathological mechanism of the injury. Single or multiple blast exposures have been commonly seen in association with chronic neurological and psychiatric sequelae including persistent cognitive impairment, post-traumatic stress disorder (PTSD), and depression. Blast injuries occur through multiple mechanisms that may be related to the effects of the primary blast wave, or by the individual being knocked down or thrown into solid objects.

3.4.1 Open-Field Exposure

This experiment determined the thresholds for mortality and injuries such as bleeding in air-filled organs such as the lungs and intestines. The potential effects on the central nervous system are, however, difficult to assess.

Advantages:

- Open-field experiments may allow for more realistic experiments with large animals that are more similar in size to humans.

Disadvantages:

- These types of experiments require large amounts of explosives and difficult to perform at every experimental laboratory.

3.4.2 Blast Tubes for Explosives

Earlier these large size blast tubes were created to study the effects of wave forms relevant to nuclear detonations. Later on, these blast tubes were modified for their utilization in rodents in around 1990s. Now presently while experimentation, the anesthetized animals are mounted in the blast tube at a distance of 1 m from the charge. Five grams of plastic explosives PETN (pentaerythritoltetranitrate) results in a peak pressure exceeding 10 bars during detonation. The animals are mounted in metallic nets or fixed to a body protection in order to limit acceleration movements. The secondary and tertiary blast effects are very difficult to perform in this model. However, smoke and gas emission contribute with quaternary blast effects.

Advantages:

- The equipment can be used in laboratory.
- The quality and quantity of blast waves can be controlled easily.

Disadvantages:

- Short duration and very simple form of the blast wave.

3.4.3 Shock Tubes with Compressed Air or Gas

Procedure:

- In this procedure, the animals are firstly anesthetized and then exposed to high-blast waves with peak reflected overpressures for inducing blast trauma.
- Helium under high pressure is utilized by placing the rats were at certain distances, i.e., 39 and 17 cm, from the shock tube opening for the low- and high-blast shockwave exposure conditions, respectively.
- Pressure sensors depict the typical reflected pressures (either 100 or 450 kPa referred to the low-blast and high-blast conditions, respectively).
- However, the exhaust gasses, or blast wind, can lead to considerable head acceleration; animals were placed 40° and 20° lateral to the shock tube axis to limit blast wind exposure for the low- and high-blast conditions, respectively (Budde et al. 2013).

Advantage and Disadvantage:

- The duration of the pulse is usually longer, and the peak pressure is much lower than in the Clemedson tube. Also, quaternary blast effects as well as other disadvantages of explosives are absent. However, this advantage can also be regarded as a disadvantage.

3.4.4 Model for Fragment Penetration

This model can be used to mimic the more severe blast TBI, where shrapnel fragments penetrate the skull and brain tissue. The velocity of such fragments is assumed to be less than 300 m/s, depending on the presence and effectiveness of helmets.

Procedure:

- Firstly, anesthetize the rats and then place in a stereotaxic frame for performing craniotomy. A lead bullet is accelerated by air pressure in a specially designed rifle and impacts a secondary projectile.
- Now, allow the pin of the secondary projectile, guide by a narrow tube penetrates the surface of the brain with a speed of 100 m/s.
- The base of the projectile is surrounded by a compressible ring that helps to control the penetration (depth) into the brain, which is usually set to 5 mm.
- The speed and shape of the penetrating secondary projectile can be changed to obtain a variation of physical properties.

Advantages:

- This penetrating TBI results in a central cavity corresponding to the actual penetration. A zone of reactive tissue, which contains a mixture of dying cells, reactive cells, and invading inflammatory cells, surrounds this cavity.
- Electron microscopy has demonstrated mainly extracellular perivascular edema.

Disadvantages:

- This method gains less success due to high mortality rate in animals.

Ethical Statement

All institutional guidelines, national guidelines, state and local laws and regulations with professional standards for the care and use of laboratory animals should be followed. Studies involving animals must state that the institutional animal ethical committee has approved the protocol. For authors using experimental animals, a statement should be made that the animals' care is in accordance with institutional guidelines and animals used have been treated humanely and with regard to the alleviation of suffering. Researchers should treat animals as sentient and must consider their proper care and use and the avoidance or minimization of discomfort, distress, or pain as imperatives. Animal experiments should be designed only after due consideration of animal health. It should be ensured that all researchers who are using animals have received instruction in research methods and in the care, maintenance, and handling of the species being used. All the surgical procedures should be performed under appropriate anesthesia and followed only those procedures which avoid infection and minimize pain during and after surgery.

References

Budde, MD et al (2013) Primary blast traumatic brain injury in the rat: relating diffusion tensor imaging and behavior. Front Neuroendocrinol 4

Feeney DM, Boyeson MG et al (1981) Responses to cortical injury: I. Methodology and local effects of contusions in the rat. Brain Res 211:67–77

Foda MA, Marmarou A (1994) A new model of diffuse brain injury in rats. Part II: morphological characterization. J Neurosurg 80:301–313

Lighthall JW (1988) Controlled cortical impact: a new experimental brain injury model. J Neurotrauma 5(1):1–15

Lindgren S, Rinder L (1969) Experimental studies in head injury II. Pressure propagation inpercussion concussion. Biophysik 3:174–180

Plantman S, Ng KC et al (2012) Characterization of a novel rat model of penetrating traumatic brain injury. J Neurotrauma 29:1219–1232. doi:10.1089/neu.2011.2182

Thompson HJ, Lifshitz J et al (2005) Lateral fluid percussion brain injury: a 15-year review and evaluation. J Neurotrauma 22:42–75

Williams AJ, Hartings JA et al (2005) Penetrating ballistic-like brain injury in the rat: differential time courses of hemorrhage, cell death, inflammation and remote degeneration. J Neurotrauma 23(12):1828–1846

Xiong Y, Asim M, Chopp M (2013) Animal models of traumatic brain injury. Nature Rev Neurosci 14(2):128–142

Animal Models of Migraine

Sumit Jamwal, Shamsher Singh and Puneet Kumar Bansal

1 Introduction

Migraine is defined as a multifactorial and episodic disorder which features unilateral, hemicranial, throbbing headache often accompanied by nausea and vomiting and aggravated by movement, sound, and light. Migraine has two major clinical subtypes: migraine without and with aura. One-third of patients with migraine experience aura, that is, transient focal neurological symptoms like sensory or motor deficits. Migraine is a chronic and disabling disorder of the brain that affects up to 15% of the population worldwide. It has a devastating consequence on the quality of a patient life and is estimated to cost \$19.6 billion and €27 billion in the USA and Europe per year, respectively.

The clinical depiction of a migraine attack suggests that a series of processes including vascular, neuronal, and biochemical elements occur at different sites. A significant contribution has been made in understanding the pathophysiology of this condition, but the exact pathophysiology of migraine is still not fully understood. It is thought to involve activation of the trigeminal afferents, which densely innervate dural structures and send projections to second-order neurons of

S. Jamwal · S. Singh · P.K. Bansal (✉)
Department of Pharmacology, ISF College of Pharmacy, Moga 142001, Punjab, India
e-mail: punnubansal79@gmail.com

S. Jamwal
e-mail: jamwalisf@gmail.com

S. Singh
e-mail: shamshersinghbajwa@gmail.com

P.K. Bansal
Department of Pharmaceutical Sciences and Technology, Maharaja Ranjit Singh Punjab Technical University (State Govt. University), Bathinda, Punjab, India

P.K. Bansal and R. Deshmukh (eds.), *Animal Models of Neurological Disorders*, https://doi.org/10.1007/978-981-10-5981-0_8

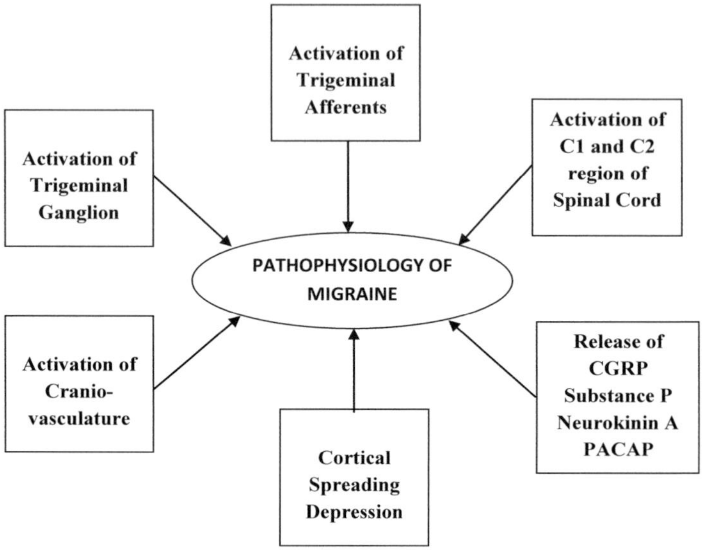

Fig. 1 Pathophysiology of migraine

trigeminal nucleus caudalis and C1–C2 region of the spinal cord (trigeminocervical complex). The trigeminovascular system includes the pseudounipolar trigeminal ganglion that projects into the trigeminal nucleus caudalis in the medullary spinal cord, and has a peripheral projection from the ophthalmic division of the trigeminal ganglion, which innervates the cranial blood vessels and pain-sensitive dura mater. Numerous potent vasodilator peptides including calcitonin gene-related peptide (CGRP), substance P, neurokinin A, and pituitary adenylate cyclase-activating peptide (PACAP) are present in the trigeminal ganglion and project nerve fibers to cranial vessels. Several lines of evidence suggest that cortical spreading depression (CSD) causes aura symptoms and impacts both short-term and long-term neurovascular function. It has been suggested that CSD is the first event identified upstream to trigeminovascular activation causing pain and characteristic blood flow changes. It has also been documented that migraine pain in humans arises due to the stimulation of blood vessels in dura mater. These points raise the concept of involvement of cerebrovascular system of dura mater in pathophysiology of migraine (Fig. 1).

1.1 Need of Animal Model

Further, it led to explosion in preclinical research of the nerve fibers that innervate the dural vasculature, and the likely involvement of the trigeminovascular system and the subsequent development of animal models of migraine. Over the past 2–3 decades, several animal models of migraine have been developed which helped out to understand the pathophysiology of migraine, identify novel drug targets and drug treatments. These animal models have provided valuable knowledge and a skeleton to understand about the effect of hormones, chemicals, and various environmental factors like sound, light on migraine pathophysiology. Although most of the researchers have suggested that these animal models have lot of shortcomings, still some promising new drugs have been developed by utilization of these preclinical models. Animal models are extremely helpful in the understanding of brain disorders and in developing new therapeutic approaches. Migraine has recently become one of the major interests to neuroscientists. Various models for studying migraine headache mechanisms have been developed and exploited efficiently, leading to better understanding of the pathophysiology of the disorder and mechanism of action of anti-migraine drugs. These model systems have primarily focused on the pain-producing cranial structures, cerebrovascular system, and dura mater, in order to provide reproducible physiological measures that can be further subjected to pharmacological investigation. A wide range of techniques (both in vivo and in vitro approaches) are now in use, which include manipulation of the mouse genome to produce animal models having similarity with human disease.

2 Classification of Animal Models of Migraine (Fig. 2)

2.1 Acetic Acid-Induced Abdominal Constriction Test

Principle:

Acetic acid-induced writhing method or abdominal constriction is the best method used for the evaluation of analgesic activity. Writhing may be defined as a stretch or tension to one side, expansion of hind legs, and contraction of the abdomen in a

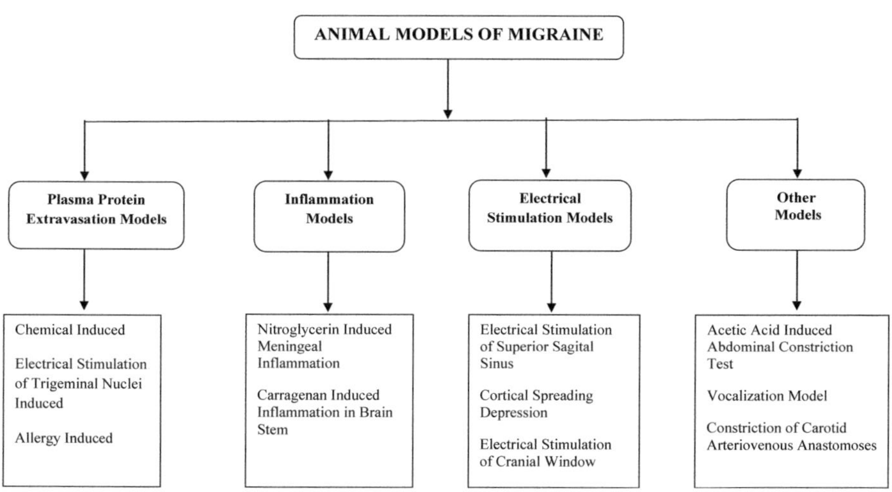

Fig. 2 Classification of animal models of migraine

way that the abdomen touches the surface and turning of trunk (twist). The anti-hyperalgesic activity of clinically used anti-migraine drugs like sumatriptan and ergotamine was first confirmed by using this model, and therefore, this test is also used as in vivo migraine model.

Procedure:

- Swiss albino mice (15–35 g) are used.
- 0.3% v/v solution of acetic acid is prepared in distilled water, and writhing is induced by the i.p. injection of 0.3% solution of acetic acid in volume of 0.1 ml/10 g body weight and is evaluated by counting the number of abdominal constrictions.
- In this model test and standard drug is administered before the acetic acid injection and number of abdominal constrictions are counted 5 min after acetic acid injection for a period of 10 min (Galeotti et al. 2002).

2.2 Nitroglycerin (GTN)-Induced Meningeal Inflammation

Principle:

Glyceryl trinitrate (GTN) infusion is the most widely used and trustworthy method to provoke migraine-like headaches in humans as well as in animals. The infusion of the nitric oxide donor nitroglycerin results in upregulation of proinflammatory mediators, macrophage activation, edema formation, and mast cell degranulation by its direct action on meningeal tissues and produces plasma protein extravasations (PPE). GTN is metabolized to nitric oxide (NO) by a combination of glutathione-S-transferase, cytochrome P450, and thiol reactions. Type II nitric oxide synthase (NOS) is expressed by macrophages and can be activated by GTN. After some hours of GTN infusion, iNOS generates NO at 100–1000 fold than its constitutive counterparts, and promotes inflammation and development of migraine. GTN activates second-order nociceptors, NF-κB, and increases the expression of c-fos, neuronal nitric oxide synthase (nNOS), and calmodulin-dependent protein kinase II in the trigeminal nucleus caudalis. Increased production of NO has been reported to upregulate COX, TNF-α, and matrix metalloprotease-9 (MMP-9).

Procedure:

- Sprague-Dawley rats (200–230 g) are used.
- Rats are infused with GTN (4 μg/kg/min, for 20 min, i.v.) a dose just 8–10 times higher than in humans (0.5 μg/kg/min, for 20 min, i.v.).
- Level of inflammatory mediators like TNF-α and mRNA and protein expression for c-fos is analyzed in the trigeminal vascular system at various time points using ELISA, RT-PCR, and immunohistochemistry.
- In other way, Sprague-Dawley rats are subcutaneously injected GTN (10 mg/kg) in the back of the neck and after 4 h rats are sacrificed for double

immunofluorescent labeling and Western blot analysis for the evaluation of nuclear NF-κB protein expression. This time point is chosen because previous studies have revealed the strongest nuclear NF-κB protein expression at this time (Bhandare et al. 2011).

Advantages:

1. It is the most widely used and reliable animal model for inducing migraine-like attacks in animals.
2. One of the most important advantages of the GTN-triggered headache model is the temporal control of events following GTN administration.

2.3 Vocalization Model in Rats

Principle:

Bradykinin (BK) is well implicated in microvasculature dilatation and in the generation of pain. It potently dilates cerebral arterial vessels and contributes to the vasodilatation, edema, and pain during migraine. BK has been reported as a possible pathogenetic factor of migraine. Taking this into consideration, the injection of a few micrograms of BK into a common carotid artery or into the cisterna magna of rabbits results in intense vocalization. Vocalization was also observed following the intra-carotid injection of BK into rabbits under general anesthesia (vocalization has long been accepted as a signal of pain in animals) (Martino et al. 2008).

Procedure:

- Wistar rats (200–250 g) are used, and BK is injected into the arterial catheter at the dose of 10 μg/kg in the volume of 10 μl/kg.
- Vocalization produced is recorded before, during, and for 5 min after BK injection with the help of microphone positioned at the snout of rat and is connected to the polygraph for recording (Ottani et al. 2004).

2.4 Plasma Protein Extravasations (PPE) Models

Principle:

Inflammation of the dura mater has been well implicated in the etiology of migraine; therefore, the blockage of PPE in the dura mater produced by trigeminal ganglion stimulation might play an important in anti-migraine activity of drug. ^{125}I-labeled bovine serum albumin or fluorescent isothiocyanate-bovine serum albumin (fluorescein) is used as marker for determining plasma protein extravasation. Plasma extravasation and vasodilation are associated with inflammatory response developed by neurogenic or non-neurogenic mechanisms. Neurogenic inflammation accompanies electrical, mechanical, or chemical stimulation of sensory nerves fibers. Capsaicin, the active ingredient of hot peppers, activates sensory axons and causes the release of vasodilators and permeability-promoting transmitters from perivascular afferent axons and subsequent development of neurogenic inflammation. Non-neurogenic plasma extravasation can be induced by the i.v. administration of 5-HT, histamine, and bradykinin (BK), which directly alter blood vessel permeability.

2.4.1 Chemical-Induced PPE

Procedure:

- Anesthetized rats are injected with any of the following chemicals (capsaicin, neuropeptides, 5-HT, bradykinin (BK), histamine, or PGE2) at dose of 1 ml/kg into the femoral vein for the development of PPE.
- 15 min after injecting tracer and 10 min after drug injection, the thorax is perfused with saline via the left ventricle.
- The brain is removed and the dura is dissected and incubated, and the amount of marker in the dura is determined by any specific methods which give amount of PPE.

2.4.2 Electrical Stimulation-Induced PPE

Procedure:

- Anesthetized animals are placed in a stereotaxic apparatus, and holes of 1–2 mm diameter are drilled on each side using following coordinates, i.e., intersection of the coronal and sagittal sutures: −1.2, 2.5, and −3.2, 2.5.
- Paired non-concentric bipolar electrodes are inserted to depth of 9.8–10.2 mm below the surface of the calvarium, ensuring bilateral placement in the trigeminal ganglia. The left or right side is arbitrarily selected for stimulation.
- Paired rectangular pulses having opposite polarity are used to stimulate test side at the two different sites for 5 min with 5 pulses/set of 5 ms duration, at an intensity of 3.0 mA.
- The potential difference and current flowing across the electrode (20–30 V) are continuously monitored using an oscilloscope. Correct electrode placement is indicated by ipsilateral contraction of the muscles of mastication during stimulation.
- The brain is removed, and the dura is dissected. The dura is further incubated, and the amount of marker in the dura is determined by any specific methods which give amount of PPE.

2.4.3 Allergy-Induced PPE

Procedure:

- Guinea pigs are used and sensitized by a single i.p. injection of ovalbumin at dose of 1 μg and 100 mg $Al(OH)_3$, in 0.5 ml saline.
- After 4–6 weeks, animals are challenged with i.v. injection of 20 μg/kg ovalbumin.
- 10 min after ovalbumin challenge, the animals are perfused with saline via left ventricle.
- The brain is then removed, and dura is dissected. The dura is further incubated, and the amount of marker in the dura is determined by any specific methods which give amount of PPE.

2.5 Carrageenan-Induced Inflammation in Brain Stem

Principle:

Iatrogenic chemical stimulation of meningeal tissues is used to produce head pain. Carrageenan is a sulfated polysaccharide which promotes inflammation via activation of proinflammatory cells. It is widely used as algesic agent in numerous experimental studies. NO can induce headache in migraine patients and often triggers a delayed migraine. The initial component of headache results involves direct action of the NO-cyclic guanosine monophosphate pathway that causes vasodilatation and vascular smooth muscle relaxation, while the delayed component of headache involves trigeminovascular activation. It has been demonstrated that the increased nNOS activity in the trigeminal system causes CGRP release and dural vessel dilation. Also, it has been shown that intracisternal injection of carrageenan is associated with a significant increase in the NOS enzymes in the brain stem of rats.

Procedure:

- Wistar rats (200–250 g) are anesthetized and injected with single intracisternal injection of carrageenan into cisterna magna.
- The brain tissues are then removed and analyzed for endothelial NOS (eNOS), neuronal NOS (nNOS), and inducible NOS (iNOS) by immunohistochemistry (Bergerot et al. 2006).

2.6 Electrical Stimulation of Superior Sagittal Sinus

Principle:

The induction of trigeminovascular nociception with the help of electrical stimulation of the dura mater surrounding the superior sagittal sinus is widely accepted model for the examination of pathophysiology of vascular headache such as migraine. The neurons of the trigeminal ganglion innervate the dural blood vessels in the periphery and trigeminal nucleus caudalis centrally (Dong et al. 2011). Pituitary adenylate cyclase-activating peptide (PACAP) is mainly present in human trigeminocervical complex and can trigger migraine. PACAP is a member of the vasoactive intestinal polypeptide/secretin/glucagon neuropeptide superfamily and acts on its specific receptor PAC1, and on two other receptors non-specifically, VPAC1 and VPAC2. PACAP may affect the paraventricular nucleus of the hypothalamus, and the hypothalamic region is well implicated in pathophysiology of migraine.

Procedure:

- Cats are anesthetized with an intraperitoneal injection of a-chloralose (70 mg/kg), with additional doses (20 mg/kg i.v.) during the experiment.
- Continuous measurement of blood pressure and heart rate is done by inserting catheters into the femoral artery and vein. A third catheter is inserted into jugular vein to permit sampling of blood.
- Cats are ventilated with 30% oxygen and are immobilized by i.v. administration of gallamine triethiodide (20 mg).
- Testing for sympathetic responses to noxious stimulation is conducted at regular intervals to assess the depth of anesthesia.
- The superior sagittal sinus is then stimulated with supramaximal square-wave stimulus-isolated shocks (100 V, 1 s duration, 10 Hz).
- Blood (5 mL) is to be taken before, and after, 7–8 min of stimulation and the volume of blood is replaced by an equivalent amount of plasma expander.

2.7 CSD Model of Migraine Aura

Principle:

Cortical spreading depression (CSD), which is known as shift in cortical steady potential, results in an increase in extracellular ions and neurotransmitter like glutamate, and sustained increase in cortical blood flow followed by transient decrease in blood flow, in occipital lobe and is responsible for generation of aura before the migraine attack. These ions and neurotransmitters cause intense vasodilatation in the cortex, pial vessel, and dura mater, and activate the trigeminal afferents and transmit impulse to the trigeminal ganglia and trigeminal nucleus caudalis (TNC) which is involved in the processing of pain. CSD is an intense depolarization of neuronal and glial membranes accompanied by a massive disruption of ionic gradients and loss of membrane resistance (Charles et al. 2013).

Procedure:

- Male Sprague-Dawley rats are anesthetized with urethane (1.8 g/kg; i.p.) and placed on a stereotaxic frame. Core temperature is maintained at 37 °C using a heating blanket.

- The dura over the left side is exposed between the bregma and 2 mm caudal to lambda. The exposed dura is kept moist throughout the experiment, with the help of modified synthetic interstitial fluid having pH 7.2. The fluid is composed of 135 mM NaCl, 5 mM KCl, 1 mM $MgCl_2$, 5 mM $CaCl_2$, 10 mM glucose, and 10 mM HEPES.
- Single waves of CSD are induced using mechanical, electrical, and chemical stimulation of the visual cortex, approx 6 mm away from the dural receptive field of the neuron under study.
- Changes in steady potential on the surface of the cortex are recorded using a glass micropipette (containing 50 mM NaCl), between the dural receptive field and the site of cortical stimulation. At a propagation rate of 3–5 mm/min, a wave of CSD enters the neuronal receptive field within 1–2 min of cortical stimulation.
- Mechanical stimulation (pin prick) is given by inserting a glass micropipette (diameter 25 μm) 1 mm into the visual cortex for 10 s.
- Electrical stimulation (cathodal pulses) is given by using a concentric bipolar electrode after every 90 min until a wave of CSD is recorded (current—0.5 or 1.0 mA; duration—1 or 3 ms; frequency—between 0.2 and 50 Hz.

2.8 Constriction of Carotid Arteriovenous Anastomoses in Anesthetised Animals

Principle:

It has been suggested that the arteriovenous oxygen difference during migraine is reduced. This finding, when corelated with clinical symptoms such as facial paleness, reduction in facial temperature, increased pulsations in temporal artery and inflammation of the frontal vein on the side of the headache lead to the implication that arteriovenous anastomoses are involved in the pathogenesis of migraine (De Vries et al. 1998).

Procedure:

- In conscious pigs, constriction of arteriovenous anastomoses is done under strong pressure of the sympathetic neuronal tone, which shunts 3% part of the total carotid blood flow.
- In contrast, under pentobarbital anesthesia, 80% of the total carotid blood flow in the pig is limited via arteriovenous anastomoses into the jugular vein.
- Radioactive microspheres are used to measure the carotid blood flow and the effects of anti-migraine drugs on this parameter.

2.9 Electrical Stimulation of Cranial Window

Principle:

One hypothesis is that nerve fibers within the trigeminal nucleus get activated during migraine headache resulting in the release of vasoactive peptides and the consequent dilation of meningeal blood vessels. Electrical stimulation of the cranial window results in activation of the trigeminal nerves via the release of CGRP from presynaptic trigeminal nerve endings ultimately leading to the dural and pial blood vessel dilation. The inhibition of dural vasodilation proved to be successful in predicting anti-migraine efficacy of triptans, 5-HT1B/1D receptor agonists, dihydroergotamine, and CGRP receptor antagonist.

Procedure:

- Male Dunkin Hartley guinea pigs (300 $\pm$ 45 g) are anesthetized throughout experiments with pentobarbitone sodium (initially 50 mg/kg i.p., then 18 mg/kg, i.v., constant infusion).
- The trachea, left carotid artery, and jugular vein are cannulated for artificial ventilation, measurement of mean arterial blood pressure (MABP), and intravenous injection of anesthetic and drugs, respectively.
- Guinea pigs are placed in a stereotaxic frame, the skull exposed and the right parietal bone is drilled until the dural blood vessels becomes clearly visible through the intact skull.
- The dural blood vessel diameter is continuously measured by a video dimension analyzer.
- In preliminary experiments, it has been observed that the dural blood vessels are observed to be maximally dilated, so that electrical stimulation of the cranial window will produce very less increase in diameter. It is therefore necessary to preconstrict the dural vessels with i.v. administration of endothelin-1 (3 mg/kg).
- After single administration of endothelin-1 (3 mg/kg, i.v.) dural vasodilation is induced approximately 3 min later by i.v. CGRP (1 mg/kg) or electrical stimulation of the cranial window (250 $\pm$ 300 mA, 5-Hz, 1 ms to 10 s) and is expressed as percentage increase in dural blood vessel diameter + SEM, from baseline (Williamson et al. 2001).

3 Conclusion

Migraine is a mysterious disorder characterized by chronic headache and is often associated with nausea, vomiting, sensitivity to light and sound. Different drugs are available to treat migraine attack, but need for novel effective therapy is insistent. Different animal models described in this chapter can be used to explore the pathophysiology of migraine and to test novel drugs. Every model described reflects some particular pathological features of the migraine like vasodilation, plasma protein extravasation, and cortical spreading depression. A specific model can be picked to induce the disease on the basis of expected mechanism of action of new drug.

Ethical Statement:

All institutional guidelines, national guidelines, state and local laws, and regulations with professional standards for the care and use of laboratory animals should be followed. Studies involving animals must state that the institutional animal ethical committee has approved the protocol. For authors using experimental animals, a statement should be made that the animals' care is in accordance with institutional guidelines and animals used have been treated humanely and with regard for the alleviation of suffering. Researchers should treat animals as sentient and must consider their proper care and use and the avoidance or minimization of discomfort, distress, or pain as imperatives. Animal experiments should be designed only after due consideration of animal health. It should be ensured that all researchers who are using animals have received instruction in research methods and in the care, maintenance, and handling of the species being used. All the surgical procedures should be performed under appropriate anesthesia and follow only those procedures which avoid infection and minimize pain during and after surgery.

References

Bergerot A, Holland PR, Akerman S et al (2006) Animal models of migraine: looking at the component parts of a complex disorder. Eur J Neurosci 24:1517–1534

Bhandare A, Kshirsagar A, Vyawahare N et al (2011) Evaluation of anti migraine potential of Areca catechu to prevent nitroglycerin-induced delayed inflammation in rat meninges: possible involvement of NOS inhibition. J Ethnopharmacol 136(1):267–270

Charles AC, Baca SM (2013) Cortical spreading depression and migraine. Nat Rev Neurol 9 (11):637–644

De Vries P, Willems EW, Heiligers JP (1998) The antimigraine agent aliniditan selectively constricts porcine carotid arteriovenous anastomoses via 5-HT1B/1D. Eur J Pharmacol 351 (2):193–201

Dong Z, Jiang L, Wang X et al (2011) Nociceptive behaviors were induced by electrical stimulation of the dura mater surrounding the superior sagittal sinus in conscious adult rats and reduced by morphine and rizatriptan benzoate. Elsevier 1368:151–158

Galeotti N, Ghelardini C, Grazioli I et al (2002) Indomethacin, caffeine and prochlorperazine alone and combined revert hyperalgesia in in vivo models of migraine. Pharmacol Res 46:245–250

Martino G, Perkins MN (2008) Tactile- induced ultrasonic vocalization in the rat: a novel assay to assess anti- migraine therapies in vivo. Cephalalgia 28(7):723–733

Ottani A, Ferraris E, Giuliani D (2004) Effect of sumatriptan in different models of pain in rats. Eur J Pharmacol 497:181–186

Williamson DJ, Hill RG, Shepheard SL et al (2001) The anti-migraine 5-HT1B/1D agonist rizatriptan inhibits neurogenic dural vasodilation in anaesthetized guinea-pigs. Br J Pharmacol 133:1029–1034

Animal Models of Schizophrenia

Onkar Bedi, Sumit Jamwal, Shamsher Singh, Ganesh Singh Bhakuni and Puneet Kumar Bansal

1 Introduction

Schizophrenia is a devastating mental sickness that impairs intellectual and social functioning and generally leads to the progress of comorbid illnesses. Schizophrenia is a syndrome: a set of indicators and signs of unknown etiology, predominantly defined by found signs of psychosis. In its most common type, schizophrenia grants with paranoid delusions and auditory hallucinations late in adolescence or in early maturity. These manifestations of the sickness have transformed little during the last century. Moreover, to these 'constructive' symptoms, schizophrenics also exhibit 'bad' and 'cognitive' symptoms.

The etiology of schizophrenia stays doubtful however tons of speculation and theories are there which offer special revilement of schizophrenia which comprise genetic, environmental, and neurochemical purpose. The most important mighty

O. Bedi · S. Jamwal · S. Singh · G.S. Bhakuni · P.K. Bansal (✉)
Department of Pharmacology, ISF College of Pharmacy, Moga 142001, Punjab, India
e-mail: punnubansal79@gmail.com

O. Bedi
e-mail: Jagrit143@gmail.com

S. Jamwal
e-mail: jamwalisf@gmail.com

S. Singh
e-mail: shamshersinghbajwa@gmail.com

G.S. Bhakuni
e-mail: Bhakuniganesh17@yahoo.com

P.K. Bansal
Department of Pharmaceutical Sciences and Technology, Maharaja Ranjit Singh Punjab Technical University (State Govt. University), Bathinda, Punjab, India

P.K. Bansal and R. Deshmukh (eds.), *Animal Models of Neurological Disorders*, https://doi.org/10.1007/978-981-10-5981-0_9

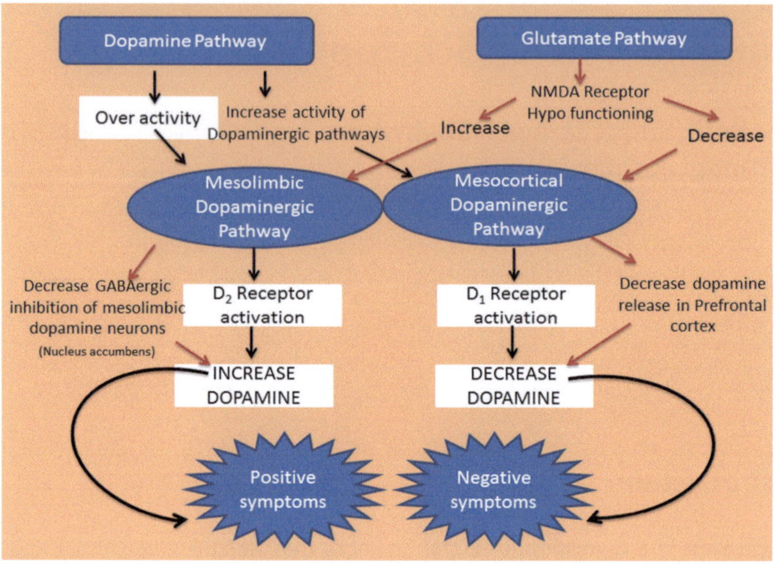

Fig. 1 Mesolimbic dopaminergic pathway in schizophrenia

association related with schizophrenia is with gene that control neuronal progress, synaptic connectivity, and glutamatergic neurotransmission. Distinct symptoms of schizophrenia appear to outcome from malfunctioning in one-of-a-kind neuronal circuits. Etiology of schizophrenia is related with mesolimbic and mesocortical pathways of mind which instantly and not directly involve dopamine and glutamate.

The role of dopamine is quite complex as the over activity of mesolimbic dopaminergic pathway leads to activation of D_2 receptor which enhance the level of dopamine and produce positive symptoms. Meanwhile the reduce activity of dopaminergic pathway activate mesocortical pathway (D_1 receptor) which produce negative symptoms as shown in the Fig. 1. However, the hypo functioning of NMDA receptor decrease mesocortical dopaminergic neuronal activity which decrease the level of dopamine in prefrontal cortex and hypo functioning of NMDA receptor also show there stimulus to increase mesolimbic pathway activation and decreasing GABAergic inhibition of mesolimbic dopamine neurons which further increase the level of dopamine release in limbic area and produce positive symptoms. So the hypo functioning of NMDA receptor also plays a vital role in the pathophysiology of schizophrenia.

The scientific evidence explores lots of psychiatric disorders and additionally numerous animal models which winning current view of neuroscience. The scientific reports advocate that disruptions in neuronal activity can have an effect on each human and animal behavior; animal models can be developed to check various predictive and causative theories that cannot be addressed in human studies. The major emphasis of current chapter is on pharmacological models and their function in focusing experimental research efforts in this area.

There are numerous problems related to modeling schizophrenia in animals, such as the faithfully reproducing the exact resemblance of cognitive modifications in cognitively advanced animals. Patients typically experience a combination of symptoms, often divided into positive (e.g., hallucinations, delusions, and thought disorganizations), negative (e.g., loss of motivation, affective blunting, alogia, and social withdrawal), and cognitive (e.g., deficits in attention, memory, and executive functions).

Many sufferers with schizophrenia additionally caused by a wide variety of catatonic phenomenon including catalepsy, stereotypies, echopraxia, and uncommon posturing and mannerisms. Similarly, the direction and final results of schizophrenia are remarkably variable, with most effective a minority of sufferers following a persistent, deteriorating route, despite enduring signs and symptoms or practical deficits in most patients.

Finally, a selection of environmental and genetic susceptibility elements has been proposed as ability causative. However, the entire modern-day animal model of schizophrenia is not supposed to function the complete animal equivalent of the human ailment for designing test specific causative or mechanistic speculation concerning schizophrenia.

1.1 Statement on the Animal Care Guidelines

The care and use of laboratory animals were considered according to the guidelines of particular country where the preclinical trail will be organized. In India, the CPCSEA Guidelines for Laboratory Animal Facility is considered for the preclinical or experimental animals studies with reference to different institutional animal societies (IAEC)

2 Classification of Animal Models (Fig. 2)

2.1 Innate Behavior Model

2.1.1 Golden Hamster Test

Principle

The ‘golden hamster-experiment’ uses the innate conduct of Mesocricetus auratus for differentiation among neuroleptic and sedative—hypnotic enterprise. The aggressive behavior of male golden hamsters is suppressed by using neuroleptics in doses which do not impair motor feature.

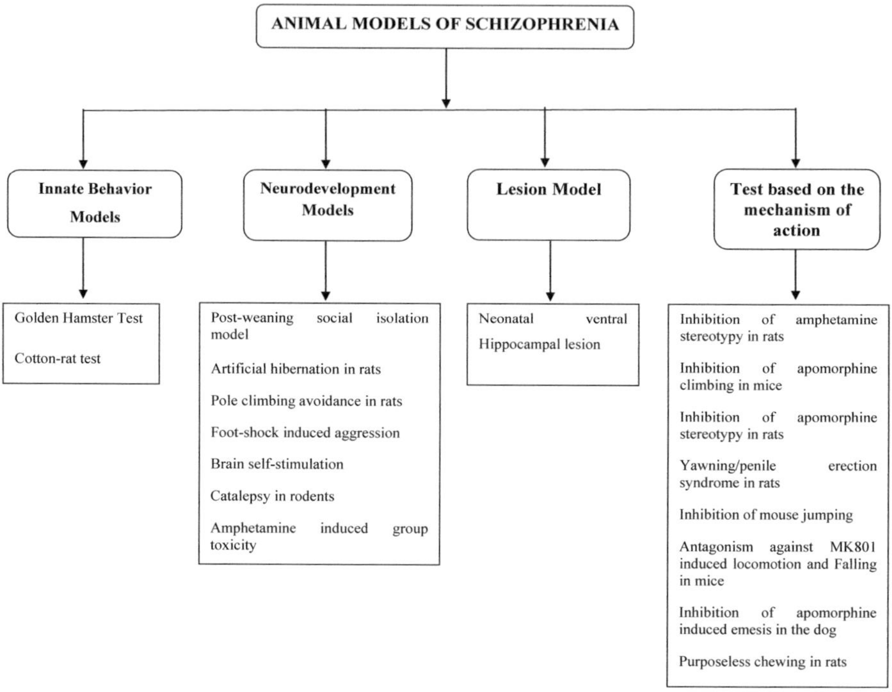

Fig. 2 Classification of animal models of schizophrenia

Procedure

- Golden hamsters (10–20) weighing 60 g are housed together in Makrolon (R) cages for at least 2 weeks which develops a characteristic fighting behavior in animals.
- Place the animals individually into glass jars of 2 L. In this situation, the hamsters assume a squatting and resting position during the day.
- If the animals are touched with a stick or a forceps, they wake up from the resting position. When held with blunted forcep the hamster throws himself onto his back, tries to bite and to push the forceps away with his legs, and utters angry shrieks.
- Touching the animals is repeated up to 6 times followed by punching with the forceps. The animals which respond to the stimulus with all three defense reactions (turning, vocalizing, and biting) are included into the test.
- The test compounds are applied s.c., i.p., or orally. Six animals are used for each dose.
- The stimuli are applied every 20 min for 3 h. The number of stimuli until response is recorded. Furthermore, the suppression of the defense reactions (turning, biting and vocalizing) is evaluated.

- An animal will be considered to be 'tamed,' if it does not defend itself even after punching with the forceps at least once during the test period. Then, the 'tamed' animal is placed on an inclined board with 20° inclination.
- Normal hamsters and hamsters tamed by neuroleptics are able to climb on the board. The animal with impaired motor function tends to slide down. This experiment is repeated thrice after each test. An animal's coordination is considered to be disturbed if it falls three times during two tests of the experiments. The number of tamed hamsters and the number of animals with impaired motor function is recorded for each dose.

Advantages

- This model is good model for experimentation point of view and used for evaluating the different antipsychotic drugs.

Disadvantages

- It needs special strain (Golden Hamster).
- It required long duration experimental study.
- It need keen observation for studying various behavioral activities.

2.1.2 Influence on Behavior of the Cotton Rat

Principle

The cotton rat (Sigmodon hispidus) is a very shy animal which hide himself at any time and 'cotton rat test' is another try to use the innate behavior as defined for numerous animal species for the differentiation of psychotropic drugs. This innate flight reflex is suppressed via centrally active drugs.

Procedure

- Cotton rats are bred in cages equipped with a clay cylinder of 20 cm length and 10 cm diameter. This cylinder is used by the animals for hiding, sleeping, and breeding.
- For the test, young animals with a body weight of 40 g are used. Normal cages (25 × 30 × 20 cm) with a wire lid are used.
- A tunnel of sheet metal (half of a cylinder) 20 cm long and 7 cm high is placed into the cage. If the tunnel is lifted and placed on another site of the cage, the cotton rats immediately hide again.
- Three rats are placed in one cage and tested for their behavior. Selective shaving of the fur enables the observer to recognize each animal. If the rats behave as described, they are then treated with the test compound s.c. or orally.

- At least 6 animals divided in two cages are used for each dose of test compound or standard. Fifteen min after application of the drug the test period of 3 h is started.
- The tunnel is lifted and placed to another site. If the animals do not show the immediate flight reflex, an airstream of short duration is blown through the wire lid. If the animal still does not respond with the flight reflex, it is considered to be positively influenced.
- Afterward, the animal is placed on an inclined board with 35° of inclination and tested for disturbance of motor coordination. A normal animal is able to climb upwards. If coordination is disturbed the rat slides down.
- The test procedure is repeated every 15 min over a period of 3 h. The animals which show at least one suppression of the flight reflex during the test period are counted as well as those who slide down on the inclined board.

Advantages

- It is used to evaluate and differentiate the psychotropic drugs.

Disadvantages

- It needs special strain (Sigmodon Hispidus).
- It is time-consuming process and test procedure is little complex.

2.2 Neurodevelopment Models

2.2.1 Post-weaning Social Isolation Model

Principle

Social deprivation of rat pups from the age of weaning (with the aid of placing them in separate cages from littermates) alters brain development and produce behavioral deficits at maturity, which are unaltered via social reintegration in later life. For example, post-weaning social isolation of rodents induces spontaneous locomotor hyperactivity, responses to novelty (neophobia), sensory-motor gating deficits, cognitive impairments, and heightened tension states and aggression. Collectively, those behavioral changes have been termed the 'isolation syndrome,' and several of these functions resemble a number of the core signs and symptoms of schizophrenia.

Procedure

- The rats with either sex are used and divided into different groups. Firstly, all rats are habituated in normal environment that is they all are in familiar environment all together. After 1 week all are separated by placing them in different cages or in the novel area.

- Social deprivation alters antioxidants level and causes behavioral deficits. The hyper activity appears within 2–3 weeks of commencing isolation.
- Data are expressed in terms of the impaired cognitive flexibility in reversal learning task and evaluated with novel object recognition test relative to the respective vehicle control data.

Advantages

- This model is less time-consuming and easy to understand.

Disadvantages

- This model practically hectic as it needs evaluation of behavioral deficits.
- Rats are not easy to handle due to social isolation and aggressiveness.

2.2.2 Artificial Hibernation in Rats

Principle

The animals are placed in closed glass vessel that is submerged in ice water. Due to the respiratory activity, the oxygen anxiety diminishes and the carbon dioxide content increases. Under the affect of cooling and of hypoxic hypercapnia, the rectal temperature falls to 15 °C, and the animal is completely anesthetized and immobilized. The rat can survive in these poikilothermic conditions for more than 20 h and whole healing happens after warming up. This sort of synthetic hibernation is augmented by means of antipsychotics.

Procedure

- Male Wistar rats (100–150 g) are deprived of food with free access to tap water overnight.
- The test compounds are injected subcutaneously 15 min prior to the start of the experiment.
- First, the rats are placed in ice-cold water to which surfactant is added in order to remove the air from the fur for 2 min.
- Then, the animals are placed into hermetically closed glass vessels of 750 ml volume which are placed into a refrigerator at 2 °C temperature.
- The vessels are opened every 10 min for exactly 10 s allowing some exchange of air and reducing the carbon dioxide accumulation.
- At each time, animals are removed from the glass vessel and observed for signs of artificial hibernation which are not shown by control animals under these conditions. Treated animals, lying on the side, are placed on the back and further examined.

- An animal is considered positive, when it remains on the back, even if the extremities are stretched out and due to which cardiac and respiration frequency are reduced and the rectal temperature has fallen to 12–15 °C.

Advantages

- This type of model used for the observation and evaluations of neuroleptics and opioid analgesics.
- It is less time-consuming.

Disadvantages

- At the time of practical, it need keen observation for exchange of air and for reducing carbon dioxide accumulation in the closed vessel (Hermetically closed glass vessel).
- A great care of animals should require due to reduction in cardiac and respiratory frequency during experiment.

2.2.3 Pole Climbs Avoidance in Rats

Principle

This is an avoidance get away method used to separate neuroleptics from sedatives and anxiolytics. The sedative compounds suppress each avoidance and get away responding at about the equal doses, whereas neuroleptic capsules reduce avoidance responding at lower doses than that affecting break out responding.

Procedure

- Male rats (200–250 g) are used. The training and testing of the rats is conducted in a 25 × 25 × 40 cm chamber that is enclosed in a dimly lit (28-V), sound-attenuating box (2.8-kHz speaker).
- The shock is delivered to the grid floor of the chamber. A smooth stainless-steel pole, 2.5 cm in diameter, is suspended by a counterbalance weight through a hole in the upper center of the chamber.
- As the animal pull down the pole to 3-mm microswitch is activated by a weight greater than 200 g and response is recorded when a rat jumps on the pole and activates the microswitch.
- The rat cannot hold the pole down while standing on the grid floor because of the counterbalance tension and cannot remain on the pole any length of time because of its smooth surface.
- The activation of the light and the speaker together are used as the conditioning stimulus. The conditioning stimulus is presented alone for 4 s and then is coincident with the unconditioned stimulus, a scrambled shock delivered for 26 s to the grid floor, and shock current is maintained at 1.5 mA.

- A pole climb response during the conditioned stimulus period terminates the conditioned stimulus and the subsequent conditioned and unconditioned stimuli. This is considered an avoidance response.
- A response during the time when both the conditioned and unconditioned stimuli are present terminates both stimuli and is considered an escape response.
- Test sessions consist of 25 trials or 60 min, whichever comes first. There is a minimum interval of 90 s. Any time remaining in the 30 s allotted to make the pole climb is added to the 90 s interval.
- Responses during this time have no scheduled consequences; however, rats having greater than 10 interval responses should not be used in the experiment.
- Before testing experimental compounds, rats are required to make at least 80% avoidance responses without any escape failures (Locke et al. 1990).
- Data are expressed in terms of the number of avoidance and escape failures relative to the respective vehicle control data.

Advantages

- Pole climbing avoidance model used to separate neuroleptics from sedative and anxiolytics in rats.
- This model is easy to understand and to be performed.

Disadvantages

- It requires specialized chamber fitted with different gazettes.
- It is more time-consuming as all animals requires training session before experimentation.
- This model is more specific to weight of the animal (micro switch are only activated due to the weight of rats which is greater than 200 g).

2.2.4 Foot-Shock-Induced Aggression

Principle

In this test, mouse fighting behavior is tested after foot-shock stimulation that is useful to detect neuroleptics but additionally shows fine results with anxiolytics and different centrally active drugs. The calming effect of BZD in monkeys and test for agents with antiaggressivity activity were developed for various animal species. These tests include foot-shock prompted aggression in mice and rats, fighting behavior of isolated mice and aggressiveness of rats which become extremely vicious after lesions in the septal area of the brain.

Procedure

- Two male mice (NMRI, Ivanovas, 20–30 g) are placed in a box with a grid floor consisting of steel rods with a distance of 6 mm.

- A constant current of 0.6 or 0.8 mA is supplied to the grid floor. A 60-Hz current is delivered for 5 s intermissions for 3 min. Each pair of mice is dosed and tested without previous exposure.
- The total number of fights is recorded for each pair during the 3-min period and fighting behavior include vocalization, leaping, running, rearing, and facing each other with some attempt to attack by hitting, biting, or boxing.
- The test compound or the standard are applied either 30 min before the test i. p. or 60 min before the test orally. The drug is given 30, 60, and 120 min prior to testing.
- A dose range is tested at the peak of drug activity. A minimum of 3 doses (10 pairs of mice/dose) is administered for a range of doses.
- The percent inhibition of aggression is calculated from the vehicle control.

Advantages

- Foot-shock-induced aggression model is used for the characterization of centrally active drugs.

Disadvantages

- It requires specialized experimental floor grid (60 Hz) and need keen observation during handling of the animals.

2.2.5 Brain Self-stimulation

Principle

Neuroleptics were proven to be potent blockers of self-stimulation. D-amphetamine and methylphenidate facilitate catecholaminergic transmission. The electric stimulation of selected brain loci produces effects in certain species which are undoubtedly reinforcing and pleasurable. Most of the facts to be received from experimental rats by using electrodes chronically implanted in the median forebrain package at the level of hypothalamus (Olds 1972).

Procedure

- Male Wistar rats (350–400 g) are anesthetized with 50 mg/kg pentobarbital i. p. and their heads placed on a level plane in a stereotactic instrument.
- A midline incision is made in the scalp and the skin held out of the way by muscle retractors and small hole is drilled in the scull with a dental burr at the point indicated by the stereotactic instrument.
- Using bregma as a reference point, the electrode is aimed at the medium forebrain bundle according to the atlas of Paxinos and Watson (2007), using the coordinates of AP = –0.8 mm, Lat = +2.8 mm, and DV = –7.2 mm below dura.

- After a minimum of 10 days for recovery, the animals are trained to bar press for electrical stimulation on a continuous reinforcement schedule in a standard operant box outfitted with a single lever.
- The reward stimulus is a train of biphasic square-wave pulses generated by a Haer stimulator. The parameters are set at pulse duration of 0.5 ms with 2.5 ms between each pulse pair.
- The train of pulses may vary between 16 and 30/s, and the intensity of the pulses that are delivered range from 0.1 to 0.5 mA using the lowest setting that will sustain maximal responding.
- The consistent baseline responding is obtained for 5 consecutive 30 min sessions. Compounds are administered 60 min prior to testing.
- The numbers of drug responses are compared to the number of responses made during each animal's 30 min control session on the preceding day, which is considered to be equal to 100%.

Advantages

- Experimentation is more précised and produce good results.

Disadvantages

- It needs good experimental handling for performing stereotactic surgery.
- This experimentation is not economical.
- It required long duration experimental study.

2.2.6 Catalepsy in Rodents

Principle

Catalepsy is a stage in which the rats fail to correct an externally imposed, unusual posture over a prolonged period of time. Neuroleptics which have an inhibitory action on the nigrostriatal dopamine system induce catalepsy while neuroleptics with little or no nigrostriatal blockade produce relatively little or no cataleptic behavior. Furthermore, cataleptic symptoms in rodents have been compared to the Parkinson-like extrapyramidal side effects seen clinically with administration of antipsychotic drugs (Honma and Fukushima 1976).

Procedure

- Wistar rats with a body weight between 200 and 250 g are used.
- Animals are dosed i.p. with the test drug or the standard and are placed individually into translucent plastic boxes with a wooden dowel mounted horizontally 10 cm from the floor and 4 cm from one end of the box.
- The floor of the box is covered with approximately 2 cm of bedding material. The animals are allowed to adapt to the box for 2 min.

- Then, each animal is grasped gently around the shoulders and under the forepaws and placed carefully on the dowel.
- The amount of time spent with at least one forepaw on the bar is determined. When the animal removes its paws, the time is recorded and the rat is repositioned on the bar and trials are conducted for each animal at 30, 60, 120, and 360 min.
- An animal is considered to be cataleptic if it remains on the bar for 60 s. Percentage of cataleptic animals is calculated. A dose of 1 mg/kg; i.p. of haloperidol was found to be effective.

Advantages

This animal model of nervous disorder used for the study of PD and epilepsy.

Disadvantages

It is time-consuming and complex.

2.2.7 Amphetamine Model of Schizophrenia

Principle

The toxicity of amphetamine greatly enhances in aggregative mice adapted in small cages. The death rate can be reduced by pretreatment with neuroleptics. This event is generally accepted as an indicator of neuroleptic activity. The increased toxicity results from increased behavioral activation due to aggregation inducing an increase of circulating catecholamines. The mechanism can be understood by the fact that amphetamine is an indirectly acting sympathomimetic amine that exerts its effects primarily by releasing norepinephrine from storage sites in the sympathetic nerves. Following administration of high doses of amphetamine, mice show signs of an elevated motor activity which is highly increased by aggregation. Neuroleptics reduce the death rate in the similar rodent animals.

Procedure

- Ten male mice of the NMRI-strain are used for each group. They are dosed with the test compound or the standard either orally or i.p. and all placed in glass jars of 18 cm diameter.
- Untreated animals serve as controls. The test has to be performed at room temperature of 24 °C.
- Thirty min after i.p. or 1 h after oral administration the mice receive 20 mg/kg d-amphetamine subcutaneously.
- The mortality is assessed 1, 4, and 24 h after dosing. The mortality of amphetamine only treated animals is at least 80%. If less than 80% die due to low ambient temperature, the test has to be repeated.
- The estimation of ED50 values for protection and their confidence limits are calculated by probit analysis of the data using the number of dosed versus the number of surviving animals.

Advantages

- It is helpful in understanding the mechanism of amphetamine induce toxicity and to study effectiveness of various anxiolytics agents.

Disadvantages

- This model requires particular mice strain (NMRI-strain).
- In this type of model, there is more chance of motility.

2.3 Lesion Models

2.3.1 Neonatal Ventral Hippocampal Lesion

Principle

Ibotenic acid (Excitotoxin) causes behavioral abnormalities that emerge after puberty and compromising the integrity of the developing behavioral changes in the neonatal by inducing neonatal and hippocampal lesions in brain. Ventral hippocampal lesions in rats cause persistent and marked impairment in several spatial working memory tasks including impaired acquisition of the T-maze delayed alteration and morris water maze tasks and retention of passive avoidance learning, even with prolong learning.

Procedure

- Wistar rats with a body weight between 200–250 g are used.
- Excitotoxin, Ibotenic acid (3–5 μg in 0.3 μl) under anesthesia with hypothermia is given by local injection.
- This produce neonatal ventral hippocampal lesion and which causes behavioral abnormalities in rats.
- These lesions subsequently produce pharmacological stress which may precipitate long-term psychiatric changes akin to those seen in schizophrenia.

Advantages

- This model greatly mimics with schizophrenic symptom due to long-term psychiatric changes.

Disadvantages

Chemical used in this model are costly.

2.4 Test Based on the Mechanism of Action

2.4.1 Inhibition of Amphetamine Stereotypy in Rats

Principle

Amphetamine is an indirect acting sympathomimetic agent which releases catecholamines from its neuronal storage pools. In rats, the drug produces a characteristic stereotypic behavior. This behavior can be prevented by neuroleptic agents.

Procedure

- Wistar rats (120 and 200 g) are injected simultaneously with d-amphetamine (10 mg/kg; s.c.) and the test compound intraperitoneally and then placed individually in stainless-steel cages (40 × 20 × 18 cm).
- The control groups receive d-amphetamine and vehicle.
- Stereotypic behavior is characterized by continuous sniffing, licking or chewing, and compulsive gnawing.
- The animals are observed 60 min after drug administration.
- An animal is considered to be protected, if the stereotypic behavior is reduced.
- The percent effectiveness of a drug is determined by the number of animals protected in each group.

Advantages

- This is very good model of choice for experimental studies due its easy experimentation and less complexity.

Disadvantages

- It need keen observation for the notification of behavioral changes.

2.4.2 Inhibition of Apomorphine Climbing in Mice

Principle

Apomorphine administration to mice results in an abnormal climbing behavior which is characterized by rearing and then full-climbing activity, predominantly mediated by the mesolimbic dopamine system. The ability of a drug to antagonize apomorphine-induced climbing behavior in the mouse has been correlated with neuroleptic potential.

Procedure

- Male mice (20–22 g) are treated i.p. or orally with the test substance or the vehicle and placed individually in wire-mesh stick cages.
- After 30 min, they are injected s.c. with 3 mg/kg apomorphine. They are observed for climbing behavior and scored as follows: 0 = four paws on the floor, 1 = forefeet holding the vertical bars, 2 = four feet holding the bars after every 10, 20 and 30 min of apomorphine administration.
- The average values of the drug-treated animals are compared with those of the controls, the decrease is expressed as percent. Three dose levels are used for each compound and the standard with a minimum of 10 animals per dose level.

Advantages

- This model is very less complex on experimentation point of view.

Disadvantages

- Apomorphine is not easily available drug in the market as this drug is covered under Controlled Substances Act. That is why it needs lots of approval process.

2.4.3 Inhibition of Apomorphine Stereotypy in Rats

Principle

Apomorphine produces a stereotyped behavior in rats, characterized by licking, sniffing, and gnawing in a repetitive, compulsive manner, which is an indication of striatal dopaminergic stimulation. Compounds which prevent apomorphine-induced stereotypy antagonize dopamine receptors in the nigrostriatal system.

Procedure

- The test drug or the standard is administered i.p. 60 min prior apomorphine dosage to Wistar rats (120–200 g).
- After administration of apomorphine HCl (s.c. at a dose of 1.5 mg/kg), all animals are placed in individual plastic cages.
- A 10 s observation period is used to measure the presence of stereotypic activity such as sniffing, licking, and chewing 10 min after apomorphine administration.
- An animal is considered protected if this behavior is reduced or abolished. The percent effectiveness of a drug is determined by the number of animals protected in each group.

Advantages

- This model is very less complex on experimentation point of view.

Disadvantages

- Apomorphine is not easily available drug in the market as this drug is covered under Controlled Substances Act. That is why it needs lots of approval process.
- It need keen observation for studying stereotypic activities.

2.4.4 Yawning/Penile Erection Syndrome in Rats

Principle

Yawning is a phylogenetically, stereotyped event that occurs singly or related to stretching and/or penile erection in humans and in animals from reptiles to birds and mammals under different situations. The yawning-penile erection syndrome may be induced in rats via apomorphine and due to dopamine autoreceptor stimulants which can be antagonized by using haloperidol and other dopamine antagonists. Antagonism against this syndrome may be appeared as indication of antipsychotic activity.

Procedure

- Naive male Wistar rats, weighing 220–280 g, are housed under controlled 12-h light–dark cycle with free access to standard food pellets and tap water.
- Rats are pretreated with subcutaneous injection of the antagonist 30 min prior to injections of the agonist, such as apomorphine (0.02–0.25 mg/kg s.c.) or physostigmine (0.02–0.3 mg/kg; s.c. or i.p.).
- After administration of the agonist, rats are placed in individual transparent Perspex cages.
- A mirror is placed behind the row of observation cages to facilitate observation of the animals for penile erections and yawns. Yawning is a fixed innate motor pattern characterized by a slow, wide opening of the mouth.
- A penile erection is considered to occur when the following behaviors are present: repeated pelvic thrusts immediately followed by an upright position, an emerging, engorged penis which the rat proceeds to lick while eating the ejaculate.
- The number of penile erections and yawns is counted for 30 min following the last injection.
- The results are expressed as the mean number of yawns and of penile erections per group ± SEM. The statistical significance is determined by comparing the results of each group with the results of the relevant control group using a nonparametric rank sum test.

Advantages

- The experimental procedure is quiet easy and not so complex.

Disadvantages

- Apomorphine is not easily available drug in the market as this drug is covered under Controlled Substances Act. That is why it needs lots of approval process.
- Observation such as yawns and penile erection is hectic and need regular observations.

2.4.5 Inhibition of Mouse Jumping

Principle

The administration of L-dopa in amphetamine pretreated animals shows jumping response where the number of jumps can be objectively counted. The jumping response is due to dopaminergic overstimulation similar to that seen in rats when stereotypy is induced by higher doses of amphetamine. The phenomenon can be blocked by neuroleptics (Lal et al. 1975, 1976).

Procedure

- Male CD-1 mice weighing 22–25 g are injected with 4 mg/kg d-amphetamine sulfate, followed 15 min later by an i.p. injection of 400 mg/kg L-dopa.
- The mice spontaneously begin to jump at a high rate. A median of 175 jumps can be observed in these mice during 60 min.
- Since mice do not show any jumping after saline administration, the responses after drug administration are specific and can be measured automatically through a pressure-sensitive switch closure or properly positioned photoelectric beam disruptions.
- Test compounds are administered 60 min prior to L-dopa injection. Jumps of mice treated with test drugs or standard are counted.
- The data is expressed as percentage of jumps in amphetamine/L-dopa treated animals. Using various doses, ED50 values with 95% confidence limits are calculated.

Advantages

- This model is very less complex on experimentation point of view.

Disadvantages

- Particular strain of mice is required (CD-1).

2.4.6 Antagonism Against MK-801 Induced Locomotion and Falling in Mice

Principle

MK-801, a noncompetitive NMDA antagonist, induce characteristic stereotypy in mice marked by locomotion and falling behavior through both dopamine-dependent and dopamine-independent mechanisms. Antipsychotic agents dose-dependent antagonize this MK-801-induced behavior.

Procedure

- Male CD-1 mice (20–30 g) are individually placed inactivity boxes lined with wire-mesh flooring and allowed to acclimate for 60 min.
- The animals are then administered with compounds 30 min prior to subcutaneous administration of MK-801 at 0.2 mg/kg.
- The mice are observed for locomotion and the presence of falling behavior 15 min following MK-801 administration.
- ED50 values and 95% confidence limits are calculated by the Litchfield and Wilcoxon method.

Advantages

- This model is very less complex on experimentation point of view.

Disadvantages

- Particular strain of mice is required (CD-1)
- It need keen observation for studying various behavioral activities.

2.4.7 Inhibition of Apomorphine-Induced Emesis in the Dog

Principle

The blockade of centrally performing dopaminergic mechanisms performs a prime role in suppression of psychotic reactions in schizophrenia. Apomorphine is seemed as a right away dopaminergic agonist, produces a said emetic effect in dogs and the blockade of apomorphine emesis which is used as an illustration of dopaminergic blockade. However, although both antiemetic activity and antipsychotic activity are thought to be due to dopaminergic blockade, the sites of action are in special brain regions and there's a lack of entire correlation of these activities.

Procedure

- Adult beagle dogs of either sex are used in treatment groups of three to nine dogs/dose.

- The dogs are given the test compounds in a gelatin capsule; they are then dosed with 0.15 mg/kg apomorphine s.c. at various intervals after administration of the test compound.
- The dogs are first observed for overt behavioral effects, e.g., pupillary response to light, changes in salivation, sedation, and tremors.
- After the administration of apomorphine, the dogs are observed for stereotypical sniffing, gnawing, and the emetic response. Emesis is defined as retching movements followed by an opening of the mouth and either attempted or successful ejection of stomach content.
- If the experimental compound is antiemetic in the primary screen, the dose is progressively lowered to obtain a minimal effective dose or an ED50 value.
- The ED50 values for haloperidol and chlorpromazine were found to be 0.06 mg/kg p.o. and 2.0 mg/kg; p.o., respectively.

Advantages

- This model is very less complex on experimentation point of view.

Disadvantages

- It is very difficult to handle dogs for experimentation.
- Particular strain of dog is required (Beagle Dogs).
- It need keen observation for studying various behavioral activities.

2.4.8 Purposeless Chewing in Rats

Principle

The cholinergic drugs or cholinesterase inhibitors induce purposeless chewing in rats via blocked of antimuscarinic agents. The chewing behavior has been proposed to be mediated through critical M2 receptors in preference to central M1 sites. Chewing can also be precipitated through persistent administration of neuroleptics in rats. Purposeless chewing is mediated via dopaminergic and nicotinic mechanisms.

Procedure

- Male albino rats are housed 10 per cage at room temperature and kept on a 12 h light-dark cycle.
- For the experiments, rats are placed individually in a large glass cylinder (height 30 cm, diameter 20 cm) at 21 ± 1 °C and allowed to habituate for 15 min before injection of drugs.
- The antagonists, e.g., sulpiride or mecamyl amine as standards, are given at different doses 30 min before treatment either with 0.01 mg/kg nicotine or 1 mg/kg pilocarpine i.p.

- Numbers of chewings are counted by direct observation immediately after drug administration. The results are presented as number of chews in a 30 min period.

Advantages

- This model is very less complex on experimentation point of view.

Disadvantages

- It need keen observation for studying various behavioral activities.

3 Conclusion

In this chapter, our main emphasize is on the important animal model of schizophrenia with reference to their importance and limitations. The use of animal models to improve the current understanding of the neurochemical and structural changes of central nervous system and their link with pathophysiology of schizophrenia. The most important advantage of neurodevelopment over pharmacological or lesion models of schizophrenia is the ability to perform behavioral, electrophysiological, and neurochemical investigation in the absence of mystifying drugs or surgical interventions and ability to detect reversal of agents operating on pharmacological mechanism (New class of antipsychotic). Ultimately, these models help to explore our knowledge in understanding this devastating disorder for the development of novel pharmacological interventions.

Ethical Statement

All institutional guidelines, national guidelines, and state and local laws and regulations with professional standards for the care and use of laboratory animals should be followed. Studies involving animals must state that the institutional animal ethical committee has approved the protocol. For authors using experimental animals, a statement should be made that the animals' care is in accordance with institutional guidelines and animals used have been treated humanely and with regard for the alleviation of suffering. Researchers should treat animals as sentient and must consider their proper care and use and the avoidance or minimization of discomfort, distress, or pain as imperatives. Animal experiments should be designed only after due consideration of animal health. It should be ensured that all researchers who are using animals have received instruction in research methods and in the care, maintenance, and handling of the species being used. All the surgical procedures should be performed under appropriate anesthesia and follow only those procedures which avoid infection and minimize pain during and after surgery.

References

Honma T, Fukushima H (1976) Correlation between catalepsy and dopamine decrease in the rat striatum induced by neuroleptics. Neuropharmacol 15:601–607

Lal H, Colpaert F, Laduron P (1975) Narcotic withdrawal-like mouse jumping produced by amphetamine and L-dopa. Eur J Pharmacol 30:113–116

Lal H, Marky M, Fielding S (1976) Effect of neuroleptic drugs on mouse jumping induced by L-dopa in amphetamine treated mice. Neuropharmacol 15:669–671

Locke KW, Dunn RW, Hubbard JW et al (1990) A centrally acting analgesic with neuroleptic properties. Drug Dev Res 19:239–256

Olds ME (1972) Alterations by centrally acting drugs of the suppression of self-stimulation behavior in the rat by tetrabenazine, physostigmine, chlorpromazine and pentobarbital. Psychopharmacology 25:299–314

Paxinos G, Watson C (2007) The rat brain in stereotaxic coordinates, 6th edn. Academic Press. ISBN: 9780125476126

Animal Model of Anxiety

Puneet Kumar Bansal, Shamsher Singh and Sumit Jamwal

1 Introduction

Anxiety is defined as state of unpleasant and uneasiness or discomfort experienced on exposure toward threat or painful stimuli both in humans and animals. It is cumulatively caused by increased activity of neuroendocrine and autonomic nervous system. Also, it is a state of behavioral disturbance, that is, sense of unrealistic worry about everyday life situations. Animal models for anxiety-related behavior are based on the assumption that anxiety in animals is comparable to anxiety in humans. Being anxious is an adaptive response to unfamiliar environmental conditions, especially during unconformity with danger or threat. Human anxiety disorders are broadly grouped according to symptomology and responsiveness to pharmacological and psychological treatment. Generalized anxiety disorder and panic disorder are the two primary classifications of pathological anxiety in humans. In generalized anxiety disorder, the peoples experience unrealistic worry about everyday life situations, which make it different from panic disorder. In contrast, panic attacks mainly indicate the primary symptoms of panic disorder with intense fear, palpitation, and sweating, etc. These events are characterized as sudden, extreme fear accompanied by autonomic nervous system arousal (Battaglia et al. 2005).

P.K. Bansal (✉)
Department of Pharmaceutical Sciences and Technology,
Maharaja Ranjit Singh Punjab Technical University (State Govt. University),
Bathinda, Punjab, India
e-mail: punnubansal79@gmail.com

P.K. Bansal · S. Singh · S. Jamwal
Department of Pharmacology, ISF College of Pharmacy, Moga 142001, Punjab, India
e-mail: shamshersinghbajwa@gmail.com

S. Jamwal
e-mail: jamwalisf@gmail.com

P.K. Bansal and R. Deshmukh (eds.), *Animal Models of Neurological Disorders*, https://doi.org/10.1007/978-981-10-5981-0_10

In response to the types of stimuli which cause anxiety, the behavioral models are classified into two types, i.e., either conditioned or unconditioned. In the conditioning models, the minor stimuli are used like deprivation of animal from food and water or giving foot electric shock, etc., but unconditioned models (spontaneous) have higher degree of ecological validity. They are also less susceptible to be arising from interference with learning/memory, hunger/thirst, or nociceptive mechanisms. These animal models provide a powerful contribution to the area of research related to anxiety at the clinical, industrial, and scientific levels. In this, the individual susceptibility difference among the animals toward anxiogenic stimuli and variable responses to different types of threats can easily be modelled in animals. It is easy to analyze the basic physiological mechanisms underlying fear in rodents because of the similar mechanisms operating in humans provide a degree of face validity for these paradigms. The rodents mainly show these responses which may be appropriate and adaptive for the current conditions, but in humans, the anxiety disorders constitute maladaptive or pathological responses to the existing situation. Further, to explore the neuroanatomy and neurochemistry involved in fear in rodents toward both conditioned and unconditional fear could offer important insights into effective targets for novel pharmacological treatment. However, it is very difficult to correlate biologically the animal studies with human behavior because: (i) the difference between human's and non-human's nervous systems; (ii) the difficulty in determining analogous behaviors among species; and (iii) the need to extrapolate the results from animals to humans.

2 Classification of Animal Models of Anxiety (Fig. 1)

2.1 Conditioned Response

2.1.1 Geller–Seifter Conflict

Principle: The Geller–Seifter conflict model is commonly used from the last few decades for the evaluation of anxiolytic drugs. In this model, multiple operant schedules are used by providing shock after the food cues (Howard et al. 1990). These food cues increase the reinforcement, and shock act as signal to confirm the behavior of the animal, i.e., if the animal is in anxiety state it does not respond to shock signals.

Procedure:

- The rats with body weight 180–250 are housed individually.
- The rats are trained in the chamber which is operated by a lever to obtain food.

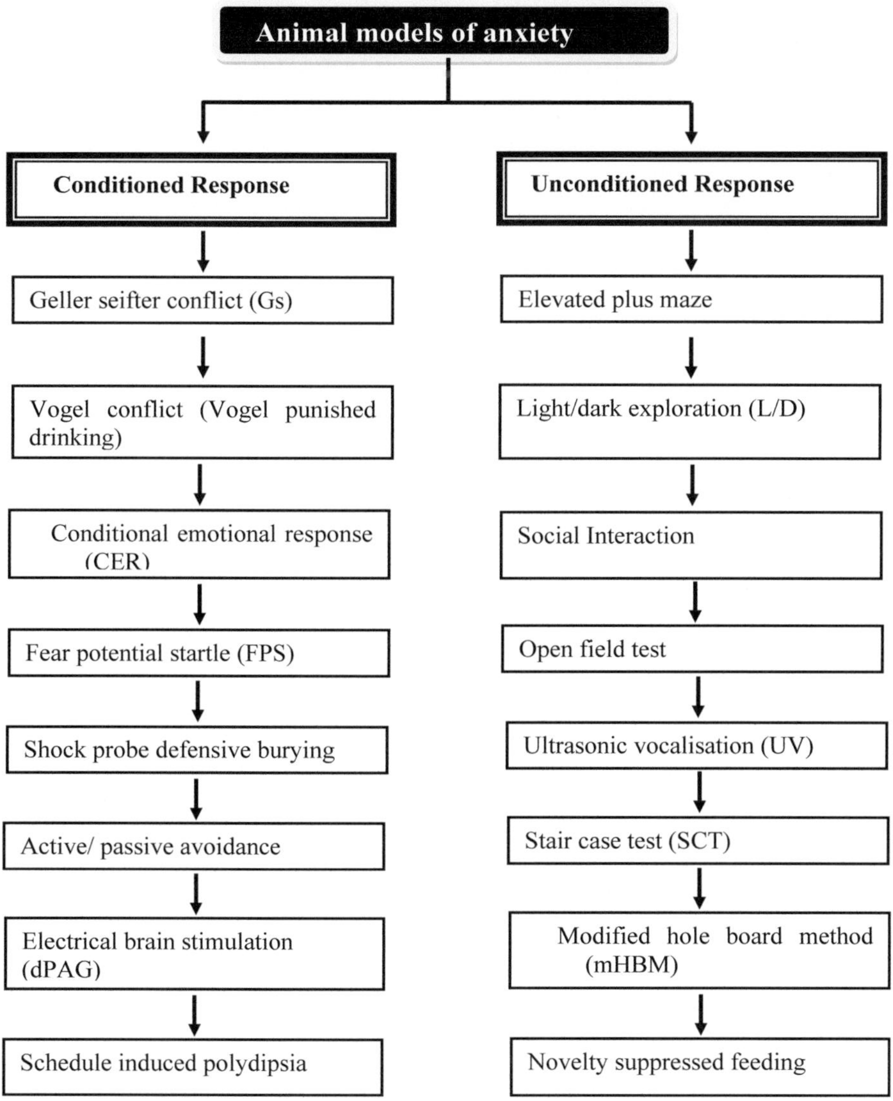

Fig. 1 Classification of animal models of anxiety

- Auditory cue in the form of signal is provided to increase the reinforcement contingencies.
- After the auditory cues, i.e., during the next session, food is available to the animals along with foot shock.
- The test procedure consists of four 15-min non-shock variable interval segments in which the reinforcement is available on a restricted basis.

- The whole test procedure consists of multiple schedule of reinforcement to evaluate the anxiolytic action of drug at different intervals.
- To analyze the drug the auditory cues: first, the response of reinforced is given at irregular intervals but afterward, every response is simultaneously reinforced (signalled by a different signal) and punished by the delivery of inescapable electro-shock.
- The response to these signals can be suppressed by administration of anxiolytics.

Advantages of G-S conflict test

1. This method has selectivity for anxiolytic drugs showing no effects of other classes of psychotropic drugs.
2. G.S method is useful for evaluation of chlordiazepoxide, diazepam, meprobamate, phenobarbital, and pentobarbital.
3. It is a suitable method for repeated drug testing.
4. Once the subjects have learned the tasks in the Geller–Seifter paradigm response rates in all operant components remain relatively stable over long periods (Willner et al. 1992). This makes the Geller–Seifter conflict a suitable test for repeated drug testing in order to demonstrate reliable and repeatable responses to anxiolytics over time in individual subjects.

Disadvantages of G-S conflict test

1. A long period of training (one to several weeks) until the animals reaches a stable baseline response to the conflict component as well as the necessity for long-term food restriction.
2. Sometime animals may die due to over electric shock.

2.1.2 Vogel Conflict

Principle: It is also called Vogel punished drinking or Vogel water-lick conflict test. The Vogel water-lick conflict is a modification of the Geller–Seifter conflict paradigm that was established to eliminate the long periods of training. It is a commonly used method to study anti-anxiety drugs in which water cues are provided for a short interval (Safi et al. 2006).

Procedure:

- Male Wistar rats of body weight 180–250 g are selected and are deprived of water for 24 h prior to the start of first training session.
- The first training session is consisted of two 3-min periods in which the number of unpunished licking spells is recorded.
- Prior to drug administration that is after the competition of first training session, the animals are placed back in the box for conflict test.

- Now the animals are administered with drug, moved to the apparatus to start the trial consisted of two 3-min periods in which rat completed 20 licks received first shock.
- After every 20 unpunished licks, 1 mA current is provided between the grid floors and drinking for the subsequent licking.
- The animals are shocked with current for fixed cycle of 3 mins. The animals which show 50% suppression of licking during second session in comparison with first trial are selected for the study.
- The drugs are to be administered after second trial competition, and again the animals are placed into their respective cages with availability of water.
- The total test time per rat is 12 min per week (Fig. 2)

Advantages of Vogel water-lick conflict

1. It is a modified form of Geller–Seifter conflict and required less training time to evaluate anti-anxiety drugs.
2. It also responds to some non-anxiolytic drugs, producing false-negative results, but antidepressants produce inconsistent results in these models.
3. By this method, we can compare the anxiolytic efficacy of different drugs.

Disadvantages of Vogel water-lick conflict

The major limitation of using the Vogel water-lick conflict is the lack of a systematic analysis of drug effects on non-conflict behavior. Later, the modified form of this method has improved replicability by preselection of subjects that lick water but are sensitive to shock induced suppression.

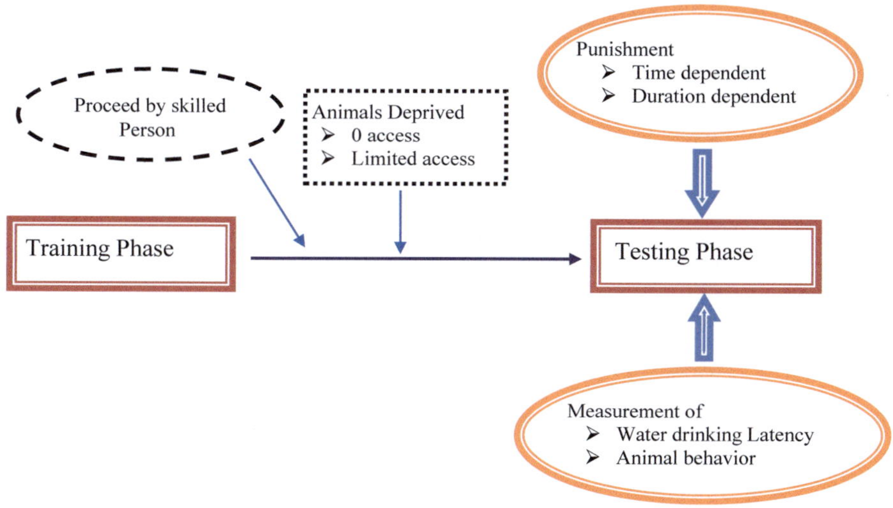

Fig. 2 Diagrammatic representation of Vogel punished drinking

2.1.3 Conditioned Emotional Response (CER)

Principal: The CER is the simplest method and was firstly discovered by Ivan Pavlov used for testing efficacy of anti-anxiety drugs. In this model, the conditional stimuli of food cues are provided with electric shock after giving training to the animals. The electric stimuli are provided to animals at different time interval of training along with food pellets.

Procedure:

- The experiment is conducted in apparatus consisted of four identical operant chambers. The floor consists of electrifiable grid, and the side walls are fitted with a single bar containing a food tray under it.
- During the preliminary training phase, each chamber is provided with 45 mg food pellets for a 1-min variable interval schedule (IV), also called magazine training.
- Now immediately after preliminary training, continuous reinforcement schedule is given with a delivery of 120 food pellets in a single session.
- At the end of training session, six daily 2-h sessions of bar pressing under a 2.5-min variable interval of food reinforcement schedule are given to the animals.
- This result in acquisition of stable bar-pressing behavior for food, and the numbers of bar presses emitted by each rat in 3-min periods on the 6th day are noted with achieving a conditional stimulus (CS).
- The conditional response consists for 3-min period of 80 dB, but noise is delivered by permanent magnet speaker placed below the floor of the experimental chamber.
- After this on dummy day conditional stimulus of 0.5 s with shock of 2-mA intensity is given at intervals 14, 48, 72, and 79 min after beginning of session.
- The procedure was repeated for three consecutive days. The magnitude of CES is measured by the "suppression ratio."

Advantages

1. The CER is a simple behavioral paradigm in which organisms learn to predict aversive events.
2. It evaluates the clinical efficacy of anxiolytic on different animals by providing different conditional stimulus.

Disadvantages

1. The main drawback of this method is that the stimuli used are painful and may also induce fear in animals.
2. Highly skilled person is required because the experiment consists of number of training phase intervals.

2.1.4 Fear Potential Startle (FPS)

Principle: In FPS, the fear reactions are used as stimuli of reflux response in animals, and the response are elicited in the form of threatening stimulus (e.g., any object, person, or situation produces feelings of fear). These can also be delivered by a neutral stimulus as a result of fear conditioning. The stimulus used is usually of auditory (e.g., loud noise) or visual (e.g., bright light) type, and startle response measures include eye blink rates and pulse/heart rate.

Procedure:

- The procedure consists of 3 days of startle acclimation in which classical fear conditioning is provided for 1 day followed by a fear-potentiated startle test session.
- In this, the animals are provided with 5-min acclimation period followed by 30 presentations of a 50 ms. After this, noise burst startle stimulus at 95, 105, or 115 dB (10 of each) are to be given in a predetermined pseudo-random order.
- Each startle consists of 15-s inter-trial interval, and these help in easy acclimation of the subjects to the experimental environment and also improve matching subjects into experimental groups.
- Measure the mean "Pre-Fear" startle scores for each subject which are formed by addition of mean startle amplitudes of all the trials over the 3 days.
- Now those animals which pass the Pre-Fear startle amplitudes are administered with different dose conditions for matching the various groups to different conditions.
- After the administration of doses, all the rats are classically fear-conditioned for the four days. During the 5-min acclimation period, foot shock is provided with five pairings of light stimuli.
- Each pairing consisted of 3-s presentation of the light, which co-terminated with the 500 ms (0.6 mA) foot shock, the inter-trial intervals ranged from 60 to 180s in a pseudo-random order.
- Now compare the results of before and after treatment for measuring the efficacy of drugs.

Advantages

1. It provides a direct correlation between the anxiety behavior of animal and anxiety disorder patients as a result of re-exposure to trauma-related stimuli or negative life events.
2. It serves as a "translational bridge" and is the first to use fear-potentiated startle to examine extinction and reinstatement in humans.

Disadvantages

1. Depending upon the signs of fair in animals, it is very difficult to correlate the model to behavior signs of anxiety in humans.
2. Sometimes, the animals may not respond to fear-induced anxiety-like state.

2.1.5 Shock Probe Defensive Burying

Principle: This model was introduced 25 years ago by Pinal and Treit. Defensive burying refers to the typical rodent behavior in which the bedding material is displaced with vigorous material. Due to this, the animals show treading-like movements of their forepaws and shovelling movements of their heads when directed toward a variety of noxious stimuli. In this, animal is exposed toward immediate threat, such as a wall-mounted electrified shock-produce.

Procedure:

- In this, the test apparatus is covered with suitable bedding material and the subjects are confronted with a wire-wrapped probe (Ø = 1 cm; 6–7 cm long). A small hole lies 2 cm above the bedding in one of the test chamber walls.
- The shock source is connected through non-insulated wires of the probe.
- Now during the teat session when the animal touches the probe, they receive an electric shock (manually operated or automatically delivered).
- Observe the animal's behavior manually or recorded on video for a 10–15-min test session.
- During this observation period, all occurring behavioral postures and the parameters are measured for maximum 15 mins.

Advantages

1. Shock probe test is helpful in detecting the neuroendocrine effect in anxiety, because noradrenaline plays a crucial role in emotional behavior in animals and humans.
2. This paradigm not only is suitable for screening potential anxiolytic properties of drugs but also seems to be especially valuable for unravelling the neural circuitry and neurochemical mechanisms involve in anxiety (Reynolds et al. 2001).

Disadvantages

1. This experiment requires a long training session for the proper acclimation of animals to evaluate anti-anxiety effect of drugs.
2. The cut-off time is too long that may increase the mortality rate because assembly is fitted with probe carrying current.

2.1.6 Active/Passive Avoidance

Passive Avoidance

Principle: Passive avoidance task is fear-aggravated test used to evaluate learning and memory in experimental animals. In this procedure, the animals are learned to avoid noxious event by suppressing a particular behavior when they are exposed to different conditions.

Procedure:

- The apparatus consists of two adjacent Plexiglas compartments of identical dimensions (27 cm × 14.5 cm × 14 cm) with grid floors.
- The floor of the two compartments has been covered with stainless steel bars (2 mm diameter) spaced 1 cm apart. The compartment is illuminated by a 5-W lamp mounted on its wall just below a movable transparent Plexiglas ceiling.
- The animals are allowed to adapt for 10 min period with free access to either the light or dark compartment of the avoidance training box after being placed in a shuttle-box.
- After the two days of adaptation period, the animals are placed into the illuminated compartment.
- To note the latency of learning phase, the sliding door is raised 30 s later.
- Close the door when the animal move into dark compartment, and a 1.5-mA constant current is applied to the fore and hind paws for 3 s.
- Again after 20 s, each animal is removed from the dark compartment and placed into the home cage.
- For the testing of short-term learning, that is, 24 h after receiving foot shock, the animals are placed in the illuminated chamber again.
- After 30 s, the sliding door is raised and latency of entering the dark compartment is recorded again constituting the step-through latency.
- The maximum cut-off time for this procedure is 5 min.

Advantages

1. Passive avoidance is a better behavioral test for learning and memory studies, because it requires little special training of the subjects and also the results are available quickly.
2. It is a simple and fast method for evaluating psychotropic and anxiolytic drugs.

Active Avoidance

Principle: The active avoidance task is a fear-motivated test in which electric current is used as a source of punishment. In this, the animals are learned to predict the occurrence of an aversive event based on the presentation of a specific stimulus in order to avoid the harmful stimuli by actively moving to a different compartment.

Procedure:

- The apparatus used for evaluating active avoidance consists of 3-equal arms like Y-maze (Narwal et al. 2012).

- Prior to the experimentation, the rats are trained in the maze for minimum 30 trials daily for 4 days.
- The conditional stimuli (CS) are provided to the animals by using a 12-W light bulb, whereas unconditioned stimuli (UCS) in the form of 3 mA electrical foot shock.
- Inter-trial interval (ITI) and inter-stimulus interval (ISI) are of 60 s and 5 s, respectively.
- Trained animals left the dark arms and entered into the light arm. If this occurred within the 5 s of ISI, the effort is counted as a conditioned response.

Advantages

1. Active avoidance is useful model for neuropharmacological and electrophysiological studies.
2. This paradigm also takes a less time to access even short-term changes in the performance of animals.

2.1.7 Electrical Brain Stimulation (dPAG)

Principle: Electrical stimulation of the dPAG has been proposed as a model of panic attacks. According to this model, a stepwise increase in the electrical current intensity to stimulate the dPAG produces alertness, then freezing, and finally the panic-like behavior characterized by running and jumping responses.

Procedure:

- In this model, the animals are placed into the experimental cage and the escape threshold is determined by applying electrical stimuli (AC, 60 Hz, 10 s) through the implanted chemitrode.
- The inter-stimulus interval is 10 s, and the current intensity is started at a level of 20 A (peak-to-peak) and is increased by steps of 4 A.
- Apply the electric stimuli until the rat started to run around the circular arena, indicating the escape behavior. Sometimes the animals also show vertical jumps as an indicative of vigorous reaction.
- After observing these behaviors, the application of electrical stimulation to the dPAG is interrupted by the experimenter person.
- The basal escape threshold is defined as the lowest current intensity that evoked escape in three successive trials of electrical stimulation. Animals with basal thresholds above 152 A are excluded from the study.

Advantages

1. This is the best model used to differentiate between panicolytic drugs like clomipramine, fluoxetine and panicogenic drugs like pentylenetetrazole.
2. The use of dPAG model helps us clearly differentiating the anxiety and panic attack.

2.1.8 Schedule-Induced Polydipsia in Rats

Principle: Schedule-induced polydipsia is a behavioral model in which the excessive drinking developed by food-deprived animals exposed to intermittent food reinforcement schedules. This short-term food exposure to the animals at different interval shows better predictivity for analyzing anti-anxiety drugs.

Procedure:

- Firstly, weigh the animals and allocate randomly to one or two groups that is the polydipsia group or the control group.
- After a 1-week acclimatization period, the animals are subjected to 15 preoperative schedule-induced polydipsia tests on weekdays.
- Place the animals in the test chamber with automatic delivery of food (45 mg) pellets on a fixed-time 60 s feeding schedule for 30 min test sessions.
- To assess schedule-induced polydipsia, water intake (g) is measured by weighing the water bottles before and after the 30 min test sessions.
- The testing of the animals is done on every day randomly. The animals which consumed 8 ml water or more are considered to be polydiptic (SIP group).
- Control animals are tested in the same environment but received all the 30 food pellets at once, and they are paired in group with an animal from the SIP group.
- Those animals which do not meet the 8 ml criterion (SIP criterion) after 15 test days are considered resistant or resistant group.

Advantages

1. It is a useful model to study those neuropsychiatric disorders characterized by the presence of compulsive behavior such as obsessive–compulsive disorder (OCD), schizophrenia, and alcohol abuse.
2. SIP provides a bitonic relationship between amount of water drinking and inter-reinforcement interval length.

2.2 Unconditioned Response

2.2.1 Elevated Plus Maze

Principal: The elevated plus maze is a widely used behavioral test for rodents, and it has been validated to assess the anti-anxiety effects of pharmacological agents and steroid hormones. Briefly, rats or mice are placed at the junction of the four arms of the maze, facing an open arm, and entries/duration in each arm is recorded manually or by a video tracking system for 5 min.

Procedure:

- The apparatus consists of two open arms (50 × 10 × 40 cm) and two enclosed arms (50 × 10 × 40 cm) with an open roof arranged, so that the two open arms are opposite to each other (Fig. 3).
- The maze lies at 50 cm height from the ground floor. The rats (200–250 g body weight) are housed in pairs for 10 days prior to testing in the apparatus.
- During this time, the rats are handled by the investigator on alternate days to reduce stress.
- The animals are divided into test and control group. Now 30 min after ip administration of the test drug or the standard, the rat is placed in the center of the maze, facing one of the enclosed arms.
- During a 5 min test period, the following measures are taken:

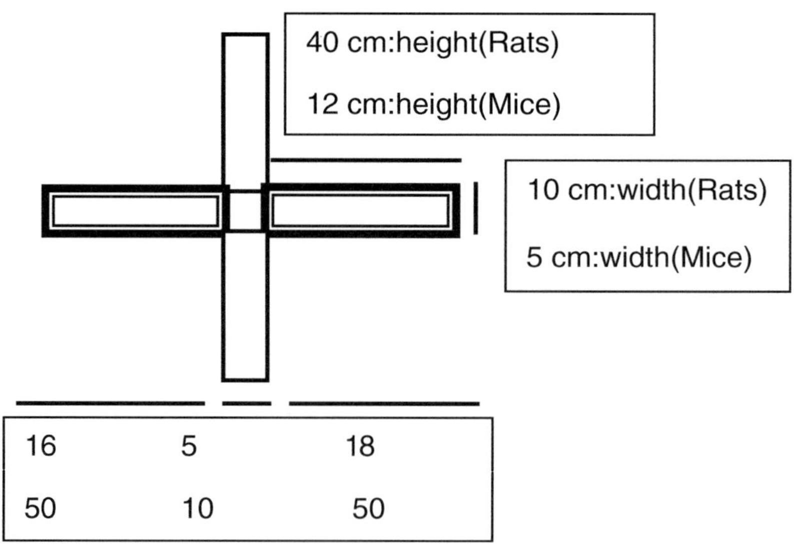

Fig. 3 Elevated plus maze

1. The number of entries into and time spent in the open and enclosed arms.
2. The total number of arm entries.

The procedure is conducted preferably in a sound attenuated area, and the observations are made from an adjacent room via a remote control TV camera or manually.

Advantages

1. Anxiolytic compounds increase open-arm activity, but anxiogenic shows opposite response.

2.2.2 Light/Dark Exploration (L/D)

Principle: The light/dark test is based on the principle that innate aversion of rodents to brightly illuminated areas and on the spontaneous exploratory behavior of rodents in response to mild stressors. The drug-induced movement of animal toward light area can be tested which indicate the efficacy of the drug.

Procedure:

- The testing apparatus consists of a light and a dark chamber divided by a photocell-equipped zone. The one-third of the animal cage is darkened with black spray.
- Both the dark one-third and the bright two-thirds of the cage are partition with a wall of 13 cm long × 5 cm height containing hole in the center.
- The cage is placed on animex activity monitor for counting the total locomotor activity of the animals under experimentation.
- An electronic system using four sets of photocells across the partition automatically counts movements through the partition. These photocells also note the time spent in the light and dark compartments.
- The animals are treated 30 min before the experiment with the test drugs or the vehicle intraperitoneally and are then observed for 10 min.

Advantages

1. The test is relatively simple with no painful stimuli to the animals.
2. This method helps in evaluating the potency of drug due to two compartment models, and also potency matches with clinical trials.

2.2.3 Social Interaction in Rats

Principle: Social interactions are a fundamental and adaptive component of the biology of numerous species. The main principle of this test is based on the free choice by a subject mouse to spend time in any of three box's compartments during two experimental sessions. It includes indirect contact with one or two animals like rat or mice with which it is unfamiliar (Stack et al. 2010).

Procedure:

- The animals are placed in apparatus made up of Plexiglas chambers fitted with clean pine shaving.
- The size of the apparatus is adjusted in such a way that the adolescent and adult animals can freely move into it (30 cm × 20 cm × 20 cm for adolescents) and (45 cm × 30 cm × 20 cm for adults).
- The test apparatus is divided along the long axis into two equally sized compartments with Plexiglas partition that contained an aperture (7 cm × 5 cm for adolescents and 9 cm × 7 cm for adults) to allow movement of the animals between compartments.
- The hole is drilled in such a way that only one animal can be move through the aperture at a time.
- The animals are marked with any color on the back before the initiation of experiment in a holding cage for 30 min.
- For reducing the bias, the animals are exposed to pretest in which baseline level is measured by depriving them in a novel environment.
- After the training or pretesting period, all the animals are then individually placed into the testing chamber having a same age and sex test partner. Also, the animals should not be familiar with both the test apparatus and the experimental animal, i.e., with the paired animals already used for testing.
- Now record the behavior of the animal manually or by a video camera during the 10-min test session.

Advantages

1. This procedure is an useful one because animals are tested at two intervals, which reduce the experimental bias.
2. Mortality rate is zero because animals are socially interacted, and no harmful stimulus (like current) is used.

2.2.4 Open-Field Test

Principle: Open-field test is a simple and novel method which provides a unique opportunity to systematically assess general locomotor activity that is to screen anxiety-related behavior in rodents. In such procedure, the anxiety behavior of animals is directly measured timely without exposing them toward noise or other stimuli. In addition, higher the level of anxiety decreases the number of entries into the various boxes in the openfield apparatus.

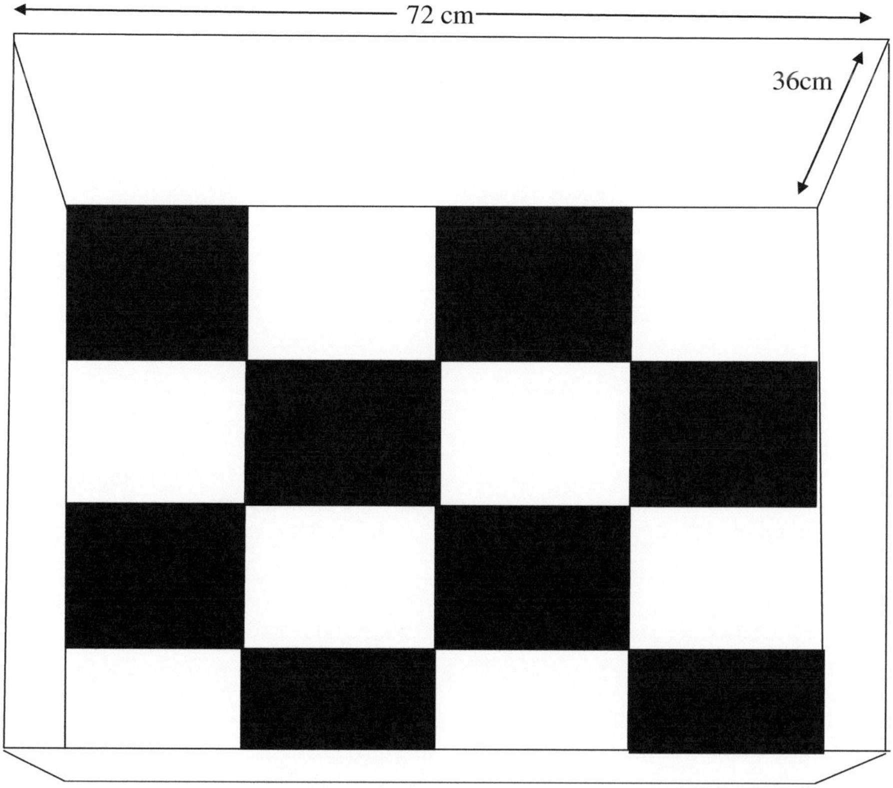

Fig. 4 Open field apparatus

Procedure:

- Open-field test is used to monitor spontaneous locomotor activity using wooden, rectangular, light brown or white black-colored open-field apparatus (100 × 100 × 40 cm) (Fig. 4).
- The floor of the apparatus is divided into 25 rectangular squares by pencil lines or with the help of marker. The experiment on the animal is performed in a room illuminated with 40 W white bulb located 150 cm above the test apparatus.
- After 2 h of first exposure of apparatus, the animal is placed in the center and number of squares cross/10 min by animal is recorded.
- Each crossing is considered only when the animal is fully moved with the four paws into the next box. Apparatus is cleaned properly after each trial and readings are taken.
- In addition to this, we can also record the horizontal units of activity, rearing behavior, defecation, and grooming activity. The maximum cut-off time provided to the animal is 5 min.

Advantages

1. It is helpful for measuring the physical motor ability of experimental animals.
2. Open-field technique is non-invasive and also do not require handling of animal's at each intervals.
3. The animal parameters are taken at stress free environment.
4. By using this test, we can check the number of behavior of animals like grooming, rearing and useful for evaluating the number of CNS disorders like Parkinson's, Huntington's, Alzheimer's, depression, and anxiety disease models.

2.2.5 Ultrasonic Vocalization (UV)

Principle: It is useful and reliable method for testing the anti-anxiety drugs in animals. In UV test, ultrasonic sound is used as indicator of the emotional and motivational status in animals. The ultrasonic vocalization directly indicates the behavioral state of animals and is suppressed by various drugs like benzodiazepines, serotonin (1A) receptor agonists, and selective serotonin reuptake inhibitors (SSRIs).

Procedure:

- The apparatus consists of Lucite box (30 × 30 × 50 cm) with two holes in which animals are trained (Knutson et al. 2002).
- Before the initiation of training session, the animals are habituated to the apparatus for 15 min. In this time, the number and duration of baseline free operant nose-pokes are recorded.
- For the smooth entry of animals, each hole is having a diameter of 3.1 cm and lies 5 cm above from on opposing walls.
- During the experimentation, the animals are placed into the apparatus and number of photo beam brooked is automatically counted in the computer along with frequency and duration of each nose-poke.
- Nose-pokes in the active hole are produced by playback of tape loop with system, recorded into a preamplifier and speaker fitted on the top of operant box.
- Animals are situated 50 cm away from the loudspeaker, and USV playback lasted as long as the animal continued to nose-poke in the active hole. However, playback is not elicited when the animals produce nose-pokes in the inactive hole.

Advantages

1. The model is useful to examine the subjective states of rats in addiction paradigms.
2. Ultrasonic vocalization is suitable method for rapid and repeated evaluation of newer anti-anxiety drugs (Knutson et al. 1999).

Disadvantages

1. It is a time-consuming procedure because in ultrasonic vocalization responding develops within five days, remains stable for at least 3 months and gives highly reproducible results later on.
2. All the animals' do not respond at same frequency of vocalization.

2.2.6 Stair Case Test (SCT)

Principle: Stair case test is used for the screening of anxiolytic and other psychopharmacological of drugs. The model is based on principle that the step—climbing is purported to reflect exploratory or locomotor activity, whereas rearing behavior is an index of anxiety state. In this the number of rearing and steps climbed latency are recorded in a 5 min period.

Procedure:

- The staircase test is carried out by the method. The apparatus is made of wood and consists of five identical steps 2.5 cm high, 10 cm wide, 7.5 cm deep surrounded by walls.
- The height of all the stairs is constant along the whole length of the staircase. On the second side of the stairs, a wooden box of dimensions (15 × 10 × 10 cm) is placed facing the staircase.
- The animal is gently placed on the floor of the box with its back to the staircase. After placing, immediately note down the number of steps climbed and rearing made for the time period of 5 min.
- The animal is considered climbed on a stem when all four paws are placed on the step.
- The number of steps climbed and the rearing responses are recorded for each animal. The apparatus is cleaned thoroughly before and after the recordings.

Advantages

1. It is a less time-consuming method because rearing index directly correlates to the anxiety state of animals.
2. The effectiveness of various anxiolytic drugs like benzodiazepines can be better evaluated by this model.
3. This model do not require food and water deprivation prior to training and also use natural stimuli.

2.2.7 Modified Hole Board Method

Principle: Modified hole board apparatus is used to explore the characteristic behavior of rodents in anxiety. The hole board setup is based on a previously modified whole board that was designed to evaluate cognitive functions. In this, the animal when dipped the head into hole in a floor is considered as valid measure of its attraction toward novelty.

Procedure:

- The apparatus is made up of opaque gray PVC (60 × 20 × 2 cm) board in which 23 holes are drilled (1.5 × 0.5 cm) in three lines.
- All holes on the board are covered by movable lids made up of the same material. The hole board is placed in the middle of a PVC box (100 × 50 50 cm), which represents the central area of an open-field.
- By using marker or by drawing white lines the outer area is divided into 12 quadrates (20 × 16 cm). The size of the PVC box is enlarged by an additional compartment (50 × 50 × 50 cm), in which the experimental animal group are placed during the test period.
- Both the compartments, that is, group compartment and experimental compartment, are separated from each other by a transparent PVC partition perforated with 120 holes (1 cm in diameter).

Advantages

1. Simple method for measuring the response of an animal to an unfamiliar environment, with advantages that several behaviors can be readily observed and quantified in this test.
2. By the use of modified hole board apparatus, we can differentiate between low and high anxiety state of animals.
3. The method is cheap and also do not require any electric shock trials.
4. Modified hole board allows the animals to maintain the visual and olfactory contact to each other, and helpful in reducing stressful conditions of social isolation.

2.2.8 Novelty Suppressed Feeding (NSF)

Principle: The novelty suppressed feeding paradigm (NSF) is a conflict test. It elicits the competing motivations between the drive to eat and the fear of moving toward food pellets placed into the center of the box. Latency to begin eating is used as an index of anxiety-like behavior, because classical anxiolytic drugs decrease this measure.

Procedure:

- The test is performed in a apparatus consisting of box with dimensions 50 × 50 cm covered with bedding and illuminated by a 70 W lamp.
- During the first day, test animals are removed from its home cage and being placed in the corner of a novel test box, containing a single pellet of food (chow)

placed in the center. In this, the latency time to approach the chow and begin eating is recorded within a 5-min period.

- If the animal is anxious, it will avoid the food and display limited exploration of the test environment, whereas if the animals are less anxious they will approach the food quickly and begin eating.
- It has been found that chronic mild stress increases the latency time in the NSF test. This effect is reversed on administration of antidepressants. The antidepressant drugs show significant reduction in the latency to NSF.

Advantages

1. Decreased latency responses to the NSF in response to antidepressants have been associated with changes in hippocampal neurogenesis—a process that is thought to be important in the recovery from depression in humans.
2. The stress employed in these models is very mild relative to most other tests because simply the animal is placed alone into the box having a food pallet.

Ethical Statement

All institutional guidelines, national guidelines, state and local laws and regulations with professional standards for the care and use of laboratory animals should be followed. Studies involving animals must state that the institutional animal ethical committee has approved the protocol. For authors using experimental animals, a statement should be made that the animals' care is in accordance with institutional guidelines, and animals used have been treated humanely and with regard for the alleviation of suffering. Researchers should treat animals as sentient and must consider their proper care and use, and the avoidance or minimization of discomfort, distress, or pain as imperatives. Animal experiments should be designed only after due consideration of animal health. It should be ensured that all researchers who are using animals have received instruction in research methods and in the care, maintenance, and handling of the species being used. All the surgical procedures should be performed under appropriate anesthesia, and follow only those procedures which avoid infection and minimize pain during and after surgery.

References

Battaglia M, Ogliari A (2005) Anxiety and panic: from human studies to animal research and back. Neurosci Biobehav Rev 29(1):169–179

Howard JL, Pollard GT (1990) Effects of buspirone in the Geller-Seifter conflict test with incremental shock. Drug Dev Res 19(1):37–49

Knutson B, Burgdorf J et al (1999) High-frequency ultrasonic vocalizations index conditioned pharmacological reward in rats. Physiol Behav 66(4):639–643

Knutson B, Burgdorf J et al (2002) Ultrasonic vocalizations as indices of affective states in rats. Psychol Bull 128(6):961

Narwal S, Sainia DR et al (2012) Behavior & pharmacological animal models for the evaluation of learning & memory condition. Indo Global J Pharm Sci 2(2):121–129

Reynolds SM, Berridge KC (2001) Fear and feeding in the nucleus accumbens shell: rostrocaudal segregation of GABA-elicited defensive behavior versus eating behavior. J Neurosci 21 (9):3261–3270

Safi K, Neuhäusser-Wespy F, Welzl H, Lipp H-P (2006) Mouse anxiety models and an example of an experimental setup using unconditioned avoidance in an automated system-intellicage. Cogn Brain Behav 10(4):475–488

Stack A, Carrier N et al (2010) Sex differences in social interaction in rats: role of the immediate-early gene zif268. Neuropsychopharmacology 35(2):570

Willner P, Muscat R, Papp M (1992) Chronic mild stress-induced anhedonia: a realistic animal model of depression. Neurosci Biobehav Rev 16(4):525–534

Animal Models of Depression and Chronic Fatigue Syndrome

Shamsher Singh, Vandana, Harsimran Kaur Gill and Puneet Kumar Bansal

1 Introduction

1.1 Depression

Major depression is one common mental illness and proliferating health problem mainly affects the person thought, behavior, feeling and emotional experiences. As per WHO, this is also responsible for morbidity with a prevalence of about 15–20% worldwide (Kulkarni et al. 2009). Common symptoms associated with the depression are irritable mood, decreased interest in pleasurable activities (anhedonia), significant weight loss or gain sometimes, insomnia, loss of energy, feeling of worthlessness or excessive guilt, decreased concentrating power, and increase in suicidal tendencies. In the old time, major depression was considered to be an old-age related mental disorder but the current trends reveal an increased percentage of younger populations being affected from its consequences (Fiske et al. 2009).

S. Singh · Vandana · P.K. Bansal (✉)
Department of Pharmacology, ISF College of Pharmacy, Moga 142001, Punjab, India
e-mail: punnubansal79@gmail.com

S. Singh
e-mail: shamshersinghbajwa@gmail.com

Vandana
e-mail: vandana2556459@gmail.com

H.K. Gill
Insectary Lab, Cornell University, 169, Helios Circle, Ithaca, NY 14853, USA
e-mail: harsimrangill.pau@gmail.com

P.K. Bansal
Department of Pharmaceutical Sciences and Technology, Maharaja Ranjit Singh Punjab Technical University (State Govt. University), Bathinda, Punjab, India

P.K. Bansal and R. Deshmukh (eds.), *Animal Models of Neurological Disorders*, https://doi.org/10.1007/978-981-10-5981-0_11

There are number of evidences about its cause but the major etiological factors include increase mental stress, disturbed sleep and reduced level of monoamines in the brain. According to monoamine hypothesis increased cholinergic and decreased adrenergic activity also increase the instance of depression. Instead of this various monoamines like norepinephrine, serotonin and dopamine help to regulate the physiological functions of brain like mood, reward processing, appetite, sleep and cognition etc. The functions like reward, emotion are regulated by several regions of the brain and these directly interconnected to limbic regions which play essential role in depression (Hasler 2010). It is confirmed by the post-mortem and neuroimaging reports that patients with low volume of grey-matter and glial cell density in the prefrontal cortex and the hippocampus regions are more prone toward the cognitive dysfunctions like depression (Drevets et al. 2008). No doubt there are plentiful medications available for the treatment of depression but the psychological counseling (psychotherapy) is also very effective for most patients. Several antidepressants are available to treat depression such as atypical antidepressants, selective serotonin reuptake inhibitors (SSRI), selective norepinephrine reuptake inhibitors (SNRI) etc. but their produce long term and dose dependent adverse effects.

The various animal models are being employed for going insight into the neurobiology of depression and to evaluate the neurochemical mechanisms involved in physiology of the antidepressants. Existing animal models of human disease have proven of significant value in elucidating basic pathophysiological mechanisms and the development of novel treatment for depression. Therefore, much of the neuroscience study of animal modeling is focused on physiological and neurobiological phenomena that may be helpful for minority of patients. These models elicit the etiological similarity between the cause and neurobehavioral abnormalities in animals which resembles to human depression.

1.2 Chronic Fatigue Syndrome

Clinically fatigue can be defined as a feeling of lack of energy resulting not only from hard work but can be raised from muscular dystrophy. Commonly it's defined as persistent and deteriorative fatigue accompanied by neuropsychiatric abnormalities which last for more than 6 or more than months (C.P.G. 2002). Chronic fatigue syndrome (CFS) is caused by number of factors but it is difficult to diagnose due to the non availability or the hidden target responsible for etiopathology. Fatigue can be associated with immunological disturbances which is well evidenced in two third of the patients. Abnormal signaling of hypothalamic–pituitary–adrenal (HPA) axis and imbalance of brain neurotransmitter levels (particularly serotonin and norepinephrine) has been reported to concern with chronic stress. However some reports generated from brain MRI and morphometric studies has showed that fatigue-related abnormalities are originated due to decreased frontal lobe activity in patients with CFS (Gomez et al. 2013; Puri et al. 2014). Selective serotonin reuptake inhibitors are the useful drugs for the treatments of CFS but long term administration is abortive to produce clinically significant effects. There are various animal models which are reliable and show better validity for research into the neurobiology of chronic fatigue syndrome also they reflect the same symptoms appeared as like humans beings (Figs. 1 and 2).

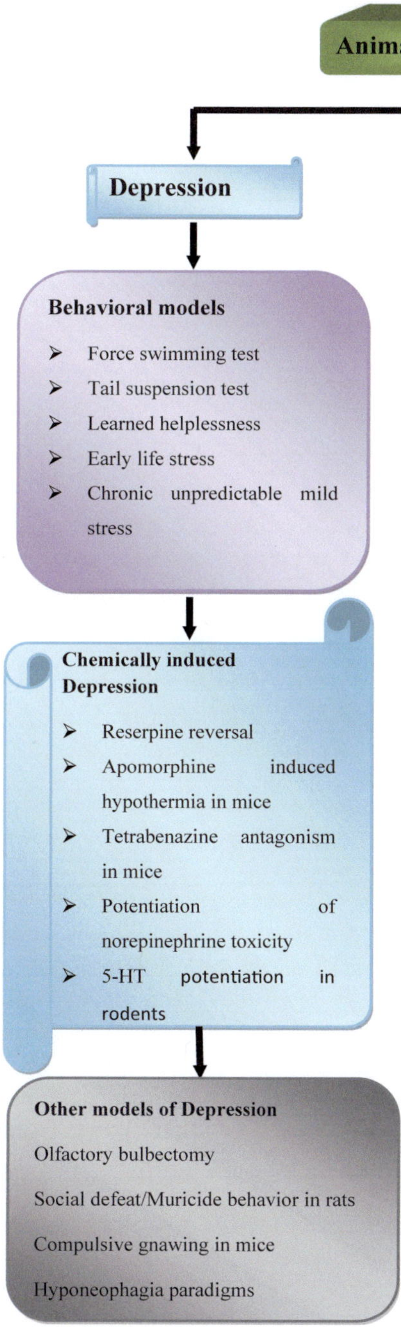

Fig. 1 HPA axis and stress response

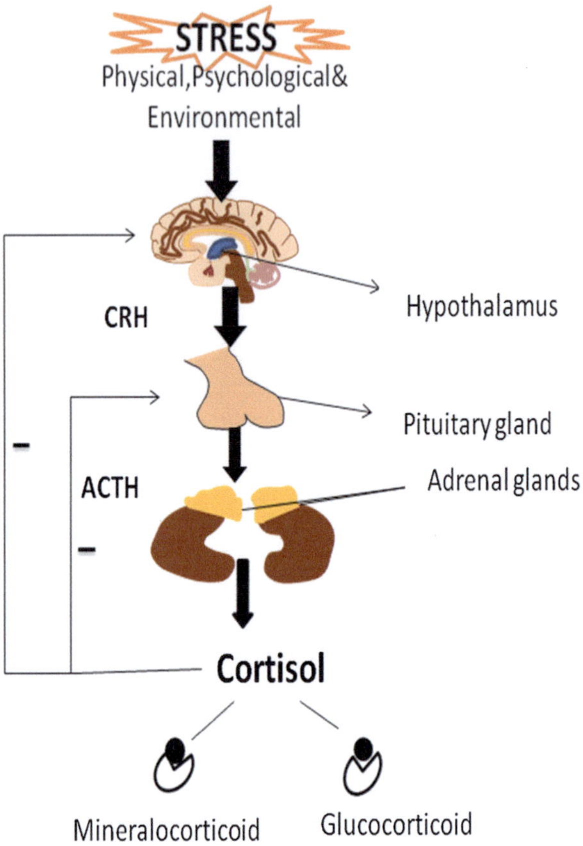

Fig. 2 Pathophysiology of depression

2 Animal Models of Depression

2.1 Force Swimming Test

Purpose: The forced-swimming test (FST) is based on the principle that animals develop an immobile posture in a nonescapable cylinder filled with water. In this test, immobility shown by the animals is interpreted as a passive stress-coping strategy or depression-like behavior (behavioral despair). Now after treatment with antidepressant, the animals will actively execute escape-directed behaviors with longer duration than the control animals administered with normal saline solution (Fig. 3).

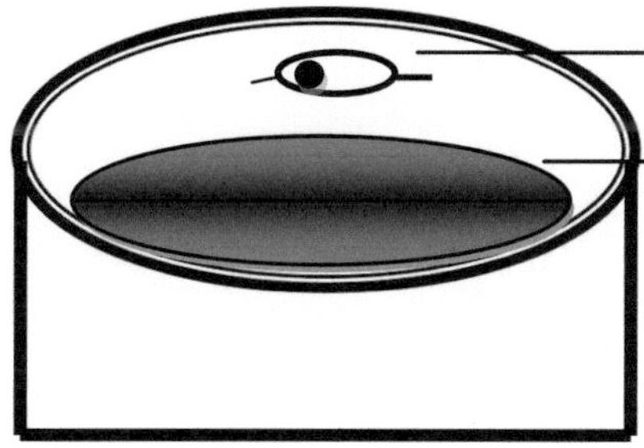

Fig. 3 Force swimming test

Procedure:

- In this model the animals are firstly forced to swim individually in a glass jar (25 × 12 × 25), containing water at room temperature (22 ± 3 °C) without any escape.
- The height of water level is adjusted to 15 cm and the level of water will be maintained during the whole experiment.
- After the completion of initial trial each animal is adapt in a typical immobile position during second time in which they remain floating in water. This is useful for the evaluation of animals during second performance but make sure to keep their head above the water level.
- The total duration of immobility will considered during a total period of 6 min. The procedure is followed for 7 days but for chronic it can be extended up to 21 days.

Merits:

1. It is a fast and reliable tool used to evaluate the potential antidepressants activities with a strong predictive validity of various drugs.
2. Animals are very easily handled and also allow rapid screening of large numbers of drugs.

Demerits:

1. It has poor face and constructs validity.
2. This test is sensitive to acute treatment only and has low validity for non-monoamine antidepressants.

2.2 Tail Suspension Test

Purpose: The "tail suspension test" is widely used method for assessing the therapeutic potential antidepressants. In this the rodents are subjected to an unavoidable and unpreventable short term stress by suspending them into water by holding their tail. Hypothetically it analyzes and reflects the behavioral despair which in turn may reflect depressive disorders in humans (Fig. 4).

Procedure:

- In the test, the animals mainly mice are hung upside down by a wire, thread or by an adhesive tape (20 mm form the tip of the tail) 50 cm above the ground.
- After an early forceful movement to escape the uncomfortable situation, it understood a motionless posture and the period of immobility during last 4 min of total 6 min interval are noted.
- Different rodent strains react differently to basal immobility in the TST, representing it is used only in mice and not in rats due to their larger size and weight.
- Acute antidepressant treatment given prior to the test reduces immobility time in the TST and it is considered to have good predictive validity.
- The TST and FST do not show the same sensitivities to pharmacologic agents or to strain differences, suggesting that responding in these tests may be determined by non identical substrates. Different mouse strains react differently to basal immobility in the TST, representing that this test is sensitive to genetic influence.

Merits:

1. Tail suspension test is unsophisticated, reliable and rapid for screening antidepressants and is able to divide locomotor stimulant dose from antidepressant dose.
2. This test does not produce hypothermia that results from immersion of animal into water in forced swimming test.
3. It provides accurate objective measurement of duration of immobility and is more sensitive to lower doses of drugs with clear dose effect relationships.
4. Its capability to detect a broad spectrum of antidepressants irrespective of their underlying mechanism.

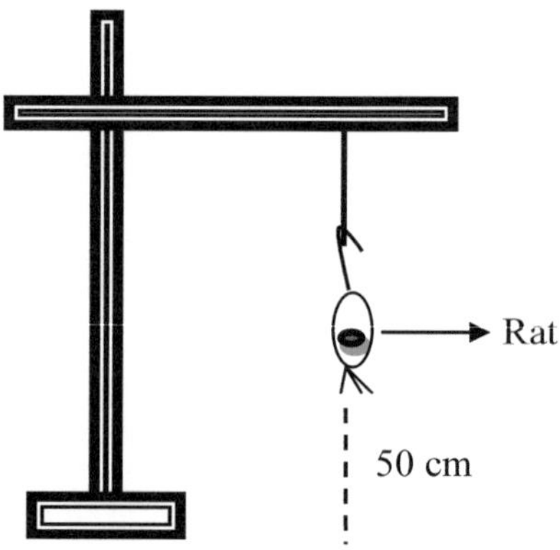

Fig. 4 Tail suspension test

5. It is economical and unsophisticated, methodologically and easily persuadable to mechanization.
6. They are useful in inducing changes that are weak to therapeutic agents in a manner projecting of their effects in humans.

Demerits:

1. The selectivity of this test for monoamine-based mechanisms may limit their ability to detect novel mechanisms.
2. TST does not reproduce the pathophysiology of depression.
3. This test is sensitive to genetic control.

2.3 Learned Helplessness

Purpose: In this animal are trained toward aversive stimuli mainly for evaluating their mental behavior. Basically animals are exposed to inescapable and unavoidable electric shocks in one situation later fail to escape shock in a different situation when getaway is possible. This phenomenon was evaluated as a potential animal model of depression (Anisman et al. 2001).

Procedure:

- In this paradigm stress-exposure phase like unavoidable stress (e.g., electrical foot shock) in one or more sessions is used for animals. During the initial session, the animals are tested for their performance in an active avoidance test.
- In a typical active avoidance test, animals are constrained to one side of a shuttle box chamber where foot shocks are dispersed but the animal has the chance of actively escaping the foot shock.
- Animals before exposing to unavoidable stress show reduced ability to escape in this model. The different forms of antidepressant treatment, including tricyclic antidepressants, selective serotonin reuptake inhibitors, monoamine oxidase inhibitors, and electroconvulsive shock therapy has the ability to restore the reduced potential to escape.
- Animals that are helpless in this model also show some features that have resemblance with human depression, including decreased motor activity, weight loss, altered sleep, decreased motivation, and increases in stress hormones.

Merits:

1. This model is reported to have a good projecting validity including alterations in hypophyseal-pituitary axis (HPA) activity and rapid eye movement (REM) sleep feature of depression and can be used as an additional screening procedure.
2. This model has been used to express the importance of controllability of the stressor as a key psychological constituent in inducing the behavioral shortage.

3. The time course for introduction and for treatment effect in LH paradigms is improved (sub-chronic) compared with the acute responsiveness of the FST and TST.
4. LH models can identify subgroups of stress exposed animals that are more liable to becoming helpless.
5. Using LH to identify endangered and unchanged subgroups can be a useful strategy for investigate mechanisms underlying differential weakness and has analogy to the human situation.

Demerits:

1. The model is time consuming and its specificity is Questionable.
2. The changes persisted for only a couple of days, reducing ease of method use and the need to repetitively administer shocks that has contributed to its unfavorable image, even unacceptable by ethics committees in some countries.

2.4 Olfactory Bulbectomy

Purpose: In rats both olfactory and limbic systems help to regulate the normal functioning as well as coupled with behavioral abnormalities. There are numeral evidences that lesions of the olfactory bulbs results in degeneration of limbic and nonlimbic regions that receive afferent projections from the bulbs. Such type of abnormalities could be responsible for behavioral changes like social behaviors deficit (e.g. increased aggression and irritability) and sexual behavior deficit (e.g., maternal and mating behaviors). Olfactory bulbectomy is reported to induce significant changes in the glutamatergic system in the olfactory cortex evidenced from decrease level of glutamate and glutamate synthesizing enzyme (like glutaminase) are observed.

Procedure:

- For the olfactory bulbectomy model, firstly the rats are anesthetized with suitable anesthesia.
- Adjust the animal in the stereotaxic apparatus and exposed the skull.
- Now drilled the holes anterior to bregma on either side of the midline at a point corresponding to the posterior margin of the orbit of the eye.
- By using suction technique, the olfactory bulbs are removed.
- To control the bleeding the holes are filled with haemostatic sponge and the scalp is sutured.
- Same surgical procedure will be given to sham operated animals, but the bulbs are left intact.

2.5 Social Defeat/Muricide Behavior in Rats

Purpose: Social stress plays an important role in the progress of depression and other psychopathology in humans. Social conflict as a stressor and social interaction can be used as valuable tool for the evaluation of depression. Social avoidance produced a phenotypic property in these models, which can be further quantify and is optional to the model like social withdrawal in human depression. Social stress models induces a depressive like state are more related to human depression as compared with models which are used as acute or severe stressors.

Procedure:

- In this model an intruder animal is placed into the home cage of another inhabitant. There are some factors according which the experiments are generally designed such as strain, body weight and social status to make sure an outcome in which a defeated animal is produced.
- Paradigms are used which vary the number of conflict sessions and the nature of the conflict (psychological vs. physical).
- Physical attack and threat of attack (exposure to sensory contact with another animal but with a barrier to physical attack) can be used separately or combined within a paradigm. In this experiment control group of animals are used which can also exposed to social contact but without any type of conflict and defeat.
- There are two important depression related features that occur in defeated animals are anhedonia, which is measured to reduce preference for sweet solutions, and social avoidance in the presence of a strange animal.
- There are some behavioral or physiologic change are observed such as, decreased sexual behavior and increased defensive behavior, increased anxiety, decreased locomotor or exploratory activity, changes in circadian rhythmicity, alterations in feeding and body weight, sleep disturbances, and impaired immune function. The HPA axis is activated in defeated animals, which is similar to other stress models.

Merits:

(1) The advantage of social defeat models in studying time-dependent neural processes relevant to depression.
(2) Social defeat has proven useful in identifying molecular mechanisms that can induce stable changes in phenotype.
(3) Animals can be identified as susceptible or resistant to the effects of social defeat, indicating further value of social defeat models for investigating substrates of individual vulnerability.

Demerits:

Social avoidance and anhedonia that result from social defeat are long lasting and are sensitive to chronic but not acute treatment with antidepressant drugs.

2.6 Chronic Unpredictable Mild Stress

The chronic unpredictable mild stress (CUS) paradigm is used to study behavioral as well as neurochemical changes that result from stress of a more chronic nature. CUS paradigms model deviate a chronic depressive like state that develops gradually over time in response to stress, and is thus considered more naturalistic in the induction (Li et al. 2009).

Procedure:

- Animals are exposed to a sequence of different stress conditions over a period of some weeks. Several stressors (6–8) are applied (1 or 2 per day) for numerous hours each day.
- Typical stressors include overnight lighting, periods of food or water control, cage angle, and isolation or crowded housing.
- The sequential and unpredictable stress exposure decreases the chances of the animals habituating to any one reoccurring condition.
- The gradual development of a decrease in reward sensitivity or anhedonia is a central focus of CUS paradigms.
- Decreased ability to experience reward is a quality which is common to all forms of depression and it is agreeable to repeated measurement as an experimental endpoint for assessing the effectiveness of CUS.
- Exposure to CUS can result in several other behavioral and physiologic changes that have similarity with symptoms of depression, such as decreased reward-related behavior, decreased self-care, and changes in sleep that respond to antidepressant treatment.
- These and other abnormalities, including increased hypothalamic–pituitary–adrenal (HPA) axis activation and immune system abnormalities, maintain face strength of this model.
- These changes develop slowly over time with CUS exposure and suggest improved face strength of this compared with the more acute stress models. Create validity for CUS is largely based on the development of reduced sucrose preference, which is interpreted to reflect anhedonia, a core symptom of depression.

Merits:

(1) Anhedonia in the CUS model responds to chronic but not acute treatment with several classes of antidepressant drugs, indicating good predictive validity.
(2) The increasing reliability and the temporal characteristics as well as the validity of using anhedonia as an endpoint has resulted in increasing use of CUS models.

Demerits:

(1) Hedonic measures are not changed by antidepressant treatment in the control animals in the model, lack of effect of antidepressants in altering hedonic response in humans.
(2) False positive responses are reported but predictive validity for CUS is strengthened by the time course and the general lack of effectiveness of non antidepressants.
(3) Reliability had been questioned for the CUS model, but is considered considerably improved.
(4) The CUS paradigm is time-consuming process.

2.7 Hyponeophagia Paradigms

Purpose: This paradigm compares feeding behavior in an anxiogenic versus non anxiogenic environment. The stress working in these models is very mild relative to most other tests for antidepressant action, and consists of placing the experimental animal in a novel environment to induce anxiety during testing. Examples of hyponeophagia tests that are used in rats and mice are novelty-induced hypophagia (NIH) and novelty-suppressed feeding (NSF) paradigms. The animal experiences difference between the desire to approach and feed or drink, and the anxiety-induced avoidance of the novel environment.

Procedure:

- Mice are adapted to drink an appetizing liquid (sweetened milk) and then given the opportunity to approach and consume it in two test sessions.
- The first session occurs in the home cage and serves as a control for potential treatment effects on appetite.
- The following test session occurs in a similar cage except that additional parameters (location, lighting) are chosen to be slightly anxiogenic.
- Consumption measures from the novel are compared with the same measures obtained in the home cage and the difference score is the measure of hyponeophagia.
- The addition of the home cage control that utilizes equal measures in a home cage detects and controls for potential alterations in consumption-related variables.
- Using palatable food or drink as the test substance avoids the use of food withdrawal, which can confuse explanation.
- These considerations suggest helplessness of common neural substrates and support the growing body of evidence for overlap in the neural circuitry that modulates anxiety and mood.

Merits:

(1) Hyponeophagia models have good predictive power.
(2) They respond to the anxiolytic effects of benzodiazepines, barbiturates and antidepressants which are anxiolytic.
(3) The hyponeophagia paradigms detect the clinically relevant acute and chronic anxiolytic effect of benzodiazepines.
(4) They detect anxiolytic effects of antidepressants only after chronic treatment, likely to depict the clinical outline for this effect in humans.
(5) Additionally hyponeophagia paradigms can detect increased anxiety, including that resulting from acute SSRI treatment, an effect that is not reliably detected in other models but is clinically relevant.

Demerits: Hyponeophagia model is sensitive to genetic influences on anxiety. The anxiety component in hyponeophagia models provides a degree of face validity. There is significant co-morbidity of major depression and anxiety disorders.

2.8 Compulsive Gnawing in Mice

Principle: Administration of apomorphine to the rodents produce compulsive gnawing instead of vomiting by stimulating the dopaminergic system. Centrally acting anticholiergic agents modify the balance between acetylcholine and dopamine resulting in an enhancement of the apomorphine effect (Randall et al. 1985). Due to this there are several compounds with psychotropic activity are reported to have an apomorphine-synergistic effect. Similarly effect is found with the administration of tricyclic antidepressants.

Procedure:

- Select the age matched animals (i.e. mice) and injects 10 mg/kg apomorphine subcutaneously (s.c).
- Now the animals are administered with test drug or the vehicle i.p. or s.c. but for testing oral activity the animals are treated 30 min prior to apomorphine injection.
- Immediately after apomorphine injection animals are placed into a cage 45 × 45 × 20 cm with a wired lid.
- The bottom of the cage is covered with corrugated paper, the corrugation facing upwards. The animals start to bite into the paper causing fine holes or tear the paper.
- This behavior is enhanced by antidepressants. The mice remain 1 h in the cage.

Merits:

1. It is simple, safe and time securable method.
2. No training is required to the animals before initiation of protocol.

Demerits:

1. It is not reliable method because of its simplicity.
2. This method is suitable only for mice not for other rodents.

2.9 Early Life Stress (ELS)

Purpose: The ELS models typically utilize stress exposure during critical periods of development and finally results into disturbed emotional, social, and cognitive performance. ELS-induced changes are mainly involved for alterations in neural systems that control or respond to stress such as the HPA axis and include endocrine, neurochemical, and behavioral alterations.

Procedure:

- Maternal paradigms are useful as developmental animal models of predisposition to emotional disorders/depression.
- Rats are exposed to daily episodes of 3–6 h separation during a significant period in the first 2 postnatal weeks and the separation can include the whole litter from the mother, or individual pups can be separated from littermates and the mother.
- Previously separated animals are then permitted to develop under normal conditions through adulthood, when phenotypic individuality are evaluated.
- As adults, previously separated rats show behavioral abnormalities, including increased anxiety, fear responses as well as reduced motor activity, social motivation, hedonic responding, endocrine and neurochemical alterations in stress-related systems.
- Licking/grooming of pups and domed back nursing are some of the important features of maternal behavior in female rats.

Prenatal stress: Parental care is ever more concerned as an important modifier of stress effects during development in humans. The impact of stress has also been modeled in prenatal stress paradigms. Maternal stress is of various types, such as, noise exposure or restraint during gestation results in alterations in the offspring, including increased anxiety, increased index of depression in depression models, and alterations in HPA axis activity.

Merits:

This paradigm represents a way to identify a population of depression/anxiety-liable individuals in an experimental setting.

Demerits:

(1) ELS paradigms are sensitive and time-consuming.
(2) The ELS paradigms produce animals with long-term depression-related features.

2.10 Chemical Induced Depression

2.10.1 Reserpine Reversal

Principle: This model analyzes the ability of antidepressants to overturn the inhibitory effects of reserpine on motility, body temperature and neuronal activity in animals. It is reported that reserpine, produce antihypertensive and antipsychotic effect by degrading the brain monoamines non-selectively has also been associated with locomotor hypomotility in rodents. The drugs which perk up the synaptic concentration of serotonin also control the behavioral syndrome induced by administration of 5-hydroxytryptophan (5-HTP). The animals with lower levels of 5-HT_{1A} autoreceptors are approved to show better results following forced swim stress compared to animals with higher levels of autoreceptors.

Procedure:

- Select the age matched animals (i.e. mice) and administered with 2 mg/kg reserpine s.c on the day of test session.
- Now eighteen hours after reserpine administration, the animals are placed into individual cages.
- Noted the initial rectal temperature by using electronic thermometer to a constant depth of 2 cm.
- Following administration of the test compound (either i.p. or p.o.), the rectal temperature is measured again at 60 min intervals for 7 h.

Merits:

These models suggest good analytical validity in term to evaluate monoamine-based antidepressant activity of compounds.

Demerits:

Skilled person is required for the animal handling and experimentation.

2.10.2 Tetrabenazine Antagonism in Mice

Principle: Tetrabenazine (TBZ) decreases the level of biogenic amines (e.g. noradrenaline, dopamine and serotonin) on the nerve terminals without disturbing their de novo synthesis. Additionally it also increase the reuptake of noradrenaline into the nerve terminals and increase the noradrenaline concentration at the receptor site which is reported to be antagonize with anti-depressants. Therefore, both MAO-inhibitors and tricyclic antidepressants are known to stop or to alienate these effects. The avoidance of TBZ induced ptosis and catalepsy can be used for evaluation of antidepressant (Shah et al. 2014).

Procedure:

- In this model the animals (mice) are treated with TBZ 40 mg/kg i.p 60 min after oral or 30 min after i.p administration of test drug.
- The animals are placed into individual cages and the test is continuing for 30 min after TBZ administration.
- Same test is repeated every 30 min up to 2 h. Catalepsy and ptosis are used as criterion which include a set of steps formed with 2 cork stoppers having 2 steps of 3 cm height.
- The animals are placed with their head downwards and hind legs upon the top cork. If the animals stay in this catatonic state indicates cataleptic effect the TBZ.
- MAO-inhibitors and tricyclic antidepressants or any other test drug antagonized cataleptic effect of the TBZ indicates its therapeutic efficacy.

Merit:

This model reflects the comparable behavior of depression likely to be seen in the human beings.

Demerit:

Mortality rate is high because so this needs a TBZ dose adjustment before administering to the animals.

2.10.3 5-Hydroxytryptophan Potentiation in Mice

Principle: It is well postulated that depressive compounds exert antidepressant activity because they are able to enhance central noradrenergic and/or serotoninergic functions. Number of antidepressant agents potentiates serotonin effects by blocking the re-uptake of serotonin. DL-5-hydroxytryptophan is used as the precursor of serotonin but the enzymatic breakdown is inhibited by the MAO-inhibitor like pargyline. In mice the distinctive symptom of head-twitches is observed.

Procedure: Male wistar rats are fasted for 3 h prior to oral administration. Now the animals are administered with vehicle/standard/test compounds orally to the respective groups. After 60 min, animals were injected with 75 mg/kg pargyline hydrochloride, subcutaneously. 90 min after pargyline, the animals were injected with 5-HTP, 10 mg/kg i.p. The number of head twitches and behavioral parameters like escape tendency, hind limb abduction, tremors, fore limb clonus and lardosis were calculated for half an hour.

Merit:

This model promises of improving and understanding the pathophysiology of depression and to evaluation the newer anti-depressant used for the treatment of depression and mood disorders.

Demerit:

Animals are exposed to severe a panic stimulus that's why technique is not reliable.

2.10.4 Apomorphine-Induced Hypothermia in Mice

Principle: Apomorphine induces hypothermia in mice which can be prohibited by antidepressants.

Procedure:

- Male NMRI mice (20–22 g) are used and randomly assigned to test groups of 6 subjects.
- One hour after oral administration of the test compounds or the vehicle a dose of 16 mg/kg apomorphine is injected s.c.
- The rectal temperature of each mouse is measured by an electronic thermometer immediately prior to apomorphine administration and 10, 20 and 30 min later.
- During the entire experiment, subjects are housed in groups of three in glass jars at room temperature.

2.10.5 Potentiation of Norepinephrine Toxicity

Principle: The antidepressant activity of some depressive compounds can be achieved because they are able to improve central noradrenergic and/or serotoninergic functions was approved by the monoamine hypothesis. Similarly the enzymatic breakdown is inhibited by the use of monoamine oxidase inhibitor pargyline. In mice the distinctive symptom of head-twitches is observed (Shank et al. 1987).

Method:

- Male mice (22–25 g) are randomly assigned to test groups of 10 subjects.
- The test drug, the standard or the vehicle is given orally 1 h prior to the s.c. injection of the sub lethal dose of 3 mg/kg noradrenaline.
- The groups of 10 mice are placed into plastic cages with free access to food and water

3 Animal Models of Chronic Fatigue Syndrome

3.1 Load Force Swimming Method (LFS)

Principle: In this the animals are forced to swim under stressful condition by giving load few times more than their body weight. Animals are provided with training in the form of swimming under competitive conditions, demanding a higher propulsive force output. The main objective is to prepare the animals for facing the

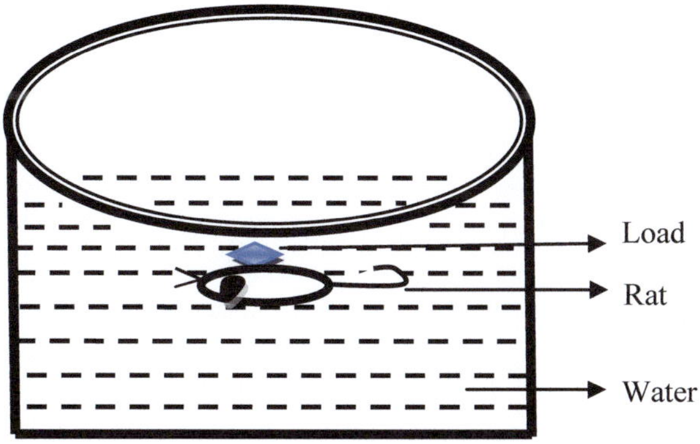

Fig. 5 Load force swimming

stressful conditions and to know about the stress induced behavioral changes (Sachdeva et al. 2011) (Fig. 5).

Procedure:

- In load force swimming test the rats are allowed to bear a load of 10 ± 2% body weight plunged individually and made to forced swim inside a rectangular glass jar (60 cm × 30 cm × 45 cm) containing 30 cm of water maintained at 22 ± 3 °C.
- The depth of the water is sufficient to prevent the animals from touching the bottom of the floor with their tails.
- The animals are forced to swim daily for 28 days in the jar till fatigue.
- When the signs of marked fatigue became clear, then animals were removed from the water and placed in clear observation chamber.
- The time onwards before rat initiate grooming (licking and rubbing of the skin/fur) was recorded to review the post exercise fatigue.
- The time to grooming was recorded in minutes after each rat was removed from the water. Survival to forced swim test was evaluated based on the rats maintained an angular position such that both the animal's nose and head were above the surface of the water.

Merits:

1. It is a suitable method for the evaluation of both muscle relaxant and anti-depressant drugs because of its similar physiology like morris water maize.
2. This method show good accuracy and reliability.

3.2 Running Wheel Activity (RWA)

Principle: CFS is a sudden flu like illness which reduces all the activities of the body. In this an apparatus consisted with a wheel rotated in a water containing animal used to measure fatigue latency. Hence the administration of antigen or drugs to the animals helps in the evaluation of protective test drug effects.

Procedure:

- The apparatus consisted of a plastic glass water tank (38 × 30 × 15 cm) with a wheel at a height of 6 cm.
- The wheel resembles a circular hollow cage with diameter of 28 cm dipped in water maintained at 15 ± 2 °C and filled in the tank.
- One hour after oral administration of drugs, the mice are placed individually in the tank at the base of wheel and removed from water after 6 min.
- After placing the mice would try to climb on the walls of wheel to escape from the cold water but it would be failed to escape from cold water due to rotation of wheel.
- The wheel would stop turning, when attempts to escape are finally abandoned.
- The assembly is fitted with a digital counting device which helps to count the number of rotations of the wheel. The wheel rotations during the complete 6 min test for mice receiving drug treatment are compared with those of control group.

Merits:

1. With the help of this paradigm we can evaluate the activity of various muscle relaxant drugs without inducing any stress.
2. It is a simple method and provides good predictive validity.

Demerits: This model is not useful for evaluation of chronic fatigue related to other disease.

3.3 Restraint Stress Model

Principle: The model depicts the stressful conditions commonly seen in human beings. In this restraint stress model we provide stressful stimuli to the animals for evaluating oxidative stress induced tissue damage likely experienced in human in the form of psychological stress (Gerecke et al. 2013).

Procedure:

- In this procedure animals are subjected to restraint stress in a glass restrainer for 21 days (3 h/day).
- To ensure the immobility restrainer apparatus size can be adjusted according to the size of the animals.
- After the restraint stress the test group animals are treated with the protective drugs.
- This chronic exposure of restraint stress produced physical and mental fatigue that represented CFS.

Merits:

1. This is time saving method and provides reliable results.
2. The stress induced in the animals commonly resembles to the human stressful conditions of human beings.

Demerits:

All the antidepressant donot effectively recover the depressive state if it is caused by some other factors.

3.4 Noise Induce Stress Model

Principle: Noise stimuli directly increase the cortical activity in animals and elevate the level of oxidative stress. Chronic exposure of animals toward noise pollution decreases the normal relaxation time and increase the muscular fatigability confirmed from the normal behavioral activity (Di et al. 2011).

Procedure:

- Animals are exposed to noise for 20 days.
- Rats are housed in rooms with controlled lightning (12 h light/dark).
- 4 h daily for 20 days rats are subjected to 100 dB SPL broadband white noise.
- Noise is produced by one loudspeaker which is (15 W), determined by a white-noise generator (80–9000 Hz), and it is installed 30 cm above the cage.
- The noise level was set at 100 dB SPL uniformly throughout the cage and it is observed by a digital sound level meter Extech instruments.
- To avoid the influence of handling-stress on evaluation of effects due to noise exposure, control rats are kept in the above-described cage during the corresponding period of time, without noise stimulation.

Merit:

It has a good predictive power for the evaluation of centrally and peripherally acting muscle relaxant.

Demerit:

In this paradigm animals need more training time and also the results obtained are with bias.

3.5 Induction of CFS by LPS and B. Abortus Antigen

Principle: In this lipopolysaccharide and B. abortus Antigen are used to induce the CFS related to fatigue likely as human beings. These antigenic substances induced fatigue either by elevating the level of immunogenic substances or by elevating the level of oxidative stress.

Procedure:

- CFS is induced by six repeated injections of original B. abortus antigen solution (0.2 ml per mouse) via the tail vein every 2 weeks.
- Treatment with the BA antigen is known to induce changes in cytokine gene expression in lymphocytes of mice.
- The interaction of lipopolysaccharide (LPS) with macrophages results in the generation of hydrogen peroxide, superoxide anion, hydroxyl radical and singlet oxygen.
- The elevated level of oxidative stress also increase tissue damage and fatigability in animals confirmed from the behavioral and biochemical estimations.

Merit:

The model is very commonly employed for evaluation of muscle relaxant because of its simplicity.

Demerit:

The rate of mortality is high because the antigenic substances are directly administered to the animals.

Ethical Statement

All institutional guidelines, national guidelines, state and local laws and regulations with professional standards for the care and use of laboratory animals should be followed. Studies involving animals must state that the institutional animal ethical committee has approved the protocol. For authors using experimental animals, a statement should be made that the animals' care is in accordance with institutional guidelines and animals used have been treated humanely and with regard for the alleviation of suffering. Researchers should treat animals as sentient and must consider their proper care and use and the avoidance or minimization of discomfort, distress, or pain as imperatives. Animal experiments should be designed only after due consideration of animal health. It should be ensured that all researchers who are using animals have received instruction in research methods and in the care, maintenance and handling of the species being used. All the surgical procedures should be performed under appropriate anesthesia and follow only those procedures which avoid infection and minimize pain during and after surgery.

References

Anisman H, Merali Z (2001) Rodent models of depression: learned helplessness induced in mice. Curr Protoc Neurosci 8.10 C. 11–18.10 C. 15

Di G-Q, Zhou B et al (2011) Aircraft noise exposure affects rat behavior, plasma norepinephrine levels, and cell morphology of the temporal lobe. J Zhejiang Univ Sci B 12(12):969–975

Drevets WC, Price JL et al (2008) Brain structural and functional abnormalities in mood disorders: implications for neurocircuitry models of depression. Brain Struct Funct 213(1–2):93–118

Fiske A, Wetherell JL, Gatz M (2009) Depression in older adults. Annu Rev Clin Psychol 27 (5):363–389

Gerecke KM, Kolobova A et al (2013) Exercise protects against chronic restraint stress-induced oxidative stress in the cortex and hippocampus. Brain Res 1509:66–78

Gómez ÁJ, Campos NV, Belenguer A, Ávila C, Forn C (2013) Regional brain atrophy and functional connectivity changes related to fatigue in multiple sclerosis. PLoS ONE 22, 8(10): e77914

Hasler G (2010) Pathophysiology of depression: do we have any solid evidence of interest to clinicians? World Psychiatry 1, 9(3):155–161

Kulkarni SK, Dhir Ashish, Akula Kiran Kumar (2009) Potentials of Curcumin as an antidepressant. Sci World J 9:1233–1241

Li Y-C, Wang F-M et al (2009) Antidepressant-like effects of curcumin on serotonergic receptor-coupled AC-cAMP pathway in chronic unpredictable mild stress of rats. Prog Neuropsychopharmacol Biol Psychiatry 33(3):435–449

Puri B, Jakeman P et al (2014) Regional grey and white matter volumetric changes in myalgic encephalomyelitis (chronic fatigue syndrome): a voxel-based morphometry 3 T MRI study. Br J Radiol

Randall PK (1985) Quantification of dopaminergic supersensitization using apomorphine-induced behavior in the mouse. Life Sci 37:1419–1423

Sachdeva AK, Kuhad A et al (2011) Epigallocatechin gallate ameliorates behavioral and biochemical deficits in rat model of load-induced chronic fatigue syndrome. Brain Res Bull 86 (3):165–172

Shah RK, Kenjale RD et al (2014) Antidepressant effect of linseed oil on various behavioral and pharmacological models of depression in Swiss albino mice. Int J Nutr Pharmacol Neurol Dis 4 (4):246

Shank RP, Vaught JL, Pelley KA et al (1987) A highly potent inhibitor of serotonin uptake. J Pharmacol Exp Ther 247:1032–1038

Working Group of the Royal Australasian College of Physicians. Chronic fatigue syndrome. Clinical practice guidelines (2002). Med J Aust 6(176):S23

Animal Models of Sleep Disorder

Shamsher Singh, Onkar Bedi, Ganesh Singh Bhakuni
and Puneet Kumar Bansal

1 Introduction

Sleep is a complicated neurological disorder on the conduct of the animals characterized by using altered consciousness with diminished sensory response. The primary feature of sleep is to provide rest and restore the body's energy levels. The duration and pattern of sleep varies considerably among individuals. Age has an important effect on quantity and depth of sleep, and it is recognized as an architecture cyclic process. The sleep cycle characterized by different phases flowed by awake, dozing, unequivocal sleep, deep sleep transitions, cerebral sleep, paradoxical sleep. The changes in normal sleep behavior affected by environmental or psychological parameters are leading causes of sleep disorder. According to American Academy of Sleep Medicine (2014), sleep disorders (Somnipathy) are classified into insomnia, dyssomnias, parasomnias, circadian rhythm sleep disorders involving the timing of sleep, and other disorders including ones caused by medical or psychological conditions and sleeping sickness. Sleep issues especially arise due to alterations in the quality, quantity, and pattern of sleep (Cappuccio et al. 2010).

S. Singh · O. Bedi · G.S. Bhakuni · P.K. Bansal (✉)
Department of Pharmacology, ISF College of Pharmacy, Moga 142001, Punjab, India
e-mail: punnubansal79@gmail.com

S. Singh
e-mail: shamshersinghbajwa@gmail.com

O. Bedi
e-mail: Jagrit143@gmail.com

G.S. Bhakuni
e-mail: Bhakuniganesh17@yahoo.com

P.K. Bansal
Department of Pharmaceutical Sciences and Technology, Maharaja Ranjit Singh Punjab
Technical University (State Govt. University), Bathinda, Punjab, India

P.K. Bansal and R. Deshmukh (eds.), *Animal Models of Neurological Disorders*, https://doi.org/10.1007/978-981-10-5981-0_12

Sleep disorders covers a huge spectrum of diseases such as inability to sleep at the desired time, excessive daytime sleepiness, abnormal movements or behavior during sleep. Sleep disorders also include sleep apnea (stops in breathing during sleep), narcolepsy and hypersomnia (excessive sleepiness at in appropriate times), cataplexy (sudden and transient loss of muscle tone while awake), sleeping sickness (disruption of sleep cycle due to infection), sleepwalking, night terrors, and bed wetting.

The pharmacological and therapeutic treatment of sleeping disorders includes barbiturates (phenobarbital, methohexitone), benzodiazepines (diazepam, lorazepam), and newer benzodiazepines (zolpidem, zopiclone). Sleep is a complicated physiological technique; this is stimulated via many factors, such as the problems associated with specific personal situations, and due to insufficient sleep. However, public and individual efforts to limit or manage sleep loss may be offset through the personal or societal desires by means of enhancing our daily recurring works or by way of adapting appropriate behavior to enhance the satisfactory of existence.

To understand this devastating disorder and to develop active therapeutic treatment, there is need to comprehend all animal models which give a wonderful step forward pathway to reveal new strategies for the diagnosis of sleep disorder. In this chapter, we reveal various animal models of narcolepsy like sleep deprivation induced changes in REM and NREM cycles. The different models helped us to reveal the pathophysiology of narcolepsy along with development of more therapeutic treatment.

2 Animal Models of Sleep Disorder

2.1 Sleep Deprivation Induced by the Modified Multiple Platform Technique

Principle

This model is based on principle that when the animals are placed on different platforms, they show variation in sleep cycles (Fig. 1). In this, small platform is used surrounded by water on which animals are placed. The animals enter the paradoxical phase of sleep, lose the postural balance partially or fully slip from the platform into the water and awaken. Now in this, large platform is also used to remove the social isolation dependency of sleep.

Procedure

- Male Wistar rats (250–280 g) are housed under a 12-h light–dark cycle. Room temperature is set at 20 °F, and animals should have free access to water and food at all times.
- The anesthesia is given with ketamine–diazepam, and two ipsilateral stainless steel screws for electroencephalogram (EEG) monitoring are implanted by

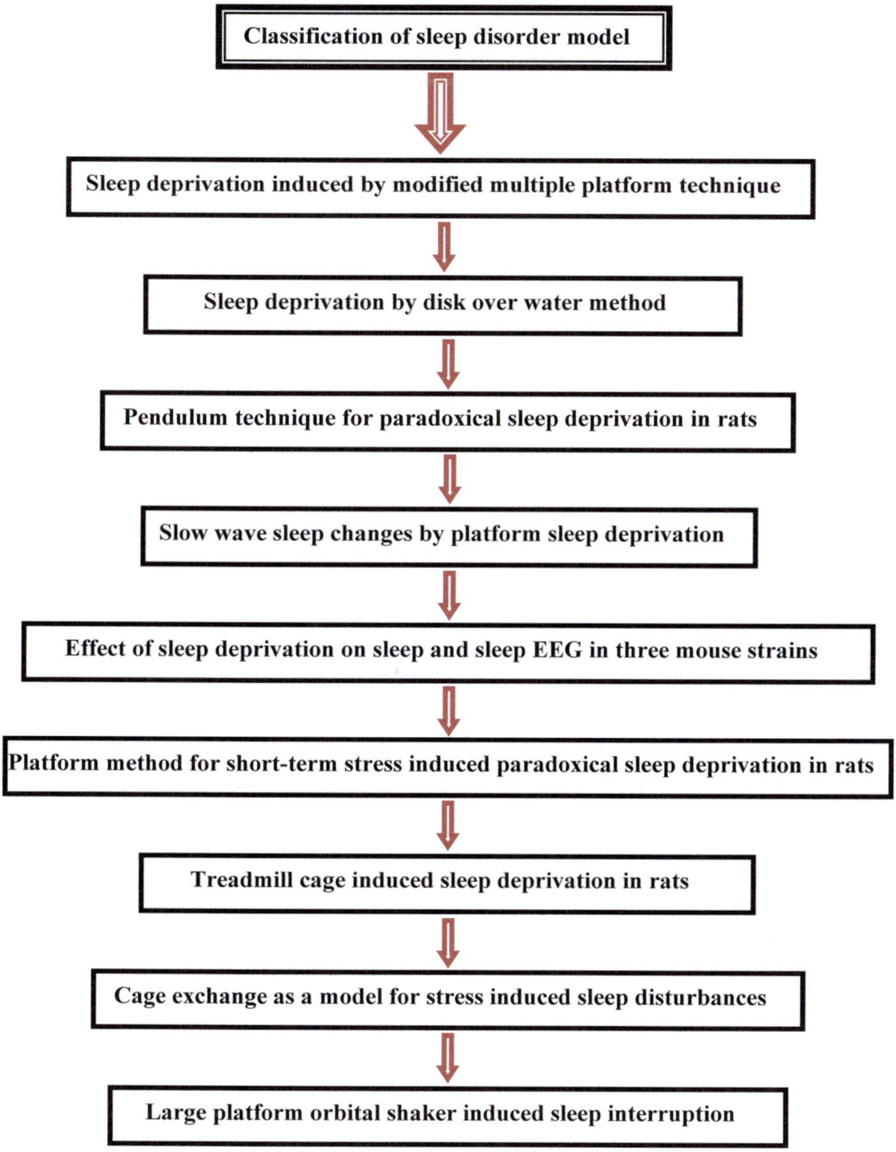

Fig. 1 Classification of animal models of sleep disorders

overlaying the right lateral fronto-parietal area and the left medial parietal cortex.

- Place additional one pair of nickel–chromium fine-wire electrodes in the dorsal neck muscle for electromyogram (EMG) recording. The electrodes are fixed firmly to the animal cranium with acrylic dental cement. After surgery, penicillin and diclofenac are administered and allow for 15 days for recovery.

- After 15 days, baselines are recorded for 3 days in a group cage (n = 5) for the animal subjected to modified multiple platform (MMP) or in individual home cages for single platform.
- To evaluate the baseline recording, animals are adapted to the sleep deprivation procedure for 30 min on 3 successive days. So, after obtaining the sleep parameters, convert these into percentage of total recording time (usually 23 h) for each day.

Experiment A: *Modified multiple platform (MMP) method*

- In MMP method, rats are tested in socially stable groups of five in separate water tanks (123 × 44 × 44 cm) containing either 18 round platforms of 6.5 cm diameter (small platforms) or 18 platform with 14-cm diameter (large platforms).
- Large platform groups are often used as a control in sleep deprivation experiments. The additional control group is placed in a third tank on a stainless steel wire mesh (2.3-mm openings), which helps in allowing rats to lie down without touching the water.
- This group is referred to as the grid control group. All the tanks are now filled with water up to a level 1 cm below the surface of the platforms or the grid. During experimentation, continuous electrophysiological monitoring is performed in one rat in each group of five.
- Now after 96 h of sleep deprivation, rats are returned to the home cage, where they are continued under recording for 4 additional days. Ten deprivation runs cycles are conducted, with different animals, to achieve a final N of 10 recorded animals per group.

Experiment B: *Single platform method*

- In the single platform procedure, animals are placed on a single small (6.5 cm diameter) platform or large (14 cm diameter) platform in an enclosed tank (23 × 23 × 35 cm).
- As similar to the group procedure, the platform was surrounded or placed above 1 cm to the level of water. Rats are again returned to the home cage after 96 h of sleep deprivation cycle, where they are continued under recording for 4 additional days.
- Twelve animals are tested in each group.

Advantages

- The model reflects the complete sleep cycle occur in the human beings or changes in sleep cycles in pathological conditions.
- This method is useful to identify the effect of social isolation on sleep paradigm.

- The model is also a useful tool for the evaluation of various sedative drugs.

Disadvantages

- Due to the fear of water present under the platform, animal may loss slow wave sleep (SWS) cycles.
- The procedure is lengthy and also needs expert person to carry it smoothly.

2.2 Sleep Deprivation by Disk-Over-Water Method

Principle

The method exhibits the physiological modifications inside the diverse effects of sleep loss by the usage of a gentle bodily stimulation.

Procedure

- The animals of 230–250 g are selected for experimentation and are divided into two groups that are sleeping deprived (SD) rats and a control rats.
- The animals are simultaneously housed each on one side of a divided 46-cm horizontal disk suspended over a shallow tray of 2–3 cm deep water.
- The recordings are monitored continuously by EEG, EMG, and theta (θ) activity to detect sleeps states. The disk is mechanically circled at the low velocity of 3.33 rpm, while the SD rat starts to sleep or enters a 'forbidden' stage. This causes awakening of the rat and forcing it to walk opposite to disk rotation to avoid being carried into the water.
- Similarly, the control rat receives the same mild physical stimulation because they are also placed on the same disk. However, the sleep or targeted sleep stages are severely reduced in the SD rats, than control rats.
- The control rats show significant sleep reduction in comparison with the SD rats because they show physiological effects in the same direction as those of experimental rats, but to a much smaller in degree.

Advantages

- It is a simple method and also uses the physical method to measure the variations in the sleep cycles.
- This model provides easy method to investigate the effect of stress over the sleep.

Disadvantages

- The sleep-deprived rats during the experimentation show reliable syndrome like body temperature changes, weight loss, and increased metabolic rate. So, these consequences can cause alteration in normal sleep paradigm.

- It is difficult to identify the animal behavior whether it is due to loss of sleep or due to change in circadian rhythm.

2.3 Pendulum Technique for Paradoxical Sleep Deprivation in Rats

Principle

The technique is based on the sound evaluation in which animals are placed on the cage which moves in forward and backward directions like pendulum. At extreme motion of cage, the animals show postural imbalance. So this causes awakeness in animals, and they starts to move in opposite direction to the movement of cage.

Procedure

- The rats with body weight 220–250 g are selected are deprived for a night and familiarized with recording conditions prior to experimentation.
- The animals are placed in apparatus consisted of a swing with room for three rats in individual cage (30 × 25 × 35) supported by frame. They are put into recording cages connected to adaptation leads nearly for one week.
- The deprivation treatments are initiated after being positioned in a desk-bound pendulum and connected to the recording leads for one to two days. Animals are divided into two groups (n = 6) and are submitted to a deprivation period of 72 h.
- Give the deprivation into two phases, that is, for one group the deprivation is started at the onset of the light phase (09:30 h) and for the second group at the onset of the dark phase (21:30 h) of the illumination cycle.
- This arrangement helps us to determining the recovery of sleep after deprivation at the beginning of the light period (Group 1) or at the beginning of the dark period (Group 2). The speed of the pendulum is adjusted very carefully to yield sufficient paradoxical sleep (PS) deprivation in the animals.
- Now, set the time of gap between two in such a way that each group takes 12-h period according to the following schedule: 45, 35, 25, 20, 17, and 15 s. Following completion of both deprivation conditions, two baseline conditions (n − 6) are conducted.
- Repeat the same for the next groups in cages when the recovery sleep was monitored in the first and second groups, respectively. Animals are weighed immediately before the deprivation period and after termination of the recovery period.

Advantages

- It is a safe method because the stimulus for the sleep deprivation does not contain nonelectric type, that is, simple cage hanged like pendulum.

- The cages for the animals are covered from all side to avoid the external disturbances.
- This method is helpful to detect the effect of light and dark cycles on the animal sleep activities.

Disadvantages

- There are maximum chances of error because animals while experimentation are judged from adjacent room and movement of animals may be affected by size of cages.

2.4 Slow Wave Sleep Changes by Platform Sleep Deprivation

Principle

The rats are sleep deprived by using different size platforms. The REM and NREM sleep paradigms are recorded in two light cycles of duration 48 h. In both platforms, first light period (0–10 h sleep deprivation) measures REM sleep and second light period (22–34 h sleep deprivation) measures NREM sleep.

Procedure

- Place the rats on inverted flowerpots surrounded by water at room temperature. The flowerpot platform is placed above 2 cm to level of water.
- All the animals are randomly assigned into two groups that is one group having (n = 7) is placed on pots with diameter 15.7 cm. Second group containing (n = 8) is placed on pot with 5.1 cm diameter. The sizes are chosen to maximize the difference between the groups. Food and water is provided continuously through disposers hanging.
- For the first 30–45 min on the platforms, rats are very active and slowly they quieted down. The response is recorded continuously during the platform sleep deprivation. Now, the baseline recordings are made on day 15 following implantation, for 10 h starting at 09.00 h.
- Again the rats are returned to their home cages after first recording until 09.00 h the next morning when they are placed on the platforms for sleep deprivation. After 48 h of platform sleep deprivation with recording, the animals are moved from the platforms to the sleep recording bins. The recovery sleep time is then recorded for 10 h from 09.00 h. The animals did not go to sleep immediately, but spent a considerable part of the first hour grooming.
- The EEG fronto-frontal and fronto-parietal and neck are recorded on paper with speed of 10 mm/s.

Advantages

- The method is a more preferable because it compares the REM and NREM sleep cycles with deep slow wave sleep (SWS) by using platform of different sizes (Alkadhi et al. 2013).
- The changes in slow wave sleep cycles depict those appeared in human beings.
- The apparatus used for measuring sleep deprivation is like animal home cage which does not disturb normal physiology of animals.

Disadvantages

- Error is more during experimentation due to external factors like noise and other environmental changes.
- The protocol is long-lasting because competition of one cycle takes 48 h.

2.5 Effects of Sleep Deprivation on Sleep and Sleep EEG in Three Mouse Strains

Principle

The transgenic animals are used to explore the contribution of genetic makeup to maintaining the sleep cycles (Deboer et al. 2003). The determination of the sleep is done by using animals of different strains to know the underlying genetic cause of variation in slow wave sleep, REM, and NREM cycles.

Procedure

- Adult male mice of three different bred strains are used, 129rOla (Ola; n = 9), 129rSvJ (SvJ; n = 6), and C57BLr6J (C57; n = 11). The animals are maintained in a 12-h light–12-h dark cycle individually, kept in Macrolon cages (36 × 20 × 35 cm), and placed in sound attenuated chambers. Food and water is provided according to need, and animals are adapted for a minimum of 3 weeks to these conditions.
- Now surgically EEG and EMG electrodes are implanted by giving barbital anesthesia with Nembutal sodium (80 mg/kg i.p., volume approximately 0.5 ml). Two gold-plated miniature crews (Ø0.9 mm) served as epidural EEG electrodes. They are placed over the right occipital cortex 2–3 mm lateral to the midline, 2 mm posterior to bregma and the cerebellum at midline, 1 mm posterior to lambda and soldered to a plug with stainless steel wires.
- The EMG was recorded with two gold wires (Ø0.2 mm) inserted into then neck muscle and soldered to the plug, which is anchored to the skull with dental cement. The animals for 3 weeks are allowed for recovery. Now, start the baseline recording at light for 18 h. At least 10 days later, the mice are subjected to 4-h sleep deprivation (SD) and recorded for remain 24 h. The SD is carried

out at light onset by introducing objects (like nesting material) into the cage and later by tapping on the cages. The tapping produces drowsiness, or the EEG exhibits slow waves.

- Again animals are placed into new cages without any disturbance during deprivation. The EEG and EMG signals are amplified by electrodes which help in differentiating the SWS, REM, and NREM movements in different strains.

Advantages

- A transgenic and knockout mouse is an important model to investigate the contribution of genes to behavior.
- It is also helpful in investigating sleep changes at genetic levels.

Disadvantages

- It is very costly method and also needs expert persons for placing electrode surgically into the animals' brain to record EEG and EMG.
- It is not easy to apply the data obtained from rodent species to human's primates.
- Transgenic animals have low survival rate.

2.6 Platform Method for Short-Term Stress-Induced Paradoxical Sleep Deprivation (PSD) in Rats

Principle

The water tank is used to induce stress stimuli to the animals. The animals when loss the consciousness, they suddenly fall into the water and become awakened (Fig. 2).

Procedure

- First rats are placed in a tiled water tank (143 × 41 × 30 cm) for 24 or 96 h. The tank contained fourteen platforms (6.5 cm in diameter) and lies above 1 cm to the water surface, this allowing the rats to move around by leaping from one platform to another.
- At the onset of each paradoxical sleep (PS) episode, the animal experiences a loss of muscle tonus and falls into water, which produces consciousness. So for the evaluation of total suppression of PS over 24 h (PSD-24 h) or 96 h (PSD-96 h) intervals, we uses the multiple platform procedure, which is well-documented to be effective in producing a total suppression of PS. Hence, it is appropriate to consider that these animals are being PS-deprived rather than being completely deprived of sleep.

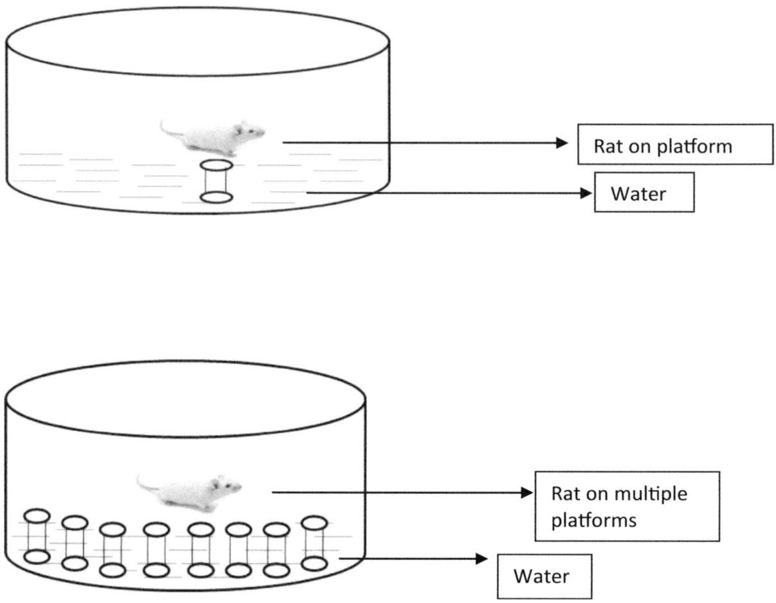

Fig. 2 Platform method for short-term stress-induced paradoxical sleep deprivation in rats

- Maintain the cage control group in the same room likely to the experimental rats for the duration of the study and showed normal sleep patterns, including PS, slow wave sleep (SWS) and wake.
- Again maintain the animal in the experimental rooms at a controlled temperature and light–dark cycle throughout the study along with availability of food and water and chow pellets located on top of the tank. The water in the tank is changed daily throughout the PSD period.

Advantages

- The number of platform used for experimentation is more, so the chances of error are less.
- The source used for stressful stimuli does not produce any harm to animals.
- It is economical, effective, and noninvasive technique to measure neurophysiology of the animals.

Disadvantages

- The aggressive behavior of animals may alter normal physiological responses.
- It does not show specificity in the results because the results vary depending on the size of platform.

2.7 Treadmill Cage Induced Sleep Deprivation in Rats

Principle

Sleep deprivation is defined as lack of restorative sleep over a cumulative period so as to cause physical or psychiatric problem in the animals induced by the physical stimuli like treadmill moving at variable rate with time which affects routine performances of tasks. This stressful movement disturbs the normal sleep regulatory systems and produce awakening by the cortical activation in the brain.

Procedure

- Firstly place the rats in a treadmill cage 50.8 cm (l) × 16.51 cm (w) × 30.48 cm (h).
- Now, switch on the apparatus during which the horizontal belt at floor automatically starts to move slowly at a rate of 0.02 m/s.
- To induce sleep fragmentation firstly, the treadmill ran at slow speed for 30 s. The total sleep deprivation is induced when the treadmill run for 4 s followed by no treadmill movement for 12 s, and these schedules ran continuously for 24 h.
- For habituations of rats in treadmill movement, the treadmills were turned on (5 min on followed by 5 min off) for one hour on each of the 2 days prior to the experiment.

Advantages

- This model is economical and noninvasively effective;
- No evidence for animal suffering from any harmful condition;
- Minimal handling is required while operation.

Disadvantages

- Treadmill procedure is very lengthy because it takes 4 week for the training session (Herman et al. 2010).
- Noise and stressful conditions may produce error in the results.

2.8 Cage Exchange as a Model for Stress-Induced Sleep Disturbances

Principle

The wakefulness in animal is promoted and maintained by multiple arousal systems (de Lecea 2012; Lee 2012). The stressful event during cage exchange paradigm produce continual wakefulness (sleep deprivation), with a pattern of neuronal hyperactivity is prominent causative factor for disturbance in sleep cycle. Specific psychological stressor or extra cellular stimuli increased immediate early genes (IEGs), i.e., c-Fos which synthesized Fos (neuronal activation) of the medial-parvicellular para-ventricular hypothalamic nucleus (mpPVH).

Procedure

Surgery (implants and lesions)

- All the animals are anesthetized and applied minor cut to explore the cranium.
- The connective tissue is cleaned and four burr holes is drilled with implanting 4 screw electrodes (two on each side) and two EMG electrodes also placed into the nuchal muscles.
- Connect all electrodes to a pedestal socket, which is fixed to the skull with dental cement. Finally, the wound is sutured and applied antibiotic ointment, and the animal is removed from the stereotaxic frame and allowed to recover from anesthesia. After 2 weeks of recovery, rats are connected to the recording apparatus during 3 d for habituation.
- Now, the baseline EEG/EMG recorded for 48 h, rats are placed into a dirty cage previously occupied by another male rat for 1 week (cage exchange). Rats were left undisturbed in the dirty cage until they were killed.
- Control rats are placed in a clean cage at the same time to synchronize the ultradian cycles of both groups.
- To examine the brain circuitry involved in stress-induced acute insomnia, the rats killed at 3:30 P.M., 90 min after the onset of this sleep-disturbed period (5.5 h after cage exchange). This time is selected because animals show sleep fragmentation and decreased sleep beginning 4 h after cage exchange, and 90 min is the optimal time to detect Fos expression associated with a specific stimulus.

Advantages

- It is a noninvasive technique used to measure narcolepsy.
- Sleep changes can easily be identified from EEG wave pattern.

Disadvantages

- Appropriate facilities and trained personnel are required;
- High mortality rate and time-consuming process;

2.9 Large Platform Orbital Shaker Induced Sleep Interruption

Principle

Stress is responsible for the number of biochemical changes at peripheral and central region of brain and also causes permanent neurological and psychological dysfunction (Moghaddam 1993). Serotonin (5-HT) is a true neuromodulator of sleep because the inhibition of 5-HT synthesis with p-chlorophenylalanine (PCPA) could inhibit tryptophan hydroxylase and impair 5-HT biosynthesis and induced a severe insomnia/sleep deprivation (Jouvet 1999).

Procedure

Surgery (implants and lesions)

- All the animals are anesthetized and applied minor cut to explore the cranium.
- The connective tissue is cleaned and four burr holes is drilled with implanting 4 screw electrodes (two on each side) and two EMG electrodes also placed into the nuchal muscles.
- All electrodes are connected to a pedestal socket, which is fixed to the skull with dental cement. Finally, the wound is sutured, antibiotic ointment is applied, and the animal is removed from the stereotaxic frame and allowed to recover from anesthesia.
- Again after 2 weeks of recovery, rats are connected to the recording apparatus during 3 d for habituation.
- Now, the recording cages are placed on the 27 × 32 cm platform of an analog orbital shaker (VWR Model OS-500; VWRL abshop, Batavia, IL) and affixed the platform, to prevent double-sided mounting tape (3 M, St. Paul, MN). Two cages could be monitored simultaneously.
- The on-off repetitive cycling of the shaker is maintained at 100 rpm and is controlled by a timer on a 120 s cycle (30 s on, 90 s off). A metal cage card holder is also suspended from the top of the cage because this creates an additional audible stimulus when the holder knocked against the side of the cage, once per cycle, during shaking.

Advantages

- It is more effective because sleep deprivation in the rodents is caused by large area platform with additional rotationary movement.

Disadvantages

- Appropriate facilities and trained personnel are required.
- It is a lengthy procedure because animals are used for experimentation after 2 weeks of surgery.

Ethical Statement

All institutional guidelines, national guidelines, state and local laws and regulations with professional standards for the care and use of laboratory animals should be followed. Studies involving animals must state that the institutional animal ethical committee has approved the protocol. For authors using experimental animals, a statement should be made that the animals' care is in accordance with institutional guidelines and animals used have been treated humanely and with regard to the alleviation of suffering. Researchers should treat animals as sentient and must consider their proper care and use and the avoidance or minimization of discomfort, distress, or pain as imperatives. Animal experiments should be designed only after due consideration of animal health. It should be ensured that all researchers who are using animals have received instruction in research methods and in the care, maintenance, and handling of the species being used. All the surgical procedures should be performed under appropriate anesthesia and follow only those procedures which avoid infection and minimize pain during and after surgery.

References

Alkadhi K, Zagaar M et al (2013) Neurobiological consequences of sleep deprivation. Curr Neuropharmacol 11(3):231

Cappuccio FP, D'Elia L et al (2010) Quantity and quality of sleep and incidence of Type 2 Diabetes: a systematic review and meta-analysis. Diabetes Care 33(2):414–420

De Lecea, Carter ME et al (2012) Shining light on wakefulness and arousal. Biol Psychiatry 71(12):1046–1052

Deboer T, Tobler I (2003) Sleep regulation in the Djungarian hamster: comparison of the dynamics leading to the slow-wave activity increase after sleep deprivation and daily torpor. Sleep 26(5):567–572

Herman T, Giladi N et al (2009) Treadmill training for the treatment of gait disturbances in people with Parkinson's disease: a mini-review. J Neural Transm 116(3):307–318

Jouvet M (1999) Sleep and serotonin: an unfinished story. Neuropsychopharmacology 21:24S–27S

Lee SH, Dan Y (2012) Neuromodulation of brain states. Neuron 76(1):209–222

Moghaddam B (1993) Stress preferentially increases extraneuronal levels of excitatory amino acids in the prefrontal cortex: comparison to hippocampus and basal ganglia. J Neurochem 60(5):1650–1657

Animal Models of Neuropathic Pain

Vanita Rani, Karamjeet Kaur, Vir Vikram Sharma
and Rahul Deshmukh

1 Introduction

In the recent history of neuromodulation period over half a century, the proof-based medical subspecialty is made. Its benefits are verified by improved pain relief, functional status, and health-related quality of life and low demand for healthcare resources. Neuromodulation is based on the innovative idea that paresthesia-inducing electrical stimulation could be analgesic. Its historic basis originates from Melzack and Wall's gate control theory of pain proposed in 1965. Neuromodulation has given us complete access to the systems of pain modulation and helped to understand the pathophysiology of pain. Neuropathic pain can be a consequence of an uncommon learning process which is associated with maladaptive plasticity of the central as well as peripheral nervous system. Various modifications of the peripheral nervous system have been defined in animal models of neuropathic pain, but their relation with human neuropathy symptoms are not fully understood. Mainly, neuropathic pain arises from injured myelinated fibers, abnormal activity in non-injured fibers, and also due to the more expression of

V. Rani · K. Kaur
Department of Pharmacology, ISF College of Pharmacy, Moga, Punjab, India
e-mail: sapanatohan@gmail.com

K. Kaur
e-mail: Karamjeet14293@gmail.com

V.V. Sharma
Dayanand Medical College and Hospital, Ludhiana, India
e-mail: virvikram76@gmail.com

R. Deshmukh (✉)
Department of Pharmacology, ISF College of Pharmacy, Moga 142001, Punjab, India
e-mail: login2rd@gmail.com

P.K. Bansal and R. Deshmukh (eds.), *Animal Models of Neurological Disorders*, https://doi.org/10.1007/978-981-10-5981-0_13

calcium channels which results in more and more release of excitatory neurotransmitters and sympathetic propagation toward the spinal ganglia. Moreover, changes in the dorsal horn alter the activity of projections toward the brain stem and enhanced spinal hyperactivity. These effects are late, signifying the maintenance of spinal sensitization. These phenomena can convince the changes in the activity of thalamo-cortical networks through which independent processes are developed and maintain the pain. The change in the cortical body areas is the demonstration after nervous lesions, and these changes may relate with the emergence of pain.

Neuropathic pain can be produced by injury or disease of the somatosensory system. The clinical expressions of neuropathic pain vary including both stimulus-evoked and non-stimulus-evoked (spontaneous) symptoms. By pharmacological intervention, the beginning for allodynia and hyperalgesia in the several pain modalities can be modulated and measured in animals and humans. Animal models have been found valuable in studies and the treatment on neuropathic pain (Fig. 1).

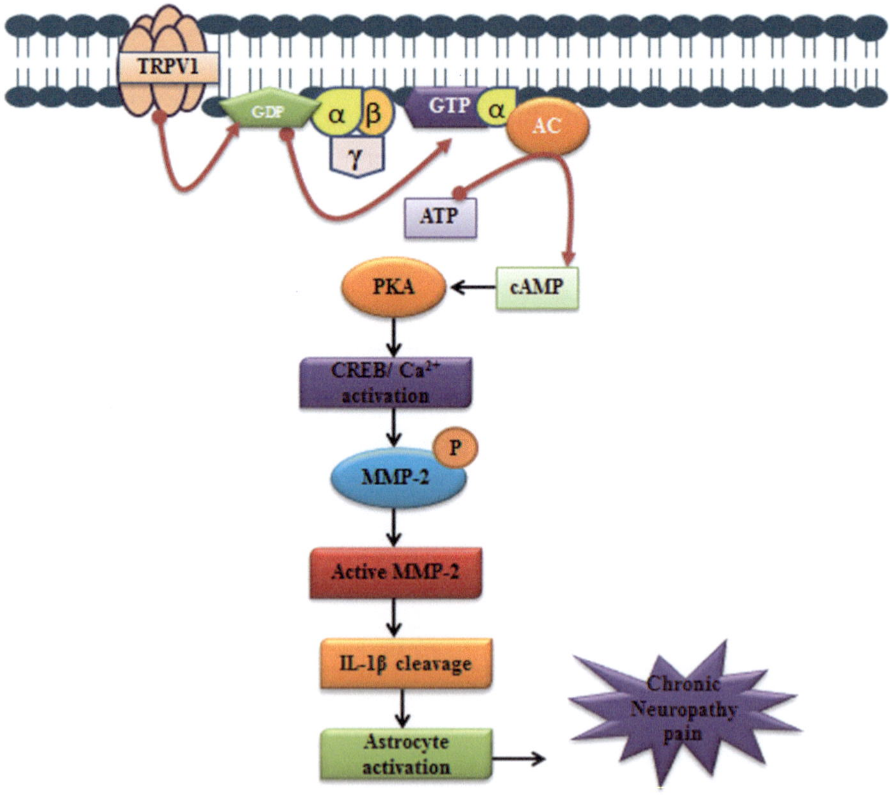

Fig. 1 Proposed mechanism of MMP transcriptional regulation through TRPV1 signaling receptors involved in the formation of chronic neuropathic pain. TRPV1 modulates Ca^{2+}-dependent intracellular signaling pathways responsible to induce neuropathic pain. *TRPV* transient receptor potential cation channel, *PKA* protein kinase A, *cAMP* cyclic adenosine monophosphate, *CREB* cAMP response element-binding protein, *MMP* matrix metalloproteinase, IL-1β *Interleukin-1 beta*

2 Pathophysiology

Neuropathic pain arises through multiple and challenging pathophysiological mechanisms. In many peripheral and CNS diseases, neuropathic pain is a common problem. The peripheral nerve diseases that most usually cause distal symmetric peripheral neuropathies and focal neuropathies are related to trauma, as well as surgical interventions. Exemplary CNS diseases include multiple scleroses, spinal cord injury, and stroke causing neuropathic pain (Fig. 2).

Neuropathic pain has plethora of different mechanisms that extend from the periphery to the central nervous system where they involve the spinal cord, supraspinal, and descending modulation systems. Lesions exist anywhere within the central nervous system (CNS) or the peripheral nervous system (PNS) that can produce neuropathic pain. Nervous system lesions have played a main role to produce NP. Many types of pathological changes in peripheral mechanism such as ectopic discharges in lesioned fibers and their corresponding ganglia, alterations in the expression and regulation of intracellular Ca^{2+} ion, sodium ion and modulatory

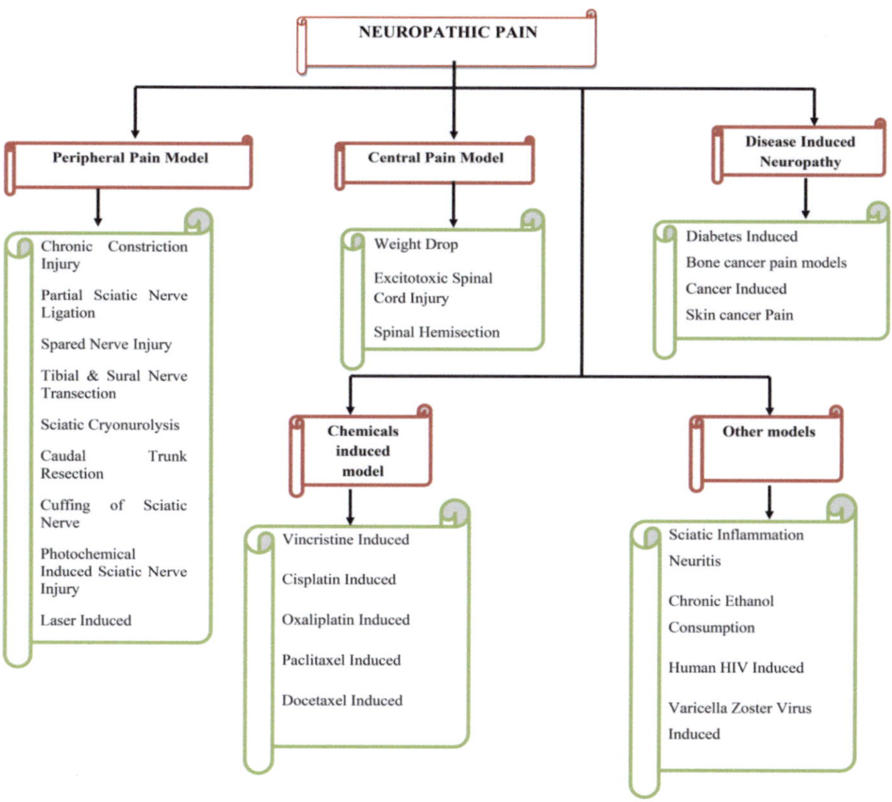

Fig. 2 Classification of animal models of neuropathic pain

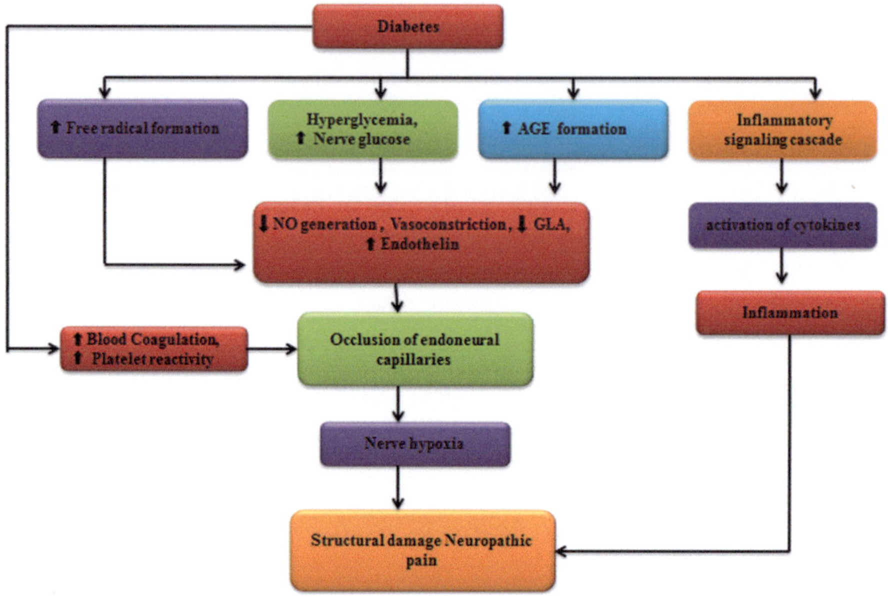

Fig. 3 Possible mechanism for neuropathic pain induced by diabetes. *AGE* advanced glycation end product, GLA *gamma linolenic acid*

receptors on primary afferent terminals occur in peripheral axons and dorsal root ganglia after nerve lesions. Neuroimmune interactions resulting in altered production of inflammatory signaling molecules, sensory-sympathetic coupling, and other alterations in receptor signaling have also been described to play a role in the pathology of NP. On the other hand, in CNS neuropathic pain involves both spinal and supraspinal mechanisms. Hyperexcitability of second-order neuron, neuroinflammatory processes, formation of a glial scar-prevented axonal regeneration, selective neuronal loss, and failure of inhibitory mechanisms have been demonstrated after lesion in spinal mechanism in NP. But in supraspinal mechanism, there is alteration in nociceptive signals (Fig. 3).

3 Animal Models of Neuropathic Pain

Several models have been developed for different pain states, and the alteration of behavior has been interpreted as a response to external stimulus or expression of pain or discomfort. Animal models are a need in the study of neuropathic pain, and greatly of what we distinguish about pain comes from the studies in mice and rats. However, very few basic findings have been rendering so far from rodent models into effective pain therapy consequently; there is still a considerable need to discover novel treatment modalities. The evaluation of neuropathic pain in humans is complex, because it is difficult to recruit the significant numbers of patients needed for a clinical trial. Thus, animal models provide advantage over humans to deveop and understand the mechanisms following neuropathic pain. The results in consistent sensory deficits (allodynia, hyperalgesia, and spontaneous pain) over a persistent period of time of animal models can be evaluated by sensory analysis. Animal models can be developed by establishing the degree of mechanical, chemical, and temperature induced allodynia and hyperalgesia, which help us to identify therapeutic potential of various pharmacotherapies (Table 1).

Table 1 List of different animal models of neuropathic pain

S. No	Name of model	Principle of injury	Species	References
1.	Chronic constriction injury	Four loose ligatures around sciatic nerve	Female Sprague–Dawley rats	Bennett and Xie (1988)
2.	Axotomy model	Complete sciatic nerve transection	Male Sprague–Dawley rats	Wall et al. (1979)
3.	Partial sciatic nerve ligation	Tight ligation of one-third to half of sciatic nerve	Male Sabra rats	Seltzer et al. (1990)
4.	Spared nerve injury	Axotomy of common peroneal and tibial nerves	Male Sprague–Dawley rats	Decosterd and Woolf (2000)
5.	Tibial and sural nerve transection	Ligation of sural, common peroneal and tibial nerves	Wistar albino rats	Jain et al. (2009)
6.	Sciatic cryoneurolysis	Freezing of the sciatic nerve	Wistar albino rats	De Leo et al. (1994)
7.	Caudal trunk resection	Resection of caudal trunk between the S3 and S4 spinal nerves	Male Sprague–Dawley rats	Na et al. (1994)
8.	Sciatic inflammatory neuritis	Applying Oxycel around the sciatic nerve	Male Sprague–Dawley rats	Eliav et al. (1999)
9.	Cuffing of sciatic nerve-induced pain	Implanting a polyethylene cuff around the common branch of the sciatic nerve	Male Sprague–Dawley rats	Mosconi and Kruger (1996)
10.	Photochemical-induced sciatic nerve injury	Thrombosis and occlusion in small vessels supplying sciatic nerve by photosensitizing dye and laser irradiation.	Male albino NMRI mice	Gazelius et al. (1996)
11.	Laser-induced sciatic nerve injury	Epineurial vessels of nerve are irradiated with laser beam results in marked reduction in the blood flow to nerve	Sprague-Dawley rats	Meyer et al. (1985)

(continued)

Table 1 (continued)

S. No	Name of model	Principle of injury	Species	References
12.	Weight drop or contusive	Constant weight is dropped over the nerve to produce an injury	Male and female Long–Evans rats	Allen (1911)
13.	Excitotoxic spinal cord injury	Intraspinal injections of excitatory amino acids	Male Sprague–Dawley rats	Brewer et al. (1999)
14.	Spinal hemisection	Laterally hemisectioning spinal cord at T11–T12 laminae	Male Sprague–Dawley rats	Hains et al. (2002)
15.	Vincristine-induced neuropathy	Direct injury of drugs to the nerves of peripheral nervous system	Male Sprague–Dawley rats	Pal (1999)
16.	STZ-induced neuropathic pain	Persistent hyperglycemia-induced changes in the nerves by administering STZ	Mice	Freireich et al. (1966)
17.	Chronic ethanol consumption-induced neuropathy	Administration of ethanol over extended period	Female Sprague–Dawley rats	Dina et al. (2007)

3.1 Peripheral Pain Model

3.1.1 Chronic Constriction Injury (CCI)

Bennett and Xie established a model of peripheral mononeuropathy in rats by CCI to the sciatic nerve, which is most commonly working animal model of neuropathic pain. The success of the model was confirmed by identifying thermal and mechanical hyperalgesia. This constriction of the sciatic nerve is related to intra neural edema, focal ischemia, and Wallerian degeneration. The behavioral signs of spontaneous pain have been reported such as tingling, burning, electric-shock-like pain, dysaesthesia, mild-to-moderate autotomy, guarding, excessive licking, limping of ipsilateral hind paw, and avoidance of placing weight on the injury side (Jaggi et al. 2011). The CCI model produces unilateral peripheral mononeuropathy, and it has been observed that symptoms in this rat model correspond to causalgia or complex regional pain syndrome in patients (Bennett and Xie 1988).

Procedure

- Rats are anesthetized with sodium pentobarbital (40 mg/kg, intraperitoneal).
- The common sciatic nerve is exposed at the level of the middle of the thigh by blunt dissection through biceps femoris; 4 ligatures (4-0 chromic gut suture) are tied loosely around it with about 1-mm spacing.
- The incision is closed in layers. The success of the modeling is confirmed by detecting thermal and mechanical hyperalgesia (Di et al. 2014).

Advantage: Chronic constriction is the easiest model to develop neuropathic pain.
Disadvantage: There has been some degree of variation observed in the animals subjected to chronic constriction injury, which may complicate quantitative analyses.

3.1.2 Axotomy Model (Complete Sciatic Nerve Transection; Neuroma Model)

The model produces anesthesia dolorosa, i.e., pain in the area which lack any sensory input in that area. Autotomy (self-attack and mutilation of the denervated limb by injured animals) is observed in this model and often measured as assigns of neuropathic pain. Clinically, the more common systems of neuropathy comprise partial lesions to peripheral nerves.

Procedure

- Rat is anesthetized, and the common sciatic nerve is exposed.
- The connective tissue attached to the sciatic nerve is cleared off, and the sciatic nerve is tightly tied by nylon suture, proximal to its divergence into the tibial and the peroneal divisions, at two locations about 1-cm apart.
- The nerve is then completely transected between the pair of ligatures, and 5 mm of the nerve between the ligatures is removed to prevent the rejoining of nerves due to regeneration.
- The lesion to other adjacent saphenous nerve is also induced entire denervation of distal hind limb. Following the complete nerve transection, a neuroma develops at the proximal nerve stump consisting of regenerative nerves sprouting in all directions (Muthuraman et al. 2008).

Advantage: This model is useful to study the spontaneous pain in the area which lacks the sensory input in that area.
Disadvantage: A complete nerve transection or lesion is relatively uncommon in patients and is usually seen only after amputation such as phantom limb pain. Moreover, ethical considerations are also the key issues in this model as animals demonstrate excessive autotomy in this model.

3.1.3 Partial Sciatic Nerve Ligation (PSNL)

This model is developed by Seltzer et al. (1990) and is one of the more often employed models of neuropathy. The partial nerve injury models are relevant for understanding neuropathic pain injury, as the partial nerve injury is the main cause of pain disorders in humans.

Procedure

- PSNL is performed under intraperitoneal ketamine/xylazine (150 and 10 mg/kg, respectively) anesthesia.
- Briefly, the right sciatic nerve is exposed after the incision of skin and separation of the muscle.

- The sciatic nerve is freed of the adhering tissue gently for about 7 mm, and one ligature is made around approximately 1/3 and 1/2 diameter of the sciatic nerve.
- Great care is taken to tie the ligatures so that the diameter of the nerve is just barely constricted.

Advantage: This model is most relevant to clinical study because partial sciatic nerve injury is the main cause of pain disorders in human.
Disadvantage: The magnitude and duration of pain responses vary considerably depending on the suture material and strains

3.1.4 Spared Nerve Injury (SNI)

The 'spared nerve injury (SNI) model' of neuropathic pain was established by Decosterd and Woolf. Two versions of SNI injury of the sciatic nerve have been developed using the same surgical procedure, but with different combination of nerve transactions. In one variant, injury is given to the common peroneal and the sural nerves, sparing the tibial nerve (t) [SNIv(t)], while in the other versions, the tibial nerve is injured leaving the sural(s) and common peroneal (cp) nerves to intact (SNIv). Within 4 days of injury, mechanical and thermal hyperalgesia and allodynia are known to occur that persist for almost 24 weeks post-injury. The substantial changes seen in mechanical and thermal sensitivities closely mimic the clinical features of neuropathic pain.

Procedure

- Surgery is to be done under suitable anesthesia. Sciatic nerve is exposed in the thigh, and a section is made directly through biceps femoris muscle and its three terminal branches: the sural, common peroneal, and tibial nerves.
- Then, the peroneal and tibial nerves are tightly ligated with 5.0 silk thread.
- The surgical procedure is to be carried out cautiously to avoid any contact with or stretching of the intact nerve.

Advantages: This model closely mimics many of the features of clinical neuropathic pain. The changes in mechanical and thermal sensitivities in this model are robust, substantial, and persistent. The surgical procedure for creating this model is relatively easy as compared to previous models.
Disadvantage: The process of separation and ligation of the sural, common peroneal, and tibial nerves is tedious. So, it is difficult to achieve success in this model.

3.1.5 Tibial and Sural Nerve Transection

This is a new model of neuropathic pain developed by Lee et al. (2000). The majorities of these behavioral changes start from the third day of surgery, reach at peak in 1–2 weeks, and persist over a period of 1½ months. Clinically in this model, profound and stable neuropathic pain symptoms with mechanical allodynia, cold allodynia, chemical allodynia, mechanical hyperalgesia, and spontaneous pain have been recognized.

Procedure

- The rat is totally anesthetized with ketamine (60 mg/kg i.p.).
- The skin of its lateral surface of the left thigh is cut through the biceps femoris muscle to uncover the sciatic nerve and its trifurcation (the sural, common peroneal, and tibial nerves).
- After that, 2 mm sections of the tibial and sural nerve (distal to the trifurcation) are ligated and cut. The common peroneal nerve is left unligated.
- The muscle and the skin are closed in two layers.

Advantage: Tibial and sural nerve transection model system is clinically relevant, as it produces profound and stable neuropathic pain symptoms with mechanical, chemical and cold allodynia, mechanical hyperalgesia and spontaneous pain in experimental animals.
Disadvantage: Heat allodynia has not been demonstrated in this model of neuropathic pain.

3.1.6 Sciatic Cryoneurolysis

This model is an interesting animal model of peripheral neuropathy, as no transection or ligation is carried out. In its place, freezing of the sciatic nerve has been used to produce nerve injury (De Leo et al. 1994).

Procedure

- Surgical procedure is performed under inhalation of anesthesia using halothane in 100% O_2, induced at 4% and maintained at 2%.
- A segment (L 1.0 cm) of the common sciatic nerve proximal to its main trifurcation is shown by blunt dissection and suspended across forceps in the surgical opening.
- The nerve is lesioned in a 3-Obsec freeze, S-set thaw, 30-set freeze cycle using a 2-mm-diameter cryoprobe cooled to −60 °C with nitrous oxide as the refrigerant.
- The nerve is allowed to defrost and then returned to its normal position. With the help of surgical staples, the wound is closed and the animal recovered in room air.

Advantage: In this model, the nerve injury is produced by freezing the sciatic nerve, no need of transection or ligation.
Disadvantage: Autotomy not occurs immediately after the freeze lesion when the limb is dysfunctional.

3.1.7 Caudal Trunk Resection

In this model, left inferior caudal trunk of the rat is resected between the 3rd and 4th spinal nerves. There is the development of mechanical, cold, and warm allodynia in the tail durable up to several weeks, and signs of mechanical and thermal allodynia appear within a day after the nerve injury (Na et al. 1994).

Procedure

- Under sodium pentobarbital anesthesia (40 mg/kg, i.p.), the right superior caudal trunk is exposed and transected at the level between the 1st and 2nd spinal nerves that innervated the rat tail.
- To avoid the probable reunite of the proximal and distal ends of the severed trunk, a portion of the trunk, approximately 2 mm inside, is removed from the proximal end.
- This surgery eliminated the 1st spinal nerve innervation to the tail via the right superior caudal trunk.
- This model is easy to apply the mechanical stimulation with von-Frey hairs and to localize the sensitive area in the tail.

Advantage: Behavioral tests have to be performed on the tail, instead of the hind paw. Testing on the tail is easier and more consistent.
Disadvantage: The proximal and distal ends of the severed trunk may get reunite and can influence the outcomes.

3.1.8 Sciatic Inflammatory Neuritis (SIN)

The SIN model is an established model of neuropathic pain and is preferred by various research groups to impose damage to nerves. Within 3 h of injection, this method has been shown to produce mechanical allodynia, but no apparent thermal hyperalgesia. The development of mirror allodynia is the characteristic trait of this type of neuropathy. Clinical studies report that the major cause of 50% of human neuropathies is inflammation or infection rather than trauma. Moreover, the traumatic nerve injuries are also accompanied by various inflammatory events (Said and Hontebeyrie 1992).

Procedure

- The surgical procedure is to be carried out under sodium pentobarbital anesthesia (45 mg/kg, i.p., supplemented as necessary). The central sciatic nerves are exposed at the mid-thigh level by blunt dissection through the biceps femoris and separated from adjacent tissue.
- The nerve is enclosed on one side, in a band (approx. 3 mm wide and 25 mm long) of sterile hemostatic oxidized cellulose.
- The Oxycel is applied by passing curved forceps beneath the nerve (cautiously avoiding stretching of nerve), then grasping one end of the band and pulling it under the nerve.
- The grasped end is gently folded over the nerve, the other end is folded over in the opposite direction, and the excess is to be trimmed away.
- The Oxycel is proposed to act as a sponge. It is wrapped around the nerve with a loose knot and does not pose any harm or injury to nerve constriction.

Advantage: This model has shown to produce mechanical allodynia within 3 h of injection.
Disadvantage: This model cannot induce apparent thermal hyperalgesia.

3.1.9 Cuffing of Sciatic Nerve-Induced Pain

In this model, neuropathic pain is induced by surgically implanting a polyethylene cuff (2 mm in length, inner diameter 0.7 mm) around the common branch of the sciatic nerve of rats. Sciatic nerve cuffs are made from thick-walled polyethylene tubing. Internal diameters of 0.026″ and 0.028″ are formed by elongating commercially available 0.023″ tubing with reaming bits. Tubing is cut into either short (<0.5 mm) segments or longer segments of 2–5 mm in length (Mosconi and Kruger 1996).

Procedure

- Sciatic nerves are exposed in the thigh using a Zeiss surgical microscope and clear from the connective tissue by dissection with least contribution of the epineurial vasculature.
- Cuffs are applied to the nerve, and after a brief blanching, the epineurial vessels filled and flow remained intact through closure of the incision.
- Two to four cuffs were put at intervals approximately 0.5 mm, and sciatic nerve was enclosed at 3–6 mm.
- Wounds are closed in two layers, and the animals are permitted to recover before being returned to cages in the vivarium (Mosconi and Kruger 1996).

Advantage: The neuropathic pain model is characterized by heat hyperalgesia lasting for 3 weeks and mechanical allodynia lasting for 2 months.
Disadvantage: This model is difficult to establish.

3.1.10 Photochemical-Induced Sciatic Nerve Injury

The aluminum foil is placed under the nerve to isolate the nearby tissue and to reflect light. It has been proved that the nerve injury is not induced by the heat generated by the laser, but is a result of the photochemical reaction forming thrombosis and occlusion in small vessels supplying the nerve (Kupers et al. 1998).

Procedure

- Under isoflurane anesthesia, the left sciatic nerve is exposed at the mid-high level and dissected free from surrounding tissue.
- A strip of aluminum foil is placed under the nerve to reflect the laser beam. Immediately after injection of erythrosin B (32.5 mg/kg in saline, iv) into the tail vein, the part of the sciatic nerve just proximal to the trifurcation is irradiated with a laser with an output power of 100 mW at 532 nm.
- The laser beam is focused into a collimated fiber delivery system. A multimode fiber with a diameter of 200/220 m is used.

- The intensity of the laser light at the tip of the probe was approximately 70 mW. In order to determine the optimal exposure time, different groups received laser irradiation of 2-, 5- or 10-min duration.
- After the surgery, the wound is closed with clips.

Advantage: In this model, highly reproducible mechanical, heat, and cold allodynia and signs of spontaneous pain have been documented.
Disadvantage: In this model, the concomitant requirement of photosensitive dyes is essential.

3.1.11 Laser-Induced Sciatic Nerve Injury

In this model, the segment of the sciatic nerve just distal to the gluteus muscle is marked with epineurial sutures, and epineurial vessels of nerve are irradiated with laser beam of 1 mm diameter for 30 s (a diode-pumped solid-state laser operating at 532 nm with an output power of 100 mW). Laser irradiation results in marked reduction in the blood flow to nerve (Chiang et al. 2005).

Procedure

- Anesthetise the animals by using 1.5% isoflurane with a mixture of 50% oxygen and 50% nitric oxide by open-mask system at 1.5 l/min and the animal could be placed on a homeothermic blanket.
- A rectal thermal probe connected to a computer ensured that animal's body temperature remained at 37 °C for the duration of the surgery.
- The rat is placed in a prone position, and the left sciatic nerve is exposed at the level of mid-thigh.
- A thin plastic film is introduced under the nerve, on top of which a strip of aluminum foil (width 5 mm) is placed for reflection of the low-energy laser beam.
- A Q-switched diode pumped a solid-state laser of the L-series with an output power of 5 mW.
- The laser light, with a wavelength of 532 nm, is focused into a transmitter fiber with a diameter of 125 mm of a PF 418 probe.
- The sciatic nerve is exposed to laser irradiation for 20 min immediately after administration of erythrosin B (32.5 mg/kg, i.v.).
- Complete hemostasis is achieved before wound closure. Each surgery carried out for less than 60 min.

Advantage: On postoperative day 2, animals develop characteristic sign of neuropathic pain including spontaneous pain behaviors such as tingling, burning, thermal hyperalgesia, and mechanical allodynia. These phenomena peak during 7–21 days after surgery and last for 3–6 weeks.
Disadvantage: In this model, nerve injury is produced by laser, which is a time-consuming procedure for surgery (60 min).

3.2 Central Pain Models

3.2.1 Weight Drop or Contusive

Injury reproducibility is a significant characteristic of experimental models of spinal cord injuries (SCIs). In this model, the spinal cord at lower thoracic–lumbar level is exposed and a constant weight is dropped over the nerve to produce an injury, which is characterized by severe paraplegia and complete segmental necrosis. Clinically, the hypersensitivity to light mechanical stimulation of the skin has been noted to develop within 1 day of injury and parallels the allodynia experienced by patients rapidly following spinal injury (Allen 1911).

Procedure

- Spinal cord contusion or transection (TX) surgeries are performed under pentobarbital (50–70 mg/kg) anesthesia and prophylactic administration of gentamycin sulfate (1 mg/kg).
- The thoracic area of the rat is properly shaved, and prodine and alcohol are applied to the skin.
- During surgery, body temperature is maintained at 37 °C using either a Harvard homeothermic feedback-controlled heating pad or heated gel packs.
- Mid-to-low thoracic regions are incised. Laminectomy of the caudal portion of T9 and all of T10 exposed the spinal cord.
- For the TX group, the spinal cord is severed with microscissors and gentle aspiration. The ends of the cord distracted and the cavity are then carefully explored with a glass probe to cut any residual fibers.
- After determining that the TX is complete, the cavity is packed with gel foam. For contusion injuries, at position T8 and T11/12 spinous processes, spinal clamps are attached, a transducer is placed over the transverse process of T9, and the impact rod is centered above T10.
- The rod contacted with dura is determined by the completion of a circuit that activated a tone. The surgical site is sutured in layers, and an antibacterial spray is applied immediately after surgery.

Advantage: Clinically, the hypersensitivity to light mechanical stimulation of the skin has been noted to develop within 1 day of injury and parallels the allodynia experienced by patients rapidly following spinal injury.
Disadvantage: The surgical procedure used in this model is very complicated to achieve nerve injury.

3.2.2 Excitotoxic Spinal Cord Injury

This model simulate the elevations of excitatory amino acids (EAAs), a well-documented neurochemical change following spinal cord injury (SCI). The continuous pathological symptoms associated with metabotropic receptor agonist quisqualic acid QUIS injections closely resemble the cascade of events, pathogenesis of ischemic, traumatic SCI, and cavities in the clinical condition of post-traumatic syringomyelia (Brewer et al. 1999).

Procedure

- Male Sprague–Dawley rats weighing 250–275 g are anesthetized with a mixture of ketamine, acepromazine, and rhombus (0.65 cc/kg, sc).
- Laminectomy-alone animals are also used as a control for the effects of surgery.
- Animals are placed in stereotaxic apparatus, and the spinal column is immobilized with a vertebral clamp.
- One intraspinal injection is made by laminectomy between spinal segments T12–L2 and the dura incised longitudinally and reflected bilaterally.
- A small hole for unilateral injections is made in the pia mater. QUIS, [125 mM], is injected unilaterally at three rostrocaudal levels (0.5 mm between tracks).
- Injections are made at a depth of 1000 mm below the surface between the dorsal root entry zone and dorsal vein.
- These parameters placed injections in the center of the gray matter between spinal laminae IV–VI.
- The total volume of QUIS or saline injected is 1.2 ml (0.4 ml per injection, delivered over a time interval of 60 s).
- A Hamilton microliter syringe with a glass micropipette extension (tip diameter 5–10 mm) is used for intraspinal injections. The syringe is positioned in a microinjector unit (Kopf 5000) attached to a micromanipulator.
- After injections, muscles are closed in layers and the skin is closed with wound clips (Brewer et al. 1999).

Advantage: This model has been shown to be a valuable model for studying the central mechanism of neuropathic pain.
Disadvantage: Dose dependently QUIS causes the toxicity of the neurons.

3.2.3 Spinal Hemisection

In this model, neuropathic pain like thermal and mechanical allodynia is established extensively not only in limbs but also in tail. Moreover, allodynia lasts for a long time following spinal cord hemisection (Kim et al. 2003).

Procedure

- Animals (175–200 g) are deeply anesthetized by intraperitoneal injection of sodium pentobarbital (40 mg/ kg), and the surgical field is shaved and prepared with Betadine.
- A longitudinal incision is made exposing several segments, and the left side of the spinal cord was laterally hemisected at T13 by the following procedure: Following the palpation of the dorsal surface to locate the rostral borders of the sacrum and dorsal spinous processes of the lower thoracic and lumbar vertebrae, the T11–T12 laminae are determined by locating the rib which attaches to the rostral end of the T13 vertebrae and counting two vertebral segments rostrally.

- The T13 spinal cord is laterally hemisected with a No. 11 scalpel blade without damage to the posterior spinal vessel or branches.
- Muscle and fascia are sutured and skin-closed with autoclips, and animals are allowed to recover on a 36.5 °C heating pad. Postoperative treatments included saline (1.0 cc sc) for rehydration and penicillin-G (0.35 ml/kg; i.m.) as a prophylactic antibiotic. Following the surgery, animals are maintained under the same preoperative conditions and the general health of the animals is carefully monitored.

Advantage: The numbers and types of effected fibers can be controlled at regular basis in each animal. The injured and intact sides are completely separated.
Disadvantage: The surgical procedure is difficult to achieve nerve injury.

3.3 Anticancer Agents-Induced Neuropathy Models

3.3.1 Vincristine-Induced Neuropathy

Vincristine is a widely employed chemotherapeutic agent for the treatment of several malignancies and primary brain tumors. It has been reported that clinical use of vincristine leads to the development of neurotoxicity of peripheral nerve fibers resulting in sensory-motor neuropathy. Among the various chemotherapeutic agents, vincristine produces consistent neurotoxicity in all the patients even at the therapeutic doses. Painful paresthesias are seen as the foremost symptom in major population of patients with this dose-dependent neuropathy.

Procedure

- Inject vincristine (0.1 mg/kg prepared with distilled water—1.0 mg/ml vincristine sulfate) intraperitoneally (0.1 mg/kg, diluted to 1 ml in saline) to rats in two 5-day cycles with 2 days break between cycles.
- In totality, 10 vincristine injections are to be given to rats on days 0–4 and 7–11 (Flatters and Bennett 2004).

Advantage: The use of vincristine has been associated with neurotoxicity of peripheral nerve fibers leading to sensory-motor neuropathy.
Disadvantage: It is difficult to develop vincristine-induced animal model of neuropathy.

3.3.2 Cisplatin-Induced Neuropathy

Cisplatin induces peripheral sensory axonal neuropathy disturbing large and small diameter sensory fibers. It generally causes clinical symptoms with a classic glove and stocking fashion after a cumulative dose of 300 mg/mm, and neuropathy may persist for long time (Markman 2003).

Procedure

- Cisplatin is administered intraperitoneally (i.p.) once a week at a dose of 5 mg/kg for 6 (36 days) weeks (cumulative dose: 30 mg/kg i.p.).
- Cisplatin is diluted in normal saline (0.9% NaCl) and delivered in a volume of 10 ml/kg bodyweight and given as i.p. injection.
- Injection is always performed after completion of mechanical and cold withdrawal testing.

Advantage: This model usually causes clinical signs and symptoms with a typical glove and stocking fashion after a cumulative dose of 300 mg/mm, and neuropathy may persist for years
Disadvantage: It has been very difficult to develop cisplatin-induced animal model of peripheral neuropathy, because of the development of nephrotoxicity before the development of neurotoxicity.

3.3.3 Oxaliplatin-Induced Neuropathy

Oxaliplatin is a key drug in the treatment of advanced metastatic colorectal cancer, but it causes acute peripheral neuropathy and chronic neuropathy. The elimination of Ca^{2+} and oxalate chelate by oxaliplatin is one of the reasons for the neuropathy, and there is little behavioral evidence (Sakurai et al. 2009).

Procedure

- Oxaliplatin (1, 2 and 4 mg/kg) or sodium oxalate (1.3 mg/kg) is administered i. p. twice a week for 4 weeks (on days 1, 2, 8, 9, 15, 16, 22 and 23).

Advantage: Oxaliplatin-induced neuropathy model is used to study both peripheral and chronic neuropathy.
Disadvantage: This method is slight difficult to set up.

3.3.4 Paclitaxel-Induced Neuropathy

Conventionally, the mechanisms underlying the pathogenesis of taxane-induced peripheral neuropathy (TIPN) include disruption of microtubule-based axonal transport, macrophage activation, and microglial activation in peripheral nerve as well as inside the spinal cord (Argyriou et al. 2008). As a result of the problematic signal transduction, there is evidence of a 'dying back' process initially affect the distal nerve endings followed by Schwann cells and neuronal bodies, or disturbed axonal transport changes in the affected neurons (Peters et al. 2007).

Procedure

- In this experimental model of neuropathy, low dose of paclitaxel (1 or 2 mg/kg i.p) has been reported to evoke pain syndrome without inducing systemic toxicity or motor impairment in rodents.

- Paclitaxel administration on four alternate days (days 0, 2, 4, and 6, with cumulative dose of 4 or 8 mg/kg) has been noted to produce peripheral neuropathy characterized by cold allodynia, long-lasting tactile (mechanical) allodynia, and endoneuronal edema of the sciatic nerve.

Advantage: TIPN model for neuropathic pain not affects the structure of internodal myelin in peripheral nerves.
Disadvantage: Sensory symptoms commonly start symmetrically in the feet, but after sometimes, it appears simultaneously in both hands and feet. With long-term dosing, the painful symptoms rise in severity and may include loss of vibratory sensation.

3.3.5 Docetaxelinduced Peripheral Neuropathy

Docetaxel-treated rats have been shown to exhibit reduced tail nerve conduction velocity, thermal threshold changes, and degeneration of skin nerves in the food pad.

Procedure

- Docetaxel-induced neuropathy model has been developed in which weekly i.v. injection of docetaxel (5; 10 or 12.5 mg/kg) for 4 weeks induces neuropathy in rats (Jaggi et al. 2011).

3.4 Disease-Induced Neuropathy Pain

3.4.1 Diabetes-Induced Neuropathy (STZ-Induced Neuropathic Pain)

Peripheral diabetic neuropathy (PDN) is a shocking complication of diabetes and a leading cause of foot amputation. Animals develop other metabolic derangements along with hyperglycemia including ketoacidosis, alterations in lipid metabolism, and general physical debility. Some of these symptoms complicate the interpretation of data in studies of nociception. It has been reported that general debility rather than peripheral neuropathy may underlie the altered measures of nociception in rats with STZ-induced diabetes. STZ has also been employed to induce diabetic neuropathy in mice (Jaggi et al. 2011) (Table 2).

Procedure

- Rats are usually made diabetic with a single dose of STZ (40–60 mg/kg, depending on bodyweight after an overnight fast to reduce competition between glucose and STZ for uptake into β-cells.
- In mice, a total dose of 180–200 mg/kgSTZ, given across 1 or 2 injections, can be used, with the higher dose reflecting correction for surface area and a degree of trial and error.

Table 2 Diabetic neuropathy

S. No	Anesthesia Dose	Route	Principle of injury	Species	References
1.	STZ (180 mg/kg)	i.p.	Persistent hyperglycemia-induced changes in the nerves	Mice	Freireich et al. (1966)
2.	STZ (40 mg/kg)	i.p.	Persistent hyperglycemia-induced changes in the nerves	Male Sprague–Dawley rats	Calcutt (2004)

- High doses of STZ can initially produce a severe hyperinsulinemia as insulin is released from degenerating β-cells, followed by marked insulinopenia.
- Both of these phenomena can cause early death in a substantial proportion of a cohort of mice unless mitigated by first sugar, then insulin, replacement.
- To counter high cohort mortalities in STZ-injected mice, lower doses of STZ given daily for anywhere from 2 to 5 days may be used to induce diabetes.
- Such multiple low-dose regimens produce a less aggressive hyperinsulinemia and subsequent insulinopenia that allow animals to survive for months.
- Nevertheless, these regimens may also produce a mild and inconveniently slowly developing neuropathy, with some animals reverting to normoglycemia over time.

Advantage: Low cost, quick induction of diabetes, conviction of duration of diabetes, and a broad background literature.
Disadvantage: Uncontrollable degree of B cell death and subsequent insulinopenia that can lead to between-study variability in the neuropathy phenotype.

One of the caveats that occasionally raised for this model has direct STZ toxicity to organs and is a particular concern for nephropathy studies. However, it is not a factor in the onset of neuropathy.

3.5 Other Models

3.5.1 Post-herpetic Neuralgia Model (Varicella Zoster Virus-Induced Neuropathy)

VZV is an infectious human alpha herpesvirus that causes chickenpox and is neurotropic, but this virus becomes dormant in neurons of the dorsal root ganglia (DRG) and trigeminal ganglia (TG). It may undergo reactivation, after a specific latent period, and induces herpes zoster (shingles) characterized by painful vesicular rash that appear along the sensory dermatomes. Post-herpetic neuralgia (PHN) is the major complication of zoster that triggers severe pain in the affected dermatome which persists for almost 3–4 months. Further, ganglionic replication is caused by processes such as focal hemorrhage, demyelination, degeneration of axons, necrosis of sensory fibers, and supporting cells. The afferent nerve fibers that innervate the damaged skin show impaired function and degeneration of posterior nerve along with sensory roots, further extending the damage to adjacent regions of spinal cord and brain stem in the central system. This damage may affect the sensory and ganglionic nerves triggering abnormal afferent pain signals that add to the nociceptive signals originating from the lesions themselves (Kennedy et al. 2013).

Procedure

- VZV (strain Dumas) is propagated on fibroblast (primary human embryonic lung) cells and harvested when cells exhibit 80% cytopathic effect (cpe) (equivalent to 10^4–10^5 plaque forming units).
- Inject 50-μl viral inoculums subcutaneously into the mid-plantar glabrous footpad of the left hind limb.

3.5.2 Human Immunodeficiency Virus (HIV)-Induced Neuropathy

HIV-1 interacts with the nervous system through binding of gp120 (external envelope protein) to CXCR4/CCR5 (chemokine receptors) that is present in neurons and glial cells resulting in peripheral axonal damage and neurotoxicity. HIV-associated sensory neuropathy is a major complication of HIV infection and highly active antiretroviral therapy (HAART), and clinical data also reveal that 50% of HIV patients are suffering from such neuropathies. It has been evidenced that HIV sensory neuropathy is triggered by pro-inflammatory molecules such as MAP kinase, tumor necrosis factor-α (TNF-α), stromal cell-derived factor 1-α (SDF1-α), and C-X-C chemokine receptor type 4 (CXCR4). However, the exact mechanistic aspect of painful HIV sensory neuropathy is not clear.

Procedure

- After anaesthetizing the rat with 1–2% isoflurane in 1% O_2 and 1% N_20, expose the left sciatic nerve to gp120 by placing a 5 mm × 2 mm piece of gel foam soaked in saline containing 200 ng HIV-1 gp120-MN in direct contact with the nerve, to form a pool of protein solution around the nerve that is left in place for 30 min.
- Then, the oxidized regenerated cellulose, previously soaked until saturation in the same saline-gp120 solution, is wrapped loosely around the sciatic nerve 2–3 mm proximal to the trifurcation so as not to cause any nerve constriction and left in situ.
- The nerve is gently manipulated back into place and skin incisions closed with 4/0 silk sutures.

Table 3 Alcohol induced hyperalgesia

S. No	Dose	Route	Principle of injury	Species	References
1	6.5% ethanol	Oral	Administration of ethanol over longer period (12 weeks)	Female Sprague–Dawley rats	Dina et al. (2007)
2	Ethanol (6.5%)	Oral	Administration of ethanol for 3 weeks	Male Sprague–Dawley rats	Ferrari and Levine (2010)

3.5.3 Chronic Ethanol Consumption/Withdrawal-Induced Neuropathy

The alcohol diet (6.5% ethanol v/v in Lieber-DeCarli formula) induced hyperalgesia with more quick onset and severity in females. Continuing alcohol consumption has been reported to induce small-fiber painful neuropathy characterized by distal axonal degeneration, i.e., dying back neuropathy. Although early use of alcohol leads to some extent of analgesia, yet over a period of time induction of pain outweighs analgesia and results in a neuropathic pain with sign and symptoms that have been described as like tearing flesh off the bones (Jaggi et al. 2011) (Table 3).

Procedure

All experimental rats are fed an ethanol-containing Lieber-DeCarli liquid diet (ED, 6.5% ethanol) daily over a period of 12 weeks (Dina et al. 2007).

4 Conclusion

Various animal models have been established and described for neuropathic pain that show predictive value for both humans and animals; still the exact pathophysiological mechanisms for the development, maintenance, and progression of the neuropathic pain syndrome are not yet clearly revealed. Although the most widely employed models still work on the principle of nerve ligation, other models featuring transaction of nerve branches of peripheral nerves are also advantageous and more frequently used currently. Models involving spinal hemisection and excitotoxin-induced SCI are preferred for conceptualizing central pain mechanisms. Furthermore, other models that make the use of chemotherapeutic agents, HIV, ethanol, and others have also been employed for the better understanding of pathogenetic mechanisms and treatment strategies for pain amelioration.

Ethical Statement

All institutional guidelines, national guidelines, state and local laws and regulations with professional standards for the care and use of laboratory animals should be followed. Studies involving animals must state that the institutional animal ethical committee has approved the protocol. For authors using experimental animals, a statement should be made that the animals' care is in accordance with institutional guidelines and animals used have been treated humanely and with regard for the alleviation of suffering. Researchers should treat animals as sentient and must consider their proper care and use and the avoidance or minimization of discomfort, distress, or pain as imperatives. Animal experiments should be designed only after due consideration of animal health. It should be ensured that all researchers who are using animals have received instruction in research methods and in the care, maintenance, and handling of the species being used. All the surgical procedures should be performed under appropriate anesthesia and follow only those procedures which avoid infection and minimize pain during and after surgery.

References

Allen AR (1911) Surgery of experimental lesion of spinal cord equivalent to crush injury of fracture dislocation of spinal column: a preliminary report. J Am Med Assoc 57:878–890

Argyriou AA, Iconomou G, Kalofonos HP (2008) Bortezomib-induced peripheral neuropathy in multiple myeloma: a comprehensive review of the literature. Blood 112(5):1593–1599

Bennett GJ, Xie YK (1988) A peripheral mononeuropathy in rat that produces disorders of pain sensation like those seen in man. Pain 33(1):87–107

Brewer KL, Bethea JR, Yezierski RP (1999) Neuroprotective effect of IL-10 following Excitotoxic Spinal Cord injury. Exp Neurol 159:484–493

Calcutt NA (2004) Modeling diabetic sensory neuropathy in rats. Pain Res: Methods Protoc 55–65

Chiang HY, Chen CT, Chien HF, Hsieh ST (2005) Skin denervation, neuropathology, and neuropathic pain in a laserinduced focal neuropathy. Neurobiol Dis 18(1):40–53

De Leo JA, Coombs DW, Willenbring S et al (1994) a neuropathic. Pain 56:9–16

Decosterd I, Woolf CJ (2000) Spared nerve injury: an animal model of persistent peripheral neuropathic pain. Pain 87:149–158

Di YX, Hong C, Jun L et al (2014) Curcumin attenuates mechanical and thermal hyperalgesia in chronic constrictive injury model of neuropathic pain. Pain Ther 3:59–69

Dina OA, Gear RW, Messing RO, Levine JD (2007) Severity of alcohol-induced painful peripheral neuropathy in female rats: role of estrogen and protein kinase (A and Cε). Neurosci 145(1):350–356

Eliav E, Herzberg U, Ruda MA, Bennett GJ (1999) Neuropathic pain from an experimental neuritis of the rat sciatic nerve. Pain 83(2):169–182

Ferrari LF, Levine JD (2010) Alcohol consumption enhances antiretroviral painful peripheral neuropathy by mitochondrial mechanisms. Eur J Neurosci 32(5):811–818

Flatters SJ, Bennett GJ (2004) Ethosuximide reverses paclitaxel-and vincristine-induced painful peripheral neuropathy. Pain 109(1):150–161

Freireich EJ, Gehan EA, Rall DP, Schmidt LH, Skipper HE (1966) Quantitative comparison of toxicity of anticancer agents in mouse, rat, hamster, dog, monkey, and man. Cancer Chemother Reports 50(4):219

Gazelius B, Cui JG, Svensson M, Meyerson B, Linderoth B (1996) Photochemically induced ischaemic lesion of the rat sciatic nerve. A novel method providing high incidence of mononeuropathy. Neuroreport 7(15–17):2619–2624

Hains BC, Everhart AW, Fullwood SD, Hulsebosch CE (2002) Changes in serotonin, serotonin transporter expression and serotonin denervation supersensitivity: involvement in chronic central pain after spinal hemisection in the rat. Exp Neurol 175:347–362

Jain V, Jaggi AS, Singh N (2009) Ameliorative potential of rosiglitazone in tibial and sural nerve transection-induced painful neuropathy in rats. Pharmacol Res 59(6):385–392

Jaggi AS, Jain V, Singh N (2011) Fundam Clin Pharmacol 25:1–28

Kennedy PGE, Montague P et al (2013) Varicella-zoster viruses associated with post-herpetic neuralgia induce sodium current density increases in the ND7-23 Nav-1.8 Neuroblastoma cell line. PLoS ONE 8(1):e51570

Kim J, Yoon YW et al (2003) Cold and mechanical allodynia in both hind paws and tail following thoracic spinal cord hemisection in rats: time courses and their correlates. Neurosci Lett 343:200–204

Kupers R, Yu W et al (1998) Photochemically-induced ischemia of the rat sciatic nerve produces a dose-dependent and highly reproducible mechanical, heat and cold allodynia, and signs of spontaneous pain. Pain 76:45–59

Lee BH, Won R, Baik EJ, Lee SH, Moon CH (2000) An animal model of neuropathic pain employing injury to the sciatic nerve branches. Neuroreport 11(4):657–661

Markman M (2003) Managing taxane toxicities. Supportive care in cancer: official journal of the Multinational Association of Supportive Care in Cancer. 11:144–147

Meyer RA, Raja SN, Campbell JN, Mackinnon SE, Dellon AL (1985) Neural activity originating from a neuroma in the baboon. Brain Res 325:255–260

Mosconi T, Kruger L (1996) Fixed-diameter polyethylene cuffs applied to the rat sciatic nerve induce a painful neuropathy: ultrastructural morphometric analysis of axonal alterations. Pain 64:37–57

Muthuraman A, Diwan V et al (2008) Ameliorative effects of Ocimum sanctum in sciatic nerve transection-induced neuropathy in rats. J Ethnopharmacol 120:56–62

Na HS, Han JS et al (1994) A behavioral model for peripheral neuropathy produced in rat's tail by inferior caudal trunk injury. Neurosci Lett 177:50–52

Pal PK (1999) Clinical and electrophysiological studies in vincristine induced neuropathy. Electromyogr Clin Neurophysiol 39:323–330

Peters CM, Jimenez-Andrade JM, Jonas BM, Sevcik MA, Koewler NJ, Ghilardi JR, Wong GY, Mantyh PW (2007) Intravenous paclitaxel administration in the rat induces a peripheral sensory neuropathy characterized by macrophage infiltration and injury to sensory neurons and their supporting cells. Exp Neurol 203(1):42–54

Said G, Hontebeyrie-Joskowicz M (1992) Nerve lesions induced by macrophage activation. Res immunol 143(6):589–599

Sakurai M, Egashira N et al (2009) Oxaliplatin-induced neuropathy in the rat: Involvement of oxalate in cold hyperalgesia but not mechanical allodynia. Pain 147:165–174

Seltzer Z, Dubner R, Shir Y (1990) A novel behavioral model of neuropathic pain disorders produced in rats by partial sciatic nerve injury. Pain 43:205–218

Wall PD, Devor M, Inbal R, Scadding JW, Schonfeld D, Seltzer Z, Tomkiewicz MM (1979) Autotomy following peripheral nerve lesions: experimental anesthesia dolorosa. Pain 7 (2):103–113

Animal Models of Attention-Deficit Hyperkinetic Disorder (ADHD)

Nidhika Sharma, Sumit Jamwal and Puneet Kumar Bansal

1 Introduction

ADHD (attention-deficit hyperkinetic disorder) is a common childhood-onset neuropsychiatric disorder characterized by attention deficit, hyperactivity, and impulsiveness, which employs to both children and adults. ADHD is a behavioral disorder and believed to affect 1 in 20 children in the USA, 3–7% of school-aged children, and persists in 30–50% of adults. ADHD was first illustrated 100 years ago as a childhood disorder, which affects boys more often than girls. In the 1930s and 1940s, children with the ADHD were called "brain damaged" or "brain injured" because their behaviors were identical with persons having brain damage. In 1960s, "hyperactive" or "hyperkinetic" became the term of choice for describing ADHD children, and in the year 1980s, the disorder was renamed to "attention deficit disorder" (ADD). This disorder was mainly divided into two types: with hyperactivity (ADD + H) and without hyperactivity (ADD-H). The first example of a disorder that appears to be related to ADHD was given by Sir Alexander Crichton in 1798. Children diagnosed with ADHD without exhibiting any signs of hyperactivity had eight of the inattention and impulsivity characteristics. In 1955, FDA

N. Sharma · S. Jamwal · P.K. Bansal (✉)
Department of Pharmacology, ISF College of Pharmacy, Moga 142001, Punjab, India
e-mail: punnubansal79@gmail.com

N. Sharma
e-mail: nidhika444@gmail.com

S. Jamwal
e-mail: jamwalisf@gmail.com

P.K. Bansal
Department of Pharmaceutical Sciences and Technology, Maharaja Ranjit Singh Punjab Technical University (State Govt. University), Bathinda, Punjab, India

P.K. Bansal and R. Deshmukh (eds.), *Animal Models of Neurological Disorders*, https://doi.org/10.1007/978-981-10-5981-0_14

approved methylphenidate (Ritalin) for the treatment of hyperactivity disorder. In 1987, DSM-III-R (revised) changed "attention deficit disorder" to "attention deficit hyperactivity disorder" (ADHD) and it accounts 14 symptoms, in which any eight symptoms were sufficient for diagnosis of ADHD.

1.1 Clinical Signs and Symptoms

Inattention, hyperactivity, and impulsivity are the major symptoms of ADHD. According to DSM-IV and ICD-10, inattention problems include: failure to give close attention to work or other activities like difficulty in sustaining attention in tasks, failure to finish schoolwork, chores, duties in the workplace, and non-willingness to engage in tasks that require sustained mental effort. Hyperactivity problems include: often fidgets or squirms on seat, often leaving seat when expected to sit, excessive inappropriate running or climbing, persistent over activity not modulated by request or context. Impulsiveness includes: often blurts out answers before the question is complete, often fails to wait in groups, games, or queues, and often talks excessively without response to social appropriateness. In adults, inattention and impulsivity seem to be more reliable markers of ADHD than hyperactivity. Daydreaming, poor concentration in work, and forgetfulness recommend inattention, although frustration with tasks, easy loss of temper, and impatience recommend inattention and impulsivity. Children and adults with ADHD frequently have comorbid antisocial, drug abuse, mood, anxiety, or learning disorders. Children with ADHD may be quite intelligent, but their lack of attention frequently results in low grades and problems in schooling. Children with ADHD continually move, run, climb, squirm, and fidget, but often have trouble with fine motor skills.

Crichton also reports that the disorder can be born with a person, and when born with a person, it becomes evident at a very early period of life. ADHD is reported to coexist with a variety of disorders. These disorders include oppositional defiant disorder 40%, developmental reading disorder, substance use disorder, anxiety disorder, mood disorder 10–20%, epilepsy, antisocial behavior, tics or tourette syndrome, 7%, obsessive compulsive disorder etc.

1.2 Pathophysiology

The exact causes of ADHD are not absolutely known. However, 80% of the disease prevalence is dependent on genetic factors, but it is not the only reason. The midbrain dopamine (DA) system, including the ventral tegmental area and the substantia nigra, is thought to play a pivotal role in the pathogenesis of ADHD and is relatively similar in different mammals. Chronic intake of dopamine agonists such as cocaine and amphetamines will produce a down-regulation of dopamine synthesis. Mesocortical DA is thought to be involved in the pathophysiology of ADHD because of its pivotal role in selective attention and working memory.

In some studies, it is well documented that the incidence of ADHD is high during short or long labor, hypoxia, fetal distress, eclampsia birth weight, and birth injuries. In 1999, various studies looked at pre- and perinatal striatal injury as a possible cause of ADHD and found that those perinatal adverse events may play a major role in the pathogenesis of some cases of ADHD. September births are significantly associated with ADHD and learning disability because, during this time, the seasonally mediated viral infections affect the both mothers and their fetus. The dysfunctioning of basal ganglia inhibitory circuits and also the shorter size on either the left or right side of caudate are make the peoples more prone to ADHD. The risk of ADHD is 2.47 times increased in the group of mothers who currently smoke than in the group who did not smoke. During pregnancy, if the mother smokes, the risk of ADHD is approximately 2.64 times higher. Organophosphate insecticide chlorpyrifos used in some fruits and vegetables causes delay in learning rates, decreases physical coordination, and causes behavioral abnormalities in children, especially ADHD. Research on several hundred children in the UK showed that 20–62.4 mg of food dyes and preservative pushes ordinary children about 10% closer to an ADHD diagnosis (Fig. 1).

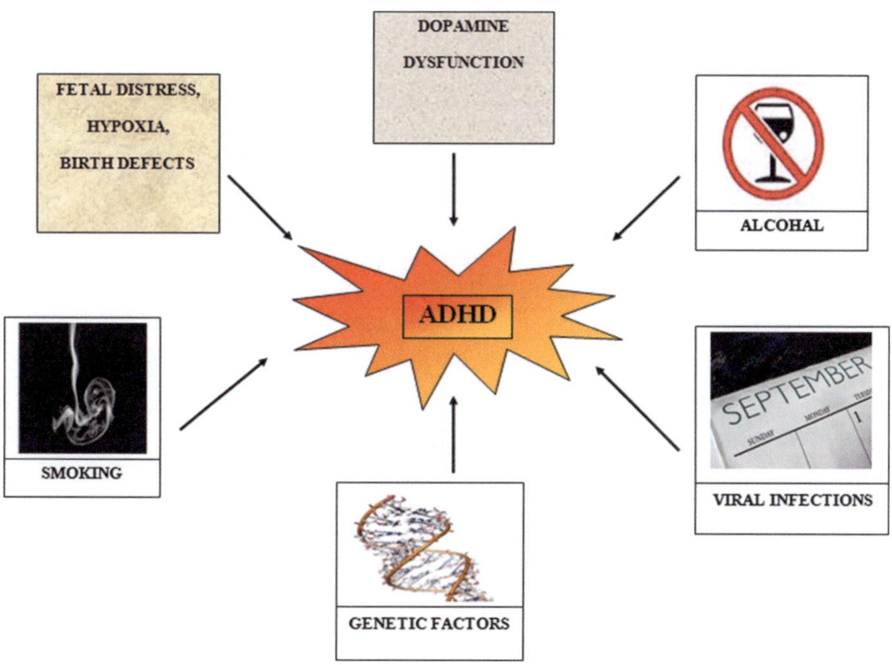

Fig. 1 Pathophysiology of ADHD

1.3 Need of Animal Model

Animal models are the best tools to evaluate any therapeutic agent. Animal models help to advance our understanding of the underlying mechanisms of disease and have proven to be invaluable in the preclinical evaluation of potential therapeutic interventions. Animal models of ADHD provide clear advantages over human studies. Animal models for ADHD with good predictive validity would allow the assessment of novel treatments and also provide tools for the study of underlying neurochemical and neuropathological alterations that lead to ADHD.

2 Classification of Animal Models of ADHD (Fig. 2)

2.1 Toxin-Induced Models

2.1.1 6-Hydroxydopamine

6-hydroxydopamine (6-OHDA) is a synthetic organic compound widely used by researchers due to its selectivity toward dopaminergic and noradrenergic neurons in the brain. 6-OHDA significantly decreases the endogenous levels of DA in the striatum, frontal cortex, and other limbic areas including the nucleus accumbens, septum, and olfactory tubercles in rats. Among these regions, striatum is mostly affected and thus dopamine dysfunction is involved in the pathogenesis of ADHD. Previous data also suggest that there is abnormality in dopamine-modulated frontal-striatal circuits, reflected by size (smaller-than-average components) and function (hypoactivation) in ADHD.

2.1.2 Dsp4

DSP4, also known as N-(2-chloroethyl)-N-ethyl-2-bromobenzylamine, can easily cross the blood–brain barrier and produce selective degeneration of noradrenergic neurons. The prefrontal cortex has been shown to be involved in the memory functions, and deficits in working memory have been observed in patients with ADHD. Literature data suggest that NA and alpha 2 agonists enhance working memory functions, whereas alpha 2 antagonists impair the memory functions. These findings suggest that depletion of NA can impair prefrontal functions, particularly in regard to memory functions.

Different doses and routes of administration of 6-hydroxydopamine

S. no.	Dose	Route	Species	Reference
1.	Single-dose administration of 6-OHDA (100 μg, 488 nmol) to 5-day male pups	Intracisternal injection	Wistar rats	Masuo et al. (2004a, b)
2.	Single-dose administration of 6-OHDA (16 mg/1 ml). Coordinates-A −1.8, L 2.4, V −7 mm	Medial forebrain bundle	Wistar rats	Erlij et al. (2012)
3.	Bilaterally injected 6-OHDA with 0.5 μg/0.5 μl at PND 11. Coordinates-AP: +2.8, ML: ±0.5, DV: 2.6	Medial prefrontal cortex (PFC)	Sprague–Dawley rats	Freund et al. (2014)

(continued)

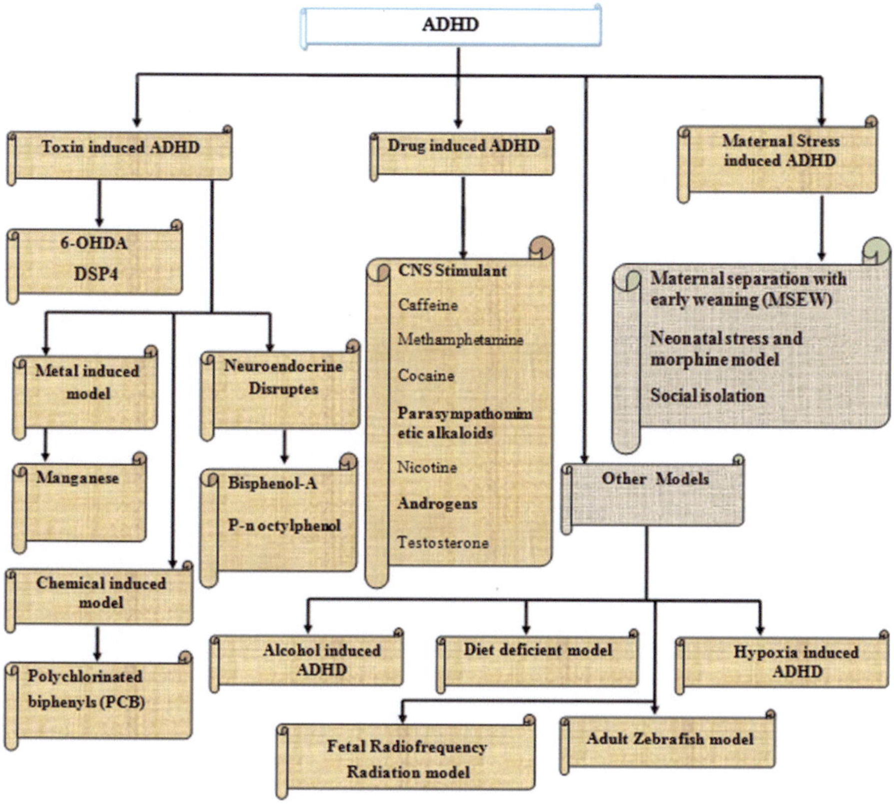

Fig. 2 Classification of animal model of ADHD

(continued)

S. no.	Dose	Route	Species	Reference
4.	Unilaterally lesion of 6-OHDA (8 μg/4 μl) on PND 7 coordinates-AP: +0.7 mm, ML: +2.2 mm, DV: −3.2 mm	Left striatum	Sprague–Dawley rats	Caballero et al. (2011)
5.	DSP4 (N-(2-chloroethyl)-N-ethyl-2-bromobenzylamine) (10, 20 or 50 mg/kg/body weight) for 5 weeks	i.p.	Wistar rats	Hauser et al. (2012)

Clinical relevance

6-OHDA produces alterations in the levels of DA in the striatum, frontal cortex, and other limbic brain regions. Similarly, disturbances in dopaminergic signaling have been reported to occur clinically in ADHD patients.

2.2 Metal-Induced Model

2.2.1 Manganese

Exposure to manganese (Mn) during neurodevelopment stage significantly alters the dopaminergic synaptic environments in the brain nuclei that control the executive functions, such as hyperactivity, impulsivity, and cognitive flexibility Kern et al. (2010).

Different doses and routes of administration of $MnCl_2$

S. no.	Dose	Route	Species	Reference
1.	$MnCl_2$ (25 μl of providing 0, 50, 145, 250, or 500 μg/day)	Orally gavaged	Sprague–Dawley rats	Tran et al. (2002)
2.	Mn doses of (0, 25, and 50 mg Mn/kg/d)	Orally	Sprague–Dawley rats	Kern et al. (2010)

Clinical relevance

Manganese toxicity has been known to be associated with hyperactivity, impulsivity, and violent behavior. Therefore, this model mimics all the behavioral symptoms as seen in ADHD patients.

2.3 Endocrine Disruptors

2.3.1 Bisphenol-A

Bisphenol-A (BPA) is an estrogenic chemical used in the production of polycarbonate and beverage cans and in dental sealants. During the development of fetal life, small changes in the levels of hormones (like estradiol) can lead to changes in the functions of brain and the behavior. Long-term exposure of BPA to female rats induces modifications in β-estrogen receptor immunoreactivity in various brain areas regulating reproductive and maternal behavior. Bisphenol-A exposure to rats has been reported to exhibit motor hyperactivity

Different doses and routes of administration of Bisphenol-A and p-n-octylphenol

S. no.	Dose	Route	Species	Reference
1.	Single-dose administration of Bisphenol-A (10, 20 and 40 μg/rat) into 5-day-old male pups	Intracisternal	Wistar rats	Kiguchi et al. (2007)
2.	Bisphenol-A (10 mg/kg/day) GD 11–PND 7	Electronic micropipette (as opposed to gavaged)	CD-1 mice	Palanza et al. (2008)

(continued)

(continued)

S. no.	Dose	Route	Species	Reference
3.	Bisphenol A (0.1 and 50 mg/l) in drinking water (unchlorinated purified water) from GD 11 to PND 21	Orally	Sprague–Dawley rats	Xu et al. (2007)
4.	Bisphenol-A(87 nmol/10 μl/rat) injected into 5-day-old pups	Intracisternal	Wistar rats	Ishido et al. (2005)
5.	Low doses of four EDCs, atrazine (10 mg/kg), perfluorooctanoic acid (PFOA 0.1 mg/kg), Bisphenol-A (50 mg/kg), 2, 3, 7, 8-tetrachlorodibenzo-p-dioxin (TCDD-0.25 mg/kg) alone, or combined in a mixture from GD 7 until weaning	Orally	C57BL/6 mice	Sobolewski et al. (2014)
6.	Single-dose administration of p-n-octylphenol (87 nmol) into 5-day-old male pups	Intracisternal	Wistar rats	Ishido et al. (2004)

Clinical relevance

According to the recent study reports, the increased level of bisphenol during early pregnancy is mainly associated with higher hyperactivity and aggression in 2-year-old girls and stayed in line with rodents study.

2.4 Chemical-Induced Model

2.4.1 Polychlorinated Biphenyls

PCBs belong to a broad family of man-made organic chemicals also known as chlorinated hydrocarbons. PCBs induce hyperactivity and motor impulsiveness and work via monoaminergic pathway. PCBs also interfere with the thyroid hormone signaling during brain development. Many of the neurodegenerative deficits linked to PCB exposures are similar to those associated with pre- and postnatal thyroid hormone insufficiency. These PCB congeners affect the intracellular regulation of calcium in rodent brain which is very important in nerve cell development and functioning. Dopamine and serotonin levels are decreased possibly via inhibition of dopamine synthesis and deficient vesicular storage or release.

Different doses and routes of administration of PCBs

S. no.	Dose	Route	Species	Reference
1.	Polychlorinated biphenyls (PCB) (1, 3 or 6 mg/kg) three times between PND 8 and 20	Orally	Spontaneously hypertensive rats (SHRs) and Wistar-Kyoto rats (WKY)	Baerland (2011)

(continued)

(continued)

S. no.	Dose	Route	Species	Reference
2.	Low combination of (1 mg/kg PCBs + 1.5 ppm MeHg) or the high combination (3 mg/kg PCBs + 4.5 ppm MeHg) throughout gestation and lactation	Orally	Nulliparous female Long-Evans rats	Sable et al. (2009)
3.	Aroclor 1254 (18 mg/kg/day) administered daily from gestational day (GD) 6–PND 21 (total 35 days)	Maternal injection	Male and female ICR mice	Nam et al. (2014)
4.	PCB 153 (5 mg/kg) or PCB 126 (2 mg/kg) of every 2 days from day 3 to 13 after delivery	Orally	DA:OLA:HSD females	Holene et al. (1998)

Clinical relevance

The number of reports in human populations represents that chronic exposures to PCBs during fetal development are linked to neurodegenerative disorders. In these reports, the populations include: the Yu-Cheng children in Taiwan (born to mothers exposed to thermally degraded PCBs between 1978 and 1979) and the Lake Michigan cohort children born to mothers who ate PCB contaminated fish. In the rodent studies, PCB exposures consistently showed the decline in serum thyroid hormone levels which are important factor for neurodevelopment.

2.5 Drug-Induced Models

2.5.1 CNS Stimulants

Cocaine is a tropane alkaloid obtained from the leaves of the coca plant. Caffeine is a bitter substance also found in coffee, tea, soft drinks, chocolate, kola nuts, and certain medicines. Dopamine agonists such cocaine and amphetamines produce down-regulation of dopamine synthesis.

Different doses and routes of administration

S. no.	Dose	Route	Species	Reference
1.	Caffeine (0.3 g/l in drinking water) to for 6 weeks	Orally	WT mice	Björklund et al. (2008)
2.	Single-dose administration of methamphetamine hydrochloride (1.5 mg/kg)	i.p.	SHR, WKY and SD rats	Womersley (2014)
3.	Single-dose administration of cocaine (20 mg/kg) during development	i.p.	SHR and WKY rats	Womersley (2014)

Clinical relevance

Dopamine agonists such as cocaine and amphetamines alter the synthesis of dopamine. The midbrain dopamine (DA) system, which includes the ventral tegmental area and the substantia nigra, is thought to play an important role in the pathogenesis of ADHD.

2.5.2 Parasympathomimetic Alkaloids

During pregnancy, if mother smokes, the risk of ADHD is approximately 2.64 times increased. Some animal experiments have shown that exposure to nicotine during the pregnancy results in hyperactivity in the child and the possible mechanisms for this may be the increase in number of nicotine receptors and an abnormality in the dopaminergic system. Smoking may reduce the amount of blood sent to the fetus which retards the supply of oxygen and nutrition to the fetus.

Different doses and routes of administration of nicotine

S. no.	Dose	Route	Species	Reference
1.	Nicotine (1.5 mg/kg) from the day 11 to 20 of gestation	s.c.	Female albino rats	Abdu (2013)
2.	Nicotine (2 mg/kg) from PND 1 to 7	Orally	Sprague–Dawley rat	Huang et al. (2006)
3	Nicotine (66 μg/kg) twice daily for 5 days starting at the age of either 3, 10, or 19 days	s.c.	Male NMRI mice	Ankarberg (2003)
4.	Nicotine (200 μg/ml) drinking solution for 30 days prior to and during pregnancy	Orally	C57Bl/6 mice	Pauly et al. (2004)

Clinical relevance

The risk of ADHD is 2.5 times higher in smoking mothers than who did not. Twenty-four studies on maternal tobacco smoking published between 1973 and 2002, indicated that there is an increased risk of ADHD in the offspring. The smoking-induced ADHD symptoms could be due to nicotine causing hypoxia to the fetus in uterus.

2.5.3 Androgens

Testosterone (a type of androgen) is a steroid hormone found in mammals, reptiles, birds, and other vertebrates. The role of androgens in the ADHD pathogens suggested that the increased levels of testosterone can diminish the brain–blood flow in the frontal cortex, via lowering of the level of estrogen receptor alpha and the vascular endothelial growth factor (VEGF)

Different doses and routes of administration of testosterone

S. no.	Dose	Route	Species	Reference
1.	Testosterone (10 ng) on PND 10 coordinates: 4.8 mm anterior to bregma, 2.3 mm lateral to the midsagittal suture, and 2.2 mm below dura	Intracranial	SHR and WKY rats	King et al. (2000)
2.	Testosterone (7.5 mg/kg) for 5 days/week	s.c.	Male Long-Evans rats	Wood et al. (2013)

Clinical relevance

It is well demonstrated that the early androgen treatment significantly reduced tyrosine hydroxylase (TH) in the specific area of brain-like caudate-putamen. This area of brain is important for behavioral inhibition, motor control, and habit teaching Therefore, this model mimics all the behavioral symptoms of ADHD.

2.5.4 Ethanol-Induced ADHD

Alcohol consumption during pregnancy is reported to increase the risk of ADHD. Exposure to alcohol consumption during pregnancy is associated with neurobehavioral disorders. The caudate has been reported to be one of the most sensitive brain areas to prenatal alcohol insult. Abnormalities in the caudate region are hypothesized to be related to defects in executive functioning, attention, and response inhibition. It is also reported that moderate levels of prenatal and postnatal alcohol exposure can have long-lasting effects on IQ and learning problems in young school children.

Different doses and routes of administration of ethanol

S. no.	Dose	Route	Species	Reference
1.	Ethanol (5 g/kg) every other day from PND 2 to 8 Ethanol (3 g/kg/day) from PND 1 to 7	i.p. oral intubation	Swiss mice Sprague–Dawley rats	Nunes et al. (2011), Smith et al. (2012)
2.	Ethanol (6 g/kg/day) from GD 7 to 16	Intragastric intubation	Sprague–Dawley rats	Choi et al. (2012)

Clinical relevance

The prevalence of ADHD is 1.55 times higher in children of mothers who consumed alcohol than in those whose mothers did not consume alcohol. The necropsy reports of an infant with fetal alcohol syndrome showed abnormalities in the basal ganglia, hippocampus, cerebellum, CNS disorganization, abnormalities in pituitary gland, and optic nerve.

2.6 Maternal Stress-Induced ADHD

2.6.1 Maternal Separation with Early Weaning (MSEW)

Early life stress has been shown to reduce glutamate and GABA transmission, particularly the $GABA_A$ receptor expression.

Species SHR and Control WKY Rats

Procedure

The day of birth is designated as PND 0. The maternal separation includes removal of the dam from the pups for 3 h daily from PNDs 2 to 14. Pups are transferred in the home cage to a different room to prevent communication with the dam by ultrasound vocalization. The temperature is maintained at 31–33 °C by infrared heating lamps so to prevent the risk of possible hypothermia. After 3 h, the pups are returned to the animal facility and the dams returned to their home cages.

Species: C57Bl/6 J Mice

Procedure

The day of birth is designated as PND 0. The maternal separation includes removal of the dam from the pups for 4 h daily from PNDs 2 to 5, and pups are transferred in a clean cage with access to food and water. The temperature is maintained by heating blanket set at 32 °C, and it is placed underneath the pups to assist with thermoregulation. From PNDs 6 to 16, the period of separation is increased to 8 h. On PND 17, MSEW animals are weaned and then left undisturbed aside from daily brief checks for dehydration on PNDs 18 to 19. They are housed in a single cage till PND 30.

2.6.2 Neonatal Stress and Morphine Model

The prolonged inescapable stress can have harmful effects on learning and development and on the immune system. This also increases the susceptibility to neurodegenerative diseases.

Species: Adult Wild-Type C57BL/6 Mice

Procedure

Morphine (2 mg/kg) is given by subcutaneous route twice daily between PNDs 5 and 9. Stressed pups are separated from the dam and isolated in individual containers within a veterinary warmer at 32 °C. To induce the stress, the pups are exposed to hypoxia (100% nitrogen 1 min followed by hyperoxia (100% oxygen 5 min) twice daily.

Clinical relevance

The study shows the absolute relation between the stress exposures during the pregnancy and ADHD symptoms in the child. Children of mothers who experienced moderate or severe stress have higher Child Behavior Checklist (CBCL) scores than the no-stressor group. This ensures that prenatal stress does cause more severe symptoms of ADHD and also shows that during the third trimester, increased stress levels correlate with greater CBCL scores.

2.6.3 Social Isolation-Induced ADHD

Species Male ICR Mice

Procedure

Male mice are socially isolated for 3–70 days before experiments in (24 cm 17 cm × 12 cm) sized cages.

2.7 Other Models

2.7.1 Hypoxia-Induced ADHD

Due to hypoxia, extracellular DA in medial PFC is reduced and transporter in PFC is increased in animals.

Species: Sprague–Dawley Rats

Procedure

Firstly unilateral ligation is performed with uterine artery supplying one uterine horn on embryonic day 17 (ED17) under the deep anesthesia (350 mg/kg chloral hydrate i.p.). Uterine artery ligation is made at the cervical level and generates the prenatal ischemia in pups located close to the site of ligation. Possibly the animals are exposed to hypoxia followed by a reperfusion phase during delivery. Prenatal ischemia at ED 17 caused a gradient of growth restriction along the uterine horn that we used to differentiate pups after delivery.

Species: Pregnant Albino Sprague–Dawley Rats

Procedure

The day of birth was designated as PND 0. Sprague–Dawley rat pups from three different litters are exposed to repeated hypoxia on PND 1–3. The rat brain during PND 1–3 is generally considered to be developmentally equivalent to the fetal human brain from mid-gestation to late in the second trimester (i.e., at 22–28) weeks of gestation. The repeated hypoxic pups are exposed to humidify 1.5% oxygen, 5% carbon dioxide, and 93.5% nitrogen at 37 °C every 2 h for 12 h daily from PND 1 to 3. The hypoxic exposure was 15 min on PND 1 and 14 min on PND 2 and PND 3.

Species: Sprague–Dawley Rats

Procedure

On the day of parturition, Sprague–Dawley dams are decapitated and hysterectomized. After this, quickly isolate entire uterus from the fetus and immediately immerse in a 37 °C saline bath for 15 min for the induction of birth hypoxia. The pups are removed from the uterus and stimulated by gentle rubs to initiate breathing.

Species: Male Wistar Rats

Procedure

In this experiment, the pups are exposed repeatedly to hypoxia during the first week of life. The mother is gently removed or separated, and the home cage containing the pups is then placed in an incubator connected to a gas supply. The temperature inside the chamber is maintained to 30 °C. Litters are assigned to be exposed to 100% nitrogen. The gas outflow is set at 40 L/min (1 bar). Exposure is to be given for 20 min on PND 1, −3, and −5 and for 10 min on PND 7.

2.7.2 N-3 PUFA-Deficient Diet Model of ADHD

The rise in consumption of processed foods and vegetable oils has markedly increased the intake of omega-6 PUFAs in Western diets, while the consumption of n-3 PUFAs in fish, nuts, and seeds has declined. Some studies have reported that children with ADHD have reduced levels of blood n-3 PUFAs as compared to healthy subjects.

S. no.	Dose	Route	Species	Reference
1.	n-3 PUFA-deficient diet to SHRs for 6 weeks	Orally	Male SHRs	Lange et al. (2013)

Clinical relevance

Some studies have reported that children with ADHD have reduced levels of blood n-3 PUFAs as compared to healthy subjects.

2.7.3 Fetal Radiofrequency Radiation Model

In utero cellular telephone, radiation exposure may lead to impairment of glutamatergic synaptic transmission onto pyramidal cells in the prefrontal cortex associated with behavioral changes.

S. no.	Dose	Route	Species	Reference
1.	Radiation from muted and silenced (800–1900 MHz) cellular phones with a SAR of 1.6 W/kg throughout GD 1–17	Radiation	Female mice	Aldad et al. (2012)

Clinical relevance

In utero exposure to radiofrequency is a potential cause of neurobehavioral disorders.

2.7.4 Adult Zebra Fish Model

The biological half-life of PCBs is long and has high liposolubility which leads to their bioaccumulation and biomagnifications through food chains over a wide range of trophic levels.

Species: Wild-Type Zebra Fish

Procedure

Eight-week-old fish are exposed to commercial food for 8 months at 26–28 °C. Commercial food consists of PCB mixtures composed of the seven indicator PCB congeners (CB-28, CB-52, CB-101, CB-118, CB-138, CB-153, and CB-180). For this, two doses are selected, i.e., intermediate and high doses. Fish exposed to an intermediate dose (equivalent to that found in the Loire Estuary, $\Sigma CB = 515$ ng/g dry weight in food) display behavioral abnormality in exploration capacities. Similarly, the fish exposed to the highest dose (equivalent to that found in the Seine Estuary, $\Sigma CB = 2302$ ng/g dry weight in food) display an increased swimming activity and larval activity.

Clinical relevance

The transparency of the zebra fish embryo allows direct and continuous visualization of tissue morphogenesis in vivo. The zebra fish also possess powerful cognitive abilities, including learning and memory. Zebra fish have been developed to be a useful tool for explaining not just the structural and chemical effects of neurotoxins, but also for assessing the behavioral impairment associated with such neurotoxins exposure.

3 Ethical Statement

All institutional guidelines, national guidelines, state and local laws, and regulations with professional standards for the care and use of laboratory animals should be followed. Studies involving animals must state that the institutional animal ethical committee has approved the protocol. For authors using experimental animals, a statement should be made that the animals' care is in accordance with institutional guidelines and animals used have been treated humanely and with regard to the alleviation of suffering. Researchers should treat animals as sentient and must consider their proper care and use and the avoidance or minimization of discomfort, distress, or pain as imperatives. Animal experiments should be designed only after due consideration of animal health. It should be ensured that all researchers who are using animals have received instruction in research methods and in the care, maintenance, and handling of the species being used. All the surgical procedures should be performed under appropriate anesthesia and follow only those procedures which avoid infection and minimize pain during and after surgery.

References

Abdu F (2013) Effect of Moringa Oleifera extract on nicotine induced neurotoxicity in female rat and their embryos

Aldad T, Gan G, Gao X et al (2012) Fetal radiofrequency radiation exposure from 800 to 1900 mhz-rated cellular telephones affects neurodevelopment and behavior in mice. Sci Rep 2

Ankarberg E (2003) Neurotoxic effects of nicotine during neonatal brain development: critical period and adult susceptibility

Baerland N (2011) Behavioral effects of PCB exposure in an animal model of attention-deficit/hyperactivity disorder: the spontaneously hypertensive rat

Björklund O, Kahlström J, Salmi P et al (2008) Perinatal caffeine, acting on maternal adenosine A 1 receptors, causes long-lasting behavioral changes in mouse offspring. PLoS ONE 3:3977

Caballero M, Núñez F, Cuffi M et al (2011) Caffeine improves attention deficit in neonatal 6-OHDA lesioned rats, an animal model of attention deficit hyperactivity disorder (ADHD). Neuroscience Lett 494:44–48

Choi I, Kim P, Joo H et al (2012) Effects of preconceptional ethanol consumption on ADHD-like symptoms in Sprague-Dawley rat offsprings. Biomol Therap 20:226

Erlij D, Acosta-García J, Floran B et al (2012) Dopamine D4 receptor stimulation in GABAergic projections of the globus pallidus to the reticular thalamic nucleus and the substantia nigra reticulata of the rat decreases locomotor activity. Neuropharmacology 62:1111–1118

Freund N, MacGillivilray H, Andersen S et al (2014) Sex-dependent changes in ADHD-like behaviors in juvenile rats following cortical dopamine depletion. Behav Brain Res 270:357–363

Hauser J, Sontag T, Tucha O et al (2012) The effects of the neurotoxin DSP4 on spatial learning and memory in Wistar rats. ADHD Atten Def Hyp Disord 4:93–99

Holene E, Nafstad I, Skaare J et al (1998) Behavioural hyperactivity in rats following postnatal exposure to sub-toxic doses of polychlorinated biphenyl congeners 153 and 126. Behav Brain Res 94:213–224

Huang L, Hsiao S, Frye G et al (2006) Chronic nicotine induces growth retardation in neonatal rat pups. Life Sci 78:1483–1493

Ishido M, Masuo Y, Oka S et al (2004) Intracisternal administration of p-n-octylphenol into neonatal rats causes hyperactivity concomitantly with the terminal deoxynucleotidyl transferase-mediated dUTP nick end-labelling (TUNEL)-positive cells in the mesencephalon where immunoreactivity for tyrosine hydroxylase is reduced by the chemical. J Health Sci 50:407–412

Ishido M, Morita M, Oka S et al (2005) Alteration of gene expression of G protein-coupled receptors in endocrine disruptors-caused hyperactive rats. Regul Pept 126:145–153

Kern C, Stanwood G, Smith D et al (2010) Preweaning manganese exposure causes hyperactivity, disinhibition, and spatial learning and memory deficits associated with altered dopamine receptor and transporter levels. Synapse 64:363–378

Kiguchi M, Fujita S, Lee J et al (2007) Behavioral responses to methylphenidate and apomorphine in rats exposed neonatally to bisphenol-A. J Oral Sci 49:311–318

King JA, Barkley RA, Delville Y, Ferris CF (2000). Early and rogen treatment decreases cognitive function and catecholamine innervation in an animal model of ADHD. Behav brain res 107:35–43

Lange K, Reisinger J, Sontag T et al (2013) Dietary omega-3 fatty acids and locomotor activity in an animal model of attention deficit hyperactivity disorder (ADHD). Funct Foods Health Dis 3:223–229

Masuo Y, Ishido M, Oka S et al (2004a) Effects of neonatal treatment with 6-hydroxydopamine and endocrine disruptors on motor activity and gene expression in rats. Neural Plast 11:59–76

Masuo Y, Morita M, Oka S et al (2004b) Motor hyperactivity caused by a deficit in dopaminergic neurons and the effects of endocrine disruptors: a study inspired by the physiological roles of PACAP in the brain. Regul Pept 123:225–234

Nam Y, Shin E, Lim Y et al (2014) YY162 prevents ADHD-like behavioral side effects and cytotoxicity induced by Aroclor1254 via interactive signaling between antioxidant potential, BDNF/TrkB, DAT and NET. Food Chem Toxicol 65:280–292

Nunes F, Kubrusly R, Manhaes A et al (2011) Acute administration of vinpocetine, a phosphodiesterase type 1 inhibitor, ameliorates hyperactivity in a mice model of fetal alcohol spectrum disorder. Drug Alcohol Depend 119:81–87

Palanza P, Gioiosa L, Saal F et al (2008) Effects of developmental exposure to bisphenol A on brain and behavior in mice. Environ Res 108:150–157

Pauly JR, Sparks JA, Hauser KF, Pauly TH (2004). In utero nicotine exposure causes persistent, gender dependant changes in locomotor activity and sensitivity to nicotine in C57Bl/6 mice. Int J Dev Neurosci 22:329–337

Sable H, Eubig P, Powers B et al (2009) Developmental exposure to PCBs and/or MeHg: effects on a differential reinforcement of low rates (DRL) operant task before and after amphetamine drug challenge. Neurotoxicol Teratol 31:149–158

Smith A, Wellmann K, Carter M et al (2012) Lobeline attenuates neonatal ethanol-mediated changes in hyperactivity and dopamine transporter function in the prefrontal cortex in rats. Neuroscience 206:245–254

Sobolewski M, Conrad K, Allen J et al (2014) Sex-specific enhanced behavioral toxicity induced by maternal exposure to a mixture of low dose endocrine-disrupting chemicals. Neurotoxicology 45:121–130

Tran T, Chowanadisai W, Lonnerdal B et al (2002) Effects of neonatal dietary manganese exposure on brain dopamine levels and neurocognitive functions. Neurotoxicology 23:645–651

Womersley J (2014) Methamphetamine and cocaine effects on dopamine neurons in a rat model of developmental stress and attention-deficit/hyperactivity disorder. University of Cape Town, Dept of Human Biology

Wood R, Armstrong A, Shah V et al (2013) ‘Roid rage in rats? testosterone effects on aggressive motivation, impulsivity and tyrosine hydroxylase. Physiol Behav 110:6–12

Xu X, Liu Y, Kato N et al (2007) Perinatal bisphenol A affects the behavior and SRC-1 expression of male pups but does not influence on the thyroid hormone receptors and its responsive gene. Neurosci Res 58:149–155

Animal Models of Autism

Nidhika Sharma, Sumit Jamwal and Puneet Kumar Bansal

1 Introduction

Autism is a neurodevelopment disorder that arises in approximately 1% of the worldwide population. Autism is a lifelong condition characterized by repetitive behavior and impaired socialization, abnormalities in communication, restricted and repetitive behavior that usually occurs in the first 3 years of life. Besides from these important symptoms, a significant number of autism individuals display higher levels of anxiety and exhibit impaired emotional learning. The most remarkable features of the social impairments in autism are defects in coordinating visual attention with others. External ear malformations are the most common physical abnormality associated with autism. About 30% individuals of autism spectrum disorder (ASD) engage in self-harming behaviors such as head banging and hand biting (Dominick et al. 2007; Johnson et al. 2007). Sir Leo Kanner in 1943 firstly described autism by as "children which are locked within them." Around 400,000 children are affected by autism in America, with 1–2 new cases per 1,000 births in the USA.

N. Sharma · S. Jamwal · P.K. Bansal (✉)
Department of Pharmacology, ISF College of Pharmacy, Moga 142001, Punjab, India
e-mail: punnubansal79@gmail.com

N. Sharma
e-mail: nidhika444@gmail.com

S. Jamwal
e-mail: jamwalisf@gmail.com

P.K. Bansal
Department of Pharmaceutical Sciences and Technology, Maharaja Ranjit Singh Punjab Technical University (State Govt. University), Bathinda, Punjab, India

P.K. Bansal and R. Deshmukh (eds.), *Animal Models of Neurological Disorders*, https://doi.org/10.1007/978-981-10-5981-0_15

Both autopsy reports and animal models of autism spectrum disorder (ASD) show impairment in different brain regions such as the frontal cortex, hippocampus, and the amygdala nucleus. Amygdala nucleus is widely considered as important for social behavior and empathy. Individuals with amygdala impairment do not process emotional information normally, and these symptoms are similar to autism individuals (Adolphs et al. 2001, 2002). It has also been suggested that early disruption in the normal functioning of frontal cortex and amygdala might be the reason for the social and emotional defects seen in autism. Many genetic studies show abnormalities in chromosomal regions such as 15q and 7q in case of autism individuals. In some cases, children with autism have impaired level of reduced glutathione and SOD both could directly lead to mitochondrial damage and dysfunction which further leads to oxidative stress. There are evidences that the blood–brain barrier function is impaired in autistic children due to neurological inflammation, immune imbalance, and increased levels of inflammatory cytokines in the brain. The anomalies in immune response are seen in the GI tract, the peripheral blood, and in the central nervous system (CNS) of the autistic patients. On the other hand, inflammation or maternal infection, and autoimmune diseases of the family of ASD children have also been shown to cause immune defects in the descendants. Abnormalities of immune function with elevated pro-inflammatory cytokine levels are commonly seen in autism individuals, most usually as overactivation or dysfunction of immune response. According to reports, the levels of IL-6 are significantly elevated in both the brain and CSF of autistics. These cytokine levels are known to play important role in neuronal development process; when the cytokine levels are chronically exalted, it can disrupt normal brain development process. The various cycles like folate, methylation, and sulfation are interrelated cycles that play important role in purine and pyrimidine synthesis, metabolism of neurotransmitters, antioxidation, and detox process. The impairment in these pathways occurs in autism individuals (Fig. 1). According to various reports, the exposure to chemicals like polychlorinated biphenyls (PCB), mercuric chloride could be a contributing factor to behavioral, neuropathological, and biochemical impairments observed in disorders like autism and attention-deficit/hyperactivity disorder (ADHD) (Winneke 2011) (Fig. 1).

A variety of animal models have been developed to understand the mechanisms that may induce one or more of the important features of autism such as socialization impairment, communication abnormalities, restrictive and behavior, interests, and activities. The information secured from these animal models is much helpful in targeting the new therapeutic approaches as well as preventive strategies for autism.

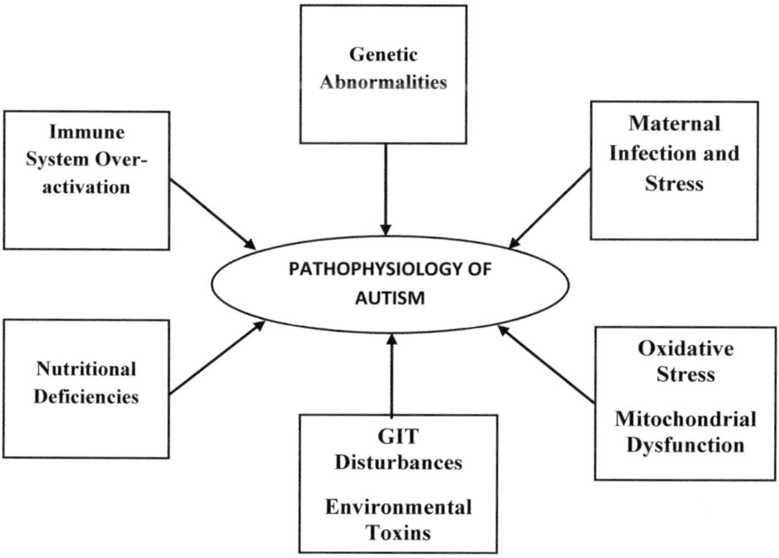

Fig. 1 Pathophysiology of autism

2 Classification of Animal Models of Autism

2.1 Toxin-Induced Autism

2.1.1 Methyl Mercury

Principle: Methyl mercury (MeHg) is a well-known and potent neurotoxin, with high-level exposures during development period leading to mental retardation, cerebral palsy, and seizures. There are evidences that mercury decreases antioxidant levels, elevated ROS generation followed by mitochondrial dysfunction. Thimerosal is firstly metabolized to ethyl mercury and then to inorganic mercury, both of which accumulate in the brain. The neonatal administration of thimerosal initiates neurodegeneration of hippocampal region and significant changes in the dopamine system with subsequent behavioral deficits. The prenatal exposure of mercury can also developmentally damage the metal detoxification system of the liver which can further lead to accumulation and toxicity of later metal exposure.

Different doses and routes of administration of methyl mercury

S. No.	Dose	Route	Species	References
1.	Methyl mercury (MeHg) (0.03 μg/kg/day) during days 8–18 of gestation	Orally	Pregnant mice	Bustamante et al. (2006)
2.	Single injection of methyl mercury (5 μg/g/body weight) on postnatal day (PND)7	s.c.	Female rats	Falluel-Morel et al. (2007)
3.	Thimerosal (1, 0.1, and 0.01 mg Hg/kg in volume of 50 μl) to on gestation day (GD) 9	Intramuscular (i.m.)	Female rats	Ida-Eto et al. (2012)

Procedure:

- Female rats are allowed to raise their own litters.
- Behavioral tests (nociception, locomotor activity, anxiety, restlessness, and social behavior) are performed in both adolescence (PND 30–40) and adulthood (PND 90–110) period in litters.
- At the end of behavioral testing, animals are killed, and brain is isolated for biochemical estimations (using serotonin, glutathione, catalase, and nitric oxide) and histopathological examination.
- Induction of autism significantly affects normal behavior, serotonin levels, glutathione levels, and oxidative stress parameters.

Clinical Relevance

US EPA estimates that approximately 25% of US kids getting mercury exposure at threatening levels. Evidence regarding this indicates that over 60,000 children are born every year with neurodevelopment disorders due to methyl mercury.

2.1.2 Methylazoxymethanol Acetate

Principle: Methylazoxymethanol acetate (MAM) is a neurotoxin which decreases DNA synthesis, and it is also used in the animal model of neurological disorders. Administration of MAM affects the brain structures by the malformations of cortical development (MCD) (Bassanini et al. 2007).

S. No.	Dose	Route	Species	Reference
1.	One or two methylazoxymethanol acetate (MAM) doses (15 mg/kg/body weight) on GD 15	i.p.	Female Sprague-Dawley rats	Bassanini et al. (2007)

Procedure: Similar procedure is performed as given in Sect. 2.1.1.

2.2 Chemicals-Induced Autism

2.2.1 Polychlorinated Biphenyls (PCB)

Principle: PCBs were primarily produced as a mixture of various congeners and widely used in many commercial and industrial applications. Proper thyroid hormone concentrations are crucial during the sensitive windows of development for normal brain. PCB has the ability to decrease thyroid hormone availability during gestational and early postgestational period.

Different doses and routes of administration of PCB

S. No.	Dose	Route	Species	Reference
1.	Equal amounts of PCB 47 and PCB 77 (12.5 mg/kg diet and 25 mg/kg diet) from the first day of pregnancy until pup weaning	Orally	Female Sprague-Dawley rats	Jolous-Jamshidi et al. (2010)

Procedure: Similar procedure is performed as given in Sect. 2.1.1.

Clinical Relevance

There are numbers of reports in human populations to conclude that exposures to PCBs during fetal development are linked to neurodegenerative disorders. Historical reports include the children born with neuronal abnormalities and showed symptoms of autism from mothers exposed to thermally degraded PCBs (In 1978 and 1979; Taiwan) (Chen et al. 1992) and the mothers who ate PCB-contaminated fish from Lake Michigan cohort (Jacobson and Jacobson 2003).

2.2.2 Mercuric Chloride

Principle: Mercury is neurotoxin and the environmental exposure of mercury contributes to neurological disorders including autism spectrum disorders and Alzheimer's disease. The mechanism for mercury-induced neurotoxicity is mitochondrial dysfunction and oxidative stress (Curtis et al. 2011). There is accumulation of mercury in the different regions of brain followed by abnormality in neuronal functioning, including hippocampus and cerebellum.

Different doses and routes of administration of mercuric chloride

S. No.	Dose	Route	Species	Reference
1.	Mercuric chloride (0, 2 and 5 mg/kg) once per week during postnatal weeks 1–3	s.c.	MT1/MT2-null mice	Eddins et al. (2008)

Procedure: Similar procedure is performed as given in Sect. 2.1.1.

Clinical Relevance

The children with autism show dysfunction in mitochondrial electron transport chain (Chauhan et al. 2011). Some autopsy reports also show mitochondrial dysfunction in autism (Kern et al. 2012). Therefore, this model is clinically relevant for better understanding of autism.

2.2.3 Lipopolysaccharide (LPS)

Principle: Microglia cells could be triggered by the Gram-negative bacteria cell wall component, lipopolysaccharide (LPS). There are numbers of evidences indicating that neurodegenerative disorders such as Parkinson's disease (PD), autism, and schizophrenia are associated with oxidative stress.

Different doses and routes of administration of LPS

S. No.	Dose	Route	Species	References
1.	LPS (100 μg/kg) on GD 9.5	i.p.	Female Wistar rats	Kirsten et al. (2013)
2.	LPS (100 μg/kg) on GD 18	i.p.	Female Wistar rats	Penteado et al. (2013)
3.	LPS (100, 300, 1000, or 2000 μg/kg) on GD 16–17	i.p.	Swiss mice	Chlodzinska et al. (2011)
5.	LPS (100 μg/kg) once daily at GD 15 and 16	i.p.	Female Sprague-Dawley rats	Baharnoori et al. (2013)
6.	LPS (50 μg/kg) for a total of 2 injections on GD 15–16	s.c.	Female Long-Evans rats	Foley et al. (2014)

Procedure: Similar procedure is performed as given in Sect. 2.1.1.

Clinical Relevance

The prenatal exposure to inflammatory agents has marked effects on several gene pathways in the brain which are directly involved in the behavioral, motor activity, and intellectual deficits observed in children exposed to inflammation during fetal life. Therefore, this model is clinically relevant to autism.

2.3 Drugs-Induced Autism

2.3.1 Valproic Acid (VPA)

Principle: VPA has ability to inhibit histone deacetylase (HDAC). HDAC plays an important role in regulating transcription during fetal development. The suppression of histone deacetylase may induce aberration in the expression of gene during fetal development and may also cause behavioral irregularities (Banerjee et al. 2014).

Different doses and routes of administration of valproic acid

S. No.	Dose	Route	Species	References
1.	Single-dose administration of VPA (350 mg/kg) on either GD 11.5 or 12.5	i.p.	Female Sprague-Dawley rats	Rodier et al. (1996)
2.	Single-dose administration of VPA (600 mg/kg) on GD 12.5	i.p.	Female Wistar rats	Foley et al. (2012)
3.	Single-dose administration of VPA (400 mg/kg) on GD 12	s.c.	Female Sprague-Dawley rats	Al-Amin et al. (2015)
4.	Single-dose administration of VPA (400 mg/kg) on PND5 or PND14	s.c.	Female BALB/c mice	Pragnya et al. (2014)
5.	Single-dose administration of VPA (400 mg/kg) on GD 12.5 or 13.5	i.p.	Female mice and Wistar albino rats	Almeida et al. (2014)
6.	Single-dose administration of VPA (500 mg/kg) on GD 11 or 12.5	i.p.	Female Wistar rats	Takuma et al. (2014)
7.	Single-dose administration of VPA (800 mg/kg) on GD 9 or 12.5	Orally	Female Wistar rats	Oyabu et al. (2013)
8.	VPA (20,100 mg/kg) daily from ED12.5 to 0ED21–22 (covering the last 9–12 days of pregnancy)	i.p.	Female barrier-raised Wistar rats	Sabers et al. (2015)
9.	Single-dose administration of VPA (400 mg/kg) on GD 7, 9.5, 12, and 15	s.c.	Female Sprague-Dawley rats	Kim et al. (2011)
9.	Single-dose administration of VPA (800 mg/kg) on ED 9	Orally	Female Sprague-Dawley rats	Narita et al. (2002)

Procedure: Similar procedure is performed as given in Sect. 2.1.1.

2.3.2 Thalidomide

Principle: Autism is a behaviorally defined disorder with impaired socialization, as well as restricted and repetitive behavior. Exposure of rat fetuses to thalidomide (THAL) on the ninth day of gestation has been reported as a useful model for human autism. It has been shown that early serotonergic neural development is disrupted in these rats.

Different doses and routes of administration of thalidomide

S. No.	Dose	Route	Species	References
1.	Thalidomide (500 mg/kg) on GD 9	Orally	Female Wistar rats	Miyazaki et al. (2005)
2.	Thalidomide (500 mg/kg) on GD 9	Orally	Female Sprague-Dawley rats	Narita et al. (2002, 2010)

Procedure: Similar procedure is performed as given in Sect. 2.1.1.

Clinical Relevance

Serotonergic system deformity has been observed in human autism. In rodents, the teratogen like thalidomide has almost same effects on the serotonergic system on GD 9, which corresponds to approximately GD 20–21 in human (Miyazaki et al. 2005).

2.3.3 Propionic Acid

Principle: Elevated levels of propionic acid can induce oxidative stress, depletion of glutathione (GSH), and elevation of lipid peroxides in various brain regions such as cortex, striatum, and hippocampus of rats (Aldbass et al. 2013).

Different doses and routes of administration of propionic acid

S. No.	Dose	Route	Species	Reference
1.	Sodium propionate (500 mg/kg) once daily on GD 12–16 for a total of 5 injections.	s.c.	Female Long-Evans rats	Foley et al. (2014)

Procedure: Similar procedure is performed as given in Sect. 2.1.1.

Clinical Relevance

Microglia activation and reactive astrocytes in hippocampus have been observed in propionic acid induced rat model having resemblance to findings from brain tissue of autistic patients (Bauman and Kemper 2005).

2.4 Maternal Stress-Induced Model

2.4.1 Prenatal Air Pollution Model of Autism

Species—C57BL/6 mice

Procedure

- Starting from ED 2, females are lightly anesthetized with 2% isoflurane for approximately a period of 1 min.

- Then, female rats are treated with diesel exhaust particles (DEP) via oropharyngeal aspiration.
- The dose is of DEP (50 μg/μl) during ED2–17 in every 3 days for a total of six doses (Bolton et al. 2013).

2.4.2 Stress-Induced Model

Species—C57BL/6 mice

Procedure

- Chronic stress variables are administered to dams on GDs 1–7.
- Every day, a novel stressor is provided to dams, i.e., on GD1 multiple cage changes, GD2 100 db white noise overnight, GD3 15-min restraint in a 50-ml conical tube, GD4 36-h constant light, GD5 novel objects in the home cage overnight (8 marbles similar in color and size), GD6 1-h exposure to fox odor, and on GD7 wet bedding overnight (Morgan and Bale 2011).
- Similar procedure is performed as given in Sect. 2.1.1.

Species—Sprague-Dawley rats

Procedure

- The repeated variable prenatal stress paradigms used in this study are adapted from Koenig and colleagues (Kinnunenet al. 2003; Koenig ret al. 2005).
- Pregnant rats are exposed to the paradigm beginning on day 14 of gestation until delivery of pups on gestational day 22 or 23.
- The stress paradigm consisted of the following: (1) restraint in Broome-style rodent restrainers; (2) exposure to a cold environment (4 ± 1 °C, 6 h); (3) overnight food privation; (4) forced swim in room temperature water (15 min); (5) reversal of the light–dark cycle; and (6) social stress induction by overpopulated housing during the dark phase of the cycle (Wilson et al. 2013).
- Similar procedure is performed as given in Sect. 2.1.1.

Species—Wistar Han rats

Procedure

- The protocol of stress during puberty (PPS) is based on exposure to fear-inducing procedures.
- The stressors are administered during PND 28−42 (a total of 7 days across postnatal day P28–P42, i.e., on P28–P30, P34, P36, P40, and P42).
- The protocol started with an exposure to an open field on PND 28 for 5 min, after which subjects experienced repeated stress exposures consisted of presenting two different fear-inducing stressors: the synthetic fox odor

trimethylthiazoline (TMT) (9 ml) which is administered in a plastic box (38 × 27.5 × 31 cm) and exposure to an elevated platform (EP) (12 × 12 cm).
- Each stressor lasted 25 min (Poirier et al. 2014; Tzanoulinou et al. 2014)
- At the end of behavioral testing, animals are killed, and brain is isolated for biochemical estimations (using serotonin, glutathione, catalase, and nitric oxide) and histopathological examination.

Clinical Relevance

Both human and animal studies have found important relationship between prenatal stress and postnatal complications in a variety of behavioral domains, such as attention, impaired socialization, abnormalities in communication language. An autism individual shows these behavioral impairments. Therefore, this animal model is very important for better understanding of behavioral deficits in autism (Kinney et al. 2008).

2.5 Maternal Immune Activation Models

2.5.1 Polyinosinic: Polycytidylic Acid

Principle: Polyinosinic: polycytidylic acid (usually abbreviated poly I:C) is an immuno-stimulant in nature. It is used in its sodium salt form to mimic viral infections. The prenatal activation of immune system in mice causes long-lasting impairments in memory (Richetto et al. 2013).

S. No.	Dose	Route	Species	References
1.	Single injection of poly(I:C) at a dose of (5 mg/kg) on GD 17	Intravenously (i.v.)	Female C57BL/6 mice	Richetto et al. (2013)
2.	Single injection of poly(I:C) at a dose of (20 mg/kg) on ED 12.5	i.p.	Female WT mice	Ehningeret al. (2014)
3.	Single injection of poly(I:C) at a dose of (20 mg/kg) on ED 12.5	i.p.	C57BL/6J mice	Shi et al. (2009)
4.	IL-6 (5 μg of carrier protein-free recombinant mouse) or 5 μg of carrier-free recombinant mouse interferon γ (IFN-γ) on ED 12.5	i.p.	Female C57BL/6J mice	Smith et al. (2007)

Procedure: Similar procedure is performed as given in Sect. 2.1.1.

Clinical Relevance

According to one study in more than 10,000 autism cases, a strong connection with viral infection in the first trimester and bacterial infection in the second trimester has been reported. Therefore, this model is important for better explanation of behavioral deficits in autism (Patterson et al. 2011).

2.5.2 Ethanol-Induced Model

Principle: The prenatal exposure to ethanol is associated with neurodevelopment disorders including autism. The abnormalities in the caudate region are hypothesized to be related to defects in attention, communication abnormalities, activities, and response inhibition (Mattson et al. 1996b).

S. No.	Dose	Route	Species	Reference
1.	Ethanol (2.9 g/kg) at time 0 (t0). Two hours later (t2), animals received a second injection (1.45 g/kg) on GD 12	i.p.	Female Long-Evans rats	Middleton et al. (2012)

Procedure: Similar procedure is performed as given in Sect. 2.1.1.

Clinical Relevance

The necropsy reports of an infant with fetal alcohol syndrome contribute abnormalities in the basal ganglia, hippocampus, cerebellum, CNS disorganization, abnormalities in pituitary gland and optic nerve (Jones and Smith 1975).

3 Conclusion

Rodents and nonhuman primates are important tools for better understanding of autism and also helpful in understanding the pathophysiology, cause, and treatment of autism. Autism is a behaviorally defined disorder. In this chapter, various animal models for autism have been described which display behavioral deficits, including impaired social behavior, repetitive and restricted behaviors and also represents an easy model to induce autism. No model is perfect, but these models demonstrate some of the pathophysiological features of autism (Fig. 2).

Ethical Statement

All institutional guidelines, national guidelines, state and local laws and regulations with professional standards for the care and use of laboratory animals should be followed. Studies involving animals must state that the institutional animal ethical committee has approved the protocol. For authors using experimental animals, a statement should be made that the animals' care is in accordance with institutional guidelines and animals used have been treated humanely and with regard to the alleviation of suffering. Researchers should treat animals as sentient and must consider their proper care and use and the avoidance or minimization of discomfort, distress, or pain as imperatives. Animal experiments should be designed only after due consideration of animal health. It should be ensured that all researchers who are

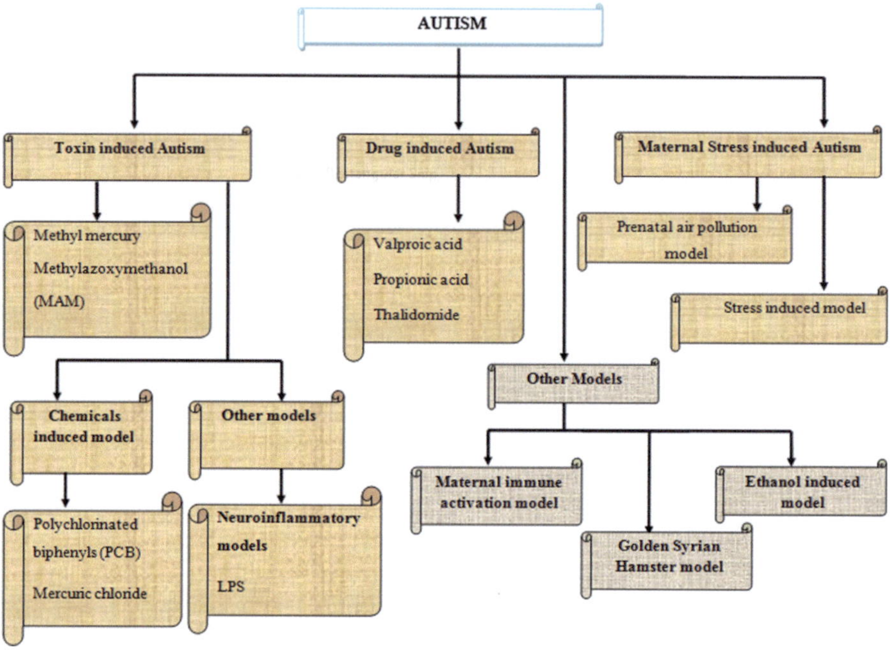

Fig. 2 Classification of animal model of autism

using animals have received instructions in research methods and in the care, maintenance, and handling of the species being used. All the surgical procedures should be performed under appropriate anesthesia and followed only those procedures which avoid infection and minimize pain during and after surgery.

References

Adolphs R, Baron-Cohen S, Tranel D (2002) Impaired recognition of social emotions following amygdala damage. J Cogn Neurosci 14:1264–1274

Adolphs R, Sears L, Piven J (2001) Abnormal processing of social information from faces in autism. J Cogn Neurosci 13:232–240

Al-Amin MM, Rahman MM, Khan FR, Zaman F, Reza HM (2015). Astaxanthin improves behavioral disorder and oxidative stress in prenatal valproic acid-induced mice model of autism. Behav Brain Res 286:112–121

Aldbass AM, Bhat RS, El-Ansary A (2013) Protective and therapeutic potency of N-acetyl-cysteine on propionic acid-induced biochemical autistic features in rats. J Neuroinflamm 10:42

Almeida LEF, Roby CD, Krueger BK (2014) Increased BDNF expression in fetal brain in the valproic acid model of autism. Mol Cell Neurosci 59:57–62

Baharnoori M, Bhardwaj SK, Srivastava LK (2013) Effect of maternal lipopolysaccharide administration on the development of dopaminergic receptors and transporter in the rat offspring. PLoS ONE 8(1):e54439

Banerjee A, Engineer C. T, Sauls et al (2014). Abnormal emotional learning in a rat model of autism exposed to valproic acid in utero Frontiers in Behavioral Neuroscience, vol. 8, article 387.doi:10.3389/fnbeh.2014.00387
Bassanini S, Hallene K, Battaglia G et al (2007). Early cerebrovascular and parenchymal events following prenatal exposure to the putative neurotoxin methylazoxymethanol. Neurobiol Dis 26(2):481–495
Bauman ML, Kemper TL (2005) Neuroanatomic observations of the brain in autism: a review and future directions. Int J Dev Neurosci 23:183–187
Bolton JL, Huff NC, Smith SH et al (2013) Maternal stress and effects of prenatal air pollution on offspring mental health outcomes in mice. Environ Health Perspect 121(9). doi:10.1289/ehp.1306560
Bustamante ND (2006) Autism and the potential role of immune function and prenatal. Methylmercury Expo
Chauhan A, Gu F, Essa MM et al (2011) Brain region-specific deficit in mitochondrial electron transport chain complexes in children with autism. J Neurochem 117:209–220
Chen YCJ, Guo YL, Hsu CC et al (1992) Cognitive-development of Yu-Cheng (oil disease) children prenatally exposed to heat-degraded PCBs. JAMA 268(22):3213–3218
Chlodzinska N, Gajerska M, Bartkowska K et al (2011) Lipopolysaccharide injected to pregnant mice affects behavior of their offspring in adulthood. Acta Neurobiol Exp 71:519–527
Curtis J. T, Chen Y, Buck DJ (2011) Chronic inorganic mercury exposure induces sex-specific changes in central TNFα expression: Importance in autism? Neurosci Lett 17, 504(1)
Dominick KC, Davis NO, Lainhart J, Tager-Flusberg H, Folstein S (2007) Atypical behaviors in children with autism and children with a history of language impairment. Res Dev Disabil 28, 2:145–162
Eddins D, Petro A, Pollard N et al (2008) Mercury-induced cognitive impairment in metallothionein-1/2 null mice. Neurotoxicol Teratol 30(2):88–95
Ehninger D (2014) Tsc2 Haploinsufficiency has limited effects on fetal brain cytokine levels during gestational immune activation. Autism Res Treat Article ID 761279. doi:10.1155/2014/761279
Falluel-Morel A, Sokolowski K, SistiH M et al (2007) Developmental mercury exposure elicits acute hippocampal cell death, reductions in neurogenesis, and severe learning deficits during puberty. J Neurochem 103(5):1968–1981
Foley AG, Gannon S, Rombach-Mullan N et al (2012) Class I histone deacetylase inhibition ameliorates social cognition and cell adhesion molecule plasticity deficits in a rodent model of autism spectrum disorder. Neuropharmacology 63(4):750–760
Foley KA, Ossenkopp K, Kavaliers M et al (2014) Pre- and neonatal exposure to lipopolysaccharide or the enteric metabolite, propionic acid, alters development and behavior in adolescent rats in a sexually dimorphic manner. PLoS ONE 9(1):e87072
Ida-Eto M, Oyabu A, Ohkawara T et al (2012) Prenatal exposure to organomercury, thimerosal, persistently impairs the serotonergic and dopaminergic systems in the rat brain: Implications for association with developmental disorders. Brain Dev. doi:10.1016/j.braindev.2012.05.004
Jacobson JL, Jacobson SW (2003) Prenatal exposure to polychlorinated biphenyls and attention at school age. J Pediatr 143(6):780–788
Johnson CP, Myers SM, Lipkin PH et al (2007) Identification and evaluation of children with autism spectrum disorders. Pediatrics 120(5):1183–1215
Jolous-Jamshidi B, Cromwell HC, McFarland AM et al (2010) Perinatal exposure to polychlorinated biphenyls alters social behaviors in rats. Toxicol Lett 199(2):136–143
Jones KL, Smith DW (1975) The fetal alcohol syndrome. Teratology 12:1–10
Kern JK, Geier DA, Audhya T et al (2012) Evidence of parallels between mercury intoxication and the brain pathology in autism. Acta Neurobiol Exp 72:113–153
Kim KC, Kim P, Go HS (2011) The critical period of valproate exposure to induce autistic symptoms in Sprague–Dawley rats. Toxicol Lett 201:137–14

Kinney DK, Munir KM, Crowley DJ et al (2008) Prenatal stress and risk for autism. Neurosci Biobehav Rev 32(8):1519–1532

Kinnunen AK, Koenig JI, Bilbe G (2003) Repeated variable prenatal stress alters pre- and postsynaptic gene expression in the rat frontal pole. J Neurochem 86:736–748

Kirsten TB, Lippi LL, Bevilacqua E et al (2013) LPS exposure increases maternal corticosterone levels, causes placental injury and increases IL-1B levels in adult rat offspring: relevance to autism. PLoS ONE 8(12):e82244

Koenig JI, Elmer GI, Shepard PD et al (2005) Prenatal exposure to a repeated variable stress paradigm elicits behavioral and neuroendocrinological changes in the adult offspring: potential relevance to schizophrenia. Behav Brain Res 156:251–261

Mattson SN, Riley EP, Delis DC et al (1996) Verbal learning and memory in children with fetal alcohol syndrome. Alcohol Clin Exp Res 20:810–816

Middleton FA, Varlinskaya EI, Mooney SM (2012) Molecular substrates of social avoidance seen following prenatal ethanol exposure and its reversal by social enrichment. Dev Neurosci 34(2–3): 115–128

Miyazaki K, Narita N, Narita M (2005) Maternal administration of thalidomide or valproic acid causes abnormal serotonergic neurons in the offspring: implication for pathogenesis of autism. Int J Dev Neurosci 23:287–297

Morgan CP, Bale TL (2011) Early prenatal stress epigenetically programs dysmasculinization in second-generation offspring via the paternal lineage. J Neurosci 17, 31(33):11748–11755

Narita M, Oyabu A, Imura Y et al (2010) Nonexploratory movement and behavioral alterations in a thalidomide or valproic acid-induced autism model rat. Neurosci Res 66(1):2–6

Narita N, Kato M, Tazoe M et al (2002) Increased monoamine concentration in the brain and blood of fetal thalidomide- and valproic acid-exposed rat: putative animal models for autism. Pediatr Res 52(4):576–579

Oyabu A, Tashiro Y, Oyama T et al (2013) Morphology of the facial motor nuclei in a rat model of autism during early development. Int J Dev Neurosci 31:138–144

Patterson PH (2011) Maternal infection and immune involvement in autism. Trends Mol Med 17(7)

Penteado S, Gomes COM, Kirsten T et al (2013) Prenatal lipopolysaccharide increases maternal behavior, decreases maternal odor preference, and induces Lipopolysaccharide hyporesponsiveness. Psychol Neurosci 6(1):31–38

Poirier GL, Imamura N, Zanoletti O et al (2014) Social deficits induced by peripubertal stress in rats are reversed by resveratrol. J Psychiatr Res 57(2014):157–164

Pragnya B, Kameshwari JSL, Veeresh B (2014) Ameliorating effect of piperine on behavioral abnormalities and oxidative markers in sodium valproate induced autism in BALB/C mice. Behav Brain Res 270:86–94

Richetto J, Calabrese F, Meye U et al (2013) Prenatal versus postnatal maternal factors in the development of infection-induced working memory impairments in mice. Brain Behav Immun 33:190–200

Rodier PM, Ingram JL, Tisdale B et al (1996) Embryological origin for autism: developmental anomalies of the cranial nerve motor nuclei. J Comp Neurol 370:247–261

Sabers A, Bertelsen FCB, Scheel-Kruger J et al (2015) Corrigendum to "Long-term valproic acid exposure increases the number of neocortical neurons in the developing rat brain" A possible new animal model of autism. Neurosci Lett 588:203–207

Shi L, Smith SEP, Malkova N et al (2009) Activation of the maternal immune system alters cerebellar development in the offspring. Brain Behav Immun 23(1):116–123

Smith SEP, Li J, Garbett K et al (2007) Maternal immune activation alters fetal brain development through Interleukin-6J. Neurosci 27(40):10695–10702

Takuma K, Hara Y, Kataoka S et al (2014) Chronic treatment with valproic acid or sodium butyrate attenuates novel object recognition deficits and hippocampal dendritic spine loss in a mouse model of autism. Pharmacol Biochem Behav 126:43–49

Tzanoulinou S, Garcia-Mompo C, Castillo-Gomez E et al (2014) Long-term behavioral programming induced by peripuberty stress in rats is accompanied by GABAergic-related alterations in the Amygdala. Plosone 9(4):e94666

Wilson CA, Vazdarjanova A, Terry AV (2013) Exposure to variable prenatal stress in rats: effects on anxietyrelated behaviors, innate and contextual fear, and fear extinction. Behav Brain Res 1(238):279–288

Winneke G (2011) Developmental aspects of environmental neurotoxicology: lessons from lead and polychlorinated biphenyls. J Neurol Sci 308(1–2):9–15

Animal Models of Tourette's Syndrome

Priya Jaswal, Harmanpreet Kaur and Rahul Deshmukh

1 Introduction

Tics are defined as non-rhythmic, involuntary, and rapid movements or sounds. Approximately 15% of children experience during their growth and development a transient tic condition, without the need of medical treatment, with one or few tics that mostly disappear within less than 1 year. The transient tic condition differs from chronic tic disorder in terms of severity and time course, as in the latter the motor or vocal tics are present for more than 1 year. Tourette's syndrome (TS), the most severe tic disorder, is a childhood-onset condition consisting of multiple motor and at least one phonic tic for duration longer than 1 year. Tourette's syndrome was first described by the French neurologist Gilles de la Tourette in 1885. TS was thought to be rare, but recent epidemiological studies have shown prevalence of about 0.3–1%. Tourette's syndrome is a neuropsychiatric disorder which is characterized by motor and vocal tics. Motor tics are sudden, repetitive, stereotyped movements involving facial twitching, eye blinking, and movements related to head and shoulder, whereas phonic tics are related to sounds generated by moving air with the help of nose, mouth, or throat (e.g., coughing and throat clearing) as well as repeating syllables, words, or phrase. Severity of tics usually peaks between

P. Jaswal · H. Kaur
Department of Pharmacology, ISF College of Pharmacy, Moga, Punjab, India
e-mail: priyajaswal138@gmail.com

H. Kaur
e-mail: Harmansaini31.hs@gmail.com

R. Deshmukh (✉)
Department of Pharmacology, ISF College of Pharmacy, Moga 142001, Punjab, India
e-mail: login2rd@gmail.com

P.K. Bansal and R. Deshmukh (eds.), *Animal Models of Neurological Disorders*, https://doi.org/10.1007/978-981-10-5981-0_16

8 and 12 years of age, but by the end of puberty, many patients show a marked reduction in severity of tics. It has also been reported that less than 20% of children with TS continue to experience a moderate level of destruction of global functioning by the age of 20 years. TS and other psychiatric disorders such as attention-deficit hyperactivity disorder (ADHD), obsessive–compulsive disorder (OCD), and impulse-control disorders (ICDs) are often found to be accompanied by TS and tic symptoms. Only 10–20% of patients suffering from TS are devoid of a co-occurring disorder. Different environmental factors may influence the waxing and waning course of tics. Specifically, this factor involves psychosocial stressors, anxiety, and emotional tension which may exacerbate the expression of tics (Robertson 2008, Hoekstra et al. 2013).

2 Pathophysiology of Tourette's Syndrome

TS is a genetic neurological disorder of disturbed neurotransmission involving basal ganglia and fronto-cortical circuits characterized by motor tics, vocal tics. Reports from the neuroimaging studies show abnormalities in various brain regions such as prefrontal cortex, paralimbic, and striatum. Dysregulation in the cortico-striato-thalamo-cortical pathways is involved in TS and its other neuropsychiatric comorbidities associated with TS. Although the etiology of tic disorders remains indefinite, several findings over the past two decades have discussed some of the key aspects related to the pathophysiology of TS. In particular, converging lines of evidence have realistically shown that tics reflect functional dysregulation within the cortico-limbic circuitry, evidenced by the imbalance of dopamine, gamma-amino-butyric acid (GABA), and other neurotransmitters (Godar et al. 2014).

2.1 Dopaminergic System Dysregulation

Dopaminergic dysfunction and TS pathophysiology originate from various clinical findings which led to the development of the 'DA hypothesis' of TS, suggesting a pathological relationship between DA abnormality and TS symptoms. According to this hypothesis, the disturbed dopaminergic activity is responsible for TS pathophysiology, either due to the super-sensitivity of dopamine receptors, hyperinnervation of dopamine, or elevated postsynaptic receptor density (Bronfeld et al. 2013). Further, the imaging studies supported this DA hypothesis in TS patients and these patients have shown increased D2 receptor densities and increased levels of the DA transporter.

2.2 Serotonergic System Dysregulation

A possible role for 5-HT in TS was initially based on the high comorbidity rates between TS and other disorders involving serotonergic dysfunction, most remarkably OCD (Zohar et al. 2000). Some findings stated that there is a decrease in levels of 5-HT, its precursors, or metabolites in urine, blood, or cerebrospinal fluid samples of TS patients when compared to healthy controls. Overall, the role of serotonin and the extent to which serotonergic dysfunction is involved in TS are still unknown.

2.3 Noradrenergic System Dysfunction

The involvement of NE in TS mainly originates from the beneficial effects of drugs which prompt the presynaptic α2-adrenergic autoreceptors in the treatment of TS patients. The mechanism of these drugs is the reduction in NE levels, suggesting that TS might be related with a hyperadrenergic dysfunction (Szabo et al. 2001). However, estimations of NE levels in TS patients have so far yielded conflicting results, with different studies reporting increased, decreased, or normal levels. Particularly, the role of NE in TS may be established by the involvement of NE in TS comorbid disorders, such as ADHD, and in the regulation of overall levels of anxiety and arousal, which are known to affect tic expression.

2.4 Environmental Factors

TS may have a multifactorial nature in which various environmental factors may contribute to the onset of the pathology. Prenatal environmental risk factors that include perinatal hypoxic/ischemic events, prenatal maternal smoking, low birthweight, and maternal stress may play a managerial role in the development and functioning of brain pathways thought to be appropriate for the emergence of tics. Beside these prenatal factors, postnatal environmental factors also supposed to affect the activity of neurons in crucial brain regions, possibly relating to the variation in tic severity. More recently, streptococcal infections have been proposed as an additional environmental factor potentially favoring the production of tics through immune-mediated mechanisms. In several cases, the beginning of tics was seen suddenly after a streptococcal infection, thus explaining pediatric autoimmune neuropsychiatric disorders associated with streptococcal infection (PANDAS). GAS-induced antibodies directed against different antigenic determinants (epitopes) in the basal ganglia have been detected in some cases of TS, thereby revealing an immune or autoantibody-mediated mechanism (Macri et al. 2013). The science of Tourette's syndrome (TS) is growing at multiple levels of analysis and will be enhanced through the use of animal models. As animal models are the best tools to evaluate any therapeutic agent. Each model has its own advantages and limitations. Particular challenges in the development of TS animal models reveal complex features related to this disorder, including its waxing and waning course of tics and sensory and psychic symptoms which are found later at stages and are invisible. Animal models can achieve face, predictive, or construct validity based on their particular features. Although the certainty of the most of these models on motor suppression may ultimately limit their utility, other models accomplish validity with organized pathogenetic mechanisms related to the immune and neural circuit-related etiologies of TS. In addition to models which will move ahead of the pharmacotherapy of TS, other animal models may enhance the utility of non-pharmacologic-related TS treatments, ranging from behavior therapy to deep brain stimulation (DBS).

3 Classification of Animal Models of Tourette's Syndrome

3.1 Models Based on Environmental Etiology

Several epidemiological studies suggest that early exposure to a number of critical environmental factors may play a role in the pathogenesis of TS and other tic disorders (Figs. 1 and 2).

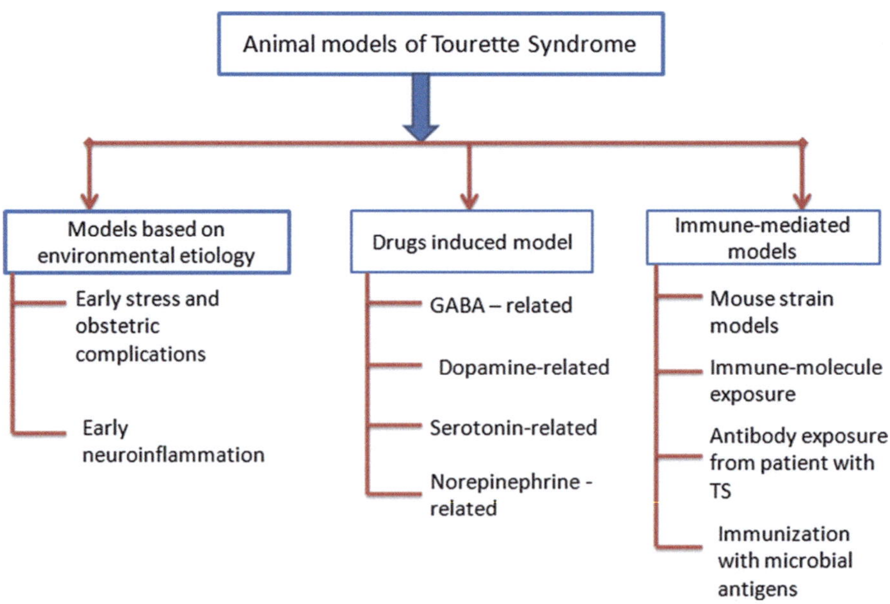

Fig. 1 Classification of animal models of Tourette's syndrome

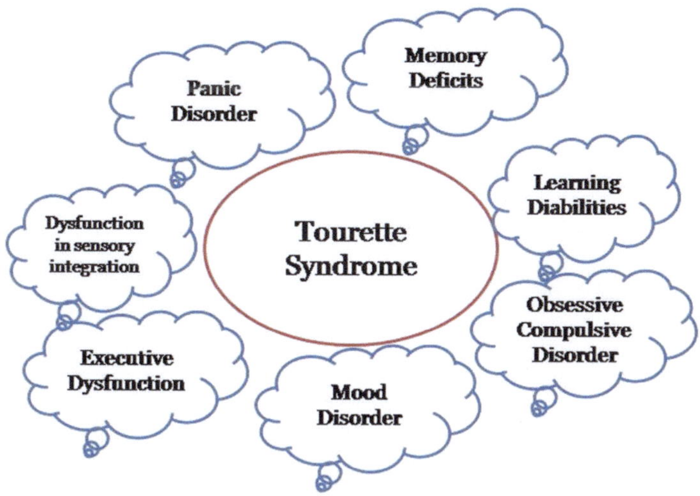

Fig. 2 Factor involved in the pathophysiology of Tourette's syndrome

3.1.1 Early Neuroinflammation Model

The best-documented environmental factor related to TS etiology that has been investigated in animal models is the implication of neuroinflammatory factors and, in particular, infections during prenatal or early postnatal life. The link between inflammatory agents and TS pathogenesis has been investigated in rodent models.

Procedure:

Immunize the animal (mice) with group A streptococci.

Changes observed:

✓ Increased grooming and rearing.

✓ Anti-brain antibodies in serum of animal.

✓ Increased IgG concentrations in numerous brain regions, including striatum, cerebellum, and hippocampus (Yaddanapudi et al. 2010).

Clinical relevance:

Several lines of evidence establish a relationship between the exposures to Group A—β hemolytic streptococci (GAHBS) and tic disorders. Using population-based data, findings suggest that streptococcal infection was associated with increased chance for obsessive–compulsive disorder (OCD), Tourette's syndrome, or tic disorder.

3.1.2 Early Stress and Obstetric Complication-Related Model

Various animal models have been developed in the attempt of reproducing the pathophysiological changes which have been related to TS. The risk of TS has been shown to be significantly enhanced in patients with a history of adverse events and stress in the prenatal and perinatal period as well as maternal smoking

(Bos-Veneman et al. 2011). Nonetheless, the majority of these models may reproduce prominent phenomenological aspects of tic disorders; however, alterations in dopaminergic signaling, as well as higher vulnerability to stereotyped behaviors and PPI deficits, were reported in these models.

Clinical evidence:

Prepulse inhibition (PPI), defined as deficits in sensorimotor gating, has been associated with subcortical dopaminergic overactivity in animal and clinical studies. Clinical data from patients' comorbidity for attention-deficit hyperactivity disorder (ADHD) and a tic disorder had significantly reduced PPI.

Advantages:

- These animal models are highly valuable to relate the immune system with TS pathophysiology.
- These studies strongly supported the contribution of environmental insults and neuroinflammatory events in the pathogenesis of TS.

Disadvantages:

- Further studies are warranted to optimize environmental manipulation procedures and elucidate the neurobiological mechanisms that underpin their role in TS.
- It has poor reproducibility.
- Predictive validity of these models is currently unknown.

3.2 Drug-Induced Models

3.2.1 $GABA_A$ Antagonism-Induced Model

GABA is an inhibitory neurotransmitter which regulates neuronal activity throughout the CNS. The GABAergic animal model of TS focuses on the disturbance of local GABAergic transmission within the BG, which is part of the neural circuit implicated in human TS, by specifically targeting the striatum. GABAa antagonists may influence multiple GABAergic transmission pathways within the striatum, including projections from MSN collaterals and local GABAergic interneurons.

Method:

Unilateral local application of $GABA_A$ antagonists such as picrotoxin or bicuculline into the rat striatum.

Changes observed:

Abnormal movements in the contralateral limbs manifested as repetitive myoclonic jerks consist of sudden, rapid, and brief flexion of the contralateral limb followed by a slower relaxation.

Abnormal movements of face include contralateral facial grimacing, teeth chattering, and tongue protrusions.

Tics usually appeared within minutes following administration of the GABAa antagonist (mostly 2–10 min, average ~5 min), and lasted for up to 2 h, in both rats and monkeys (Bronfeld et al. 2013).

3.2.2 Dopamine-Agonist-Induced Model

Dopamine receptor stimulation through non-selective indirect (i.e., d-amphetamine) or direct (apomorphine) agonists is known to produce stereotypies, as well as PPI deficits (Godar et al. 2014). Recent neuroimaging and anatomical studies have provided evidence for abnormal basal ganglia and dopaminergic function in TS.

Method:
Dopaminergic agonist (amphetamine) administered systemically to rodents.

Doses of amphetamine

Low dose: 0.2–5.0 mg/kg, i.v.,
Moderate dose: 0.2 and 0.5 mg/kg,
High dose: 5.0 mg/kg (Porrino et al. 1984).

Changes observed:
Stereotypic behaviors are sniffing, licking, or biting.

3.2.3 Serotonin-Related Models

The animal model of 5-HT-induced behaviors has also highlighted the complex relationship between different neuromodulator systems in relation to TS psychopharmacology.

Method:
5-HT2A receptor agonists [2,5-dimethoxy-4-iodophenyl-2-aminopropane (DOI)] administered systemically in rats.

Changes observed:
- Head shakes in mice and rats as well as PPI deficits in rats.
- Wet dog shakes in rats observed which are single- or multiple-rotation movements involving parts such as head, neck, shoulders, and upper trunk, reminiscent of the shaking movements seen in dogs emerging from water (Bronfeld et al. 2013).

3.2.4 Norepinephrine-Related Models

Pharmacological manipulations of the NE system do not directly induce tic-like behaviors in animals. However, this system is involved in modulating the PPI response in animals, which is hypothesized to model SMG deficits in TS patients.

Method:
Adrenergic agonist, Cirazoline, an α-1 adrenergic agonist administered systemically in rats.

Changes observed:
PPI deficits in rats and mice—this effect is shown to be mediated by the activation of central rather than peripheral adrenergic receptors (Sallinen et al. 1998b).

Advantages:

- The systemic or intracerebral administration of drugs can reproduce some of the neurochemical alterations observed in TS and provide a rapid, cost-effective method to understand the pathophysiology behind tic disorders and screen for novel therapeutic targets.

Disadvantages:

- Dopaminergic animal model clearly demonstrates the involvement of a central DA dysfunction in motor and sensorimotor abnormalities, but its relation to TS is not defined yet.
- Limited construct validity of these models for TS raises doubts about their translational relevance.
- Further studies are needed to test the predictive validity of these behaviors.

3.3 Immune-Mediated Models

3.3.1 Mouse Strain Models of Immune-Mediated

Some of mouse strains and transgenic models with altered immune function or sensitivity to the development of autoimmune disease have been shown to exhibit repetitive behavior patterns. The BTBR T + tf/J (BTBR) mouse strain was developed primarily as a model of ASD, but this strain shares some common characteristics with the immune and behavioral disturbances as present in TS. It has been reported that BTBR mice had increased serum IgG and IgE, as similar to reports related to TS, even in the absence of a specific exposure (Amodeo et al. 2012). Along with this, BTBR mice found to increase anti-brain antibodies of the IgG isotype in serum, reliable with their high levels of IgG-secreting B cells in peripheral blood and of IgG and IgE deposits in brain.

3.3.2 Immune Molecule Exposure

Cytokines and chemokines
Acute disruption of cytokines can modify CNS development and lead to constant behavioral disturbances similar to TS, tic disorder, and OCD. Different proinflammatory cytokines have been found to affect behavior in a different manner

in a pattern equivalent to the regionally specific changes these cytokines induce in brain monoamines and the distribution of their cognate cytokine receptors in brain.

Methods used:

1. Introduction of IL-2 mid-gestationally to pregnant autoimmune disease-sensitive SJL/J mice.

   Changes observed:
   Immune and behavioral disturbances in the offspring of mice
2. Pregnant female mice are treated with IL-2 during mid-gestation period.
   Changes observed:
   ✓ Behavioral alterations, including self-grooming and rearing
   ✓ Stereotyped movements and multiple alterations of locomotor activity (Ponzio et al. 2007).
3. Intraperitoneal (i.p.) administration of IL-2 or IL-6 to male BALB/c mice.
   Changes observed:
   ✓ Increase in dopamine (DA) turnover in prefrontal cortex
   ✓ Increased norepinephrine (NE) utilization in hippocampus and hypothalamus
   ✓ More digging and rearing behavior.
   ✓ IL-6-treated mice showed increased locomotor and grooming behaviors.
4. Peripheral administration of TGF-β1 in mice.
   Changes observed:
   ✓ Increase self-grooming and other stereotypic behaviors.
   ✓ Overexpression of TGF-β1 in the dentate gyrus of hippocampus during early postnatal development can lead to persistently increase repetitive (self-grooming) behavior and social interaction deficits.

- **Soluble cytokine receptors**
  Peripheral administration of immune mediators such as cytokines and soluble cytokine receptors can affect the development of the CNS leading to persistent behavioral changes, such as motor stereotypies and abnormal repetitive behaviors (Macri et al. 2013).

Method:
Single subcutaneous (sc) injections of soluble IL-2 receptors (sIL-2R) α or β and sIL-6R will be given in Balb/c mice.

Changes observed:
✓ Increased repetitive head bobbing.
✓ Increased grooming and rearing/sniffing behaviors as well as other stereotypic behaviors (Zalcman et al. 2012).

3.3.3 Antibodies Exposure from Patients with TS

TS pathogenesis may be linked to the presence of autoantibodies. Although these models do not clearly explain the important factors required for the production of pathogenic antibodies or scrutinize how circulating antibodies enter brain and bind to their specific antigens, they do have the potential to describe the patient phenotypes and antibody distinctiveness that are most tightly associated with behavioral disturbances similar to TS and/or OCD.

Method:

1. **Direct CNS administration of antibodies**
   Inject serum from TS patients into the striatum of rats.

   Changes observed:
   ✓ Motor and oral stereotypies, as well as episodic vocalizations and also increased genital grooming in association with IgG deposits.
   ✓ Increased levels of dopamine and reduction in DAT expression (Singer et al. 2005).
2. **Peripheral administration of antibodies**
   Two studies employing peripheral injection strategies for the introduction of antibodies show symptoms related to TS.
   ✓ In rhesus macaque monkeys, IgG antibodies obtained from mothers of children with ASD induced stereotypic and hyperactive symptoms in association with stressful contexts (Martin et al. 2008).
   ✓ Reports from another study showed repetitive movements in mice injected peripherally with commercial anti-GAS antibodies, but in association with IgM only.

3.3.4 Immunization with Microbial Antigens

The peripheral exposure of rodents to microbial antigens provides basis for the study of the mechanisms involved in the induction of autoantibodies supposed to be associated with the etiology of TS.

Method:

1. Active immunization of female SJL/J mice (a strain having high tendency to show autoimmune responses) with GAS homogenate.

   Changes observed:
   ✓ Increased rearing behavior, compulsive grooming, and reduced motor coordination.
2. Passive immunization of naive mice with GAS sera derived from actively immunized mice.
   Changes observed:
   ✓ Behavioral dysfunctions similar to those exhibited by donor mice (increased rearing and passive social behaviors.

✓ IgG deposits in the brains of donor mice are present within the cerebellum, globus pallidus, and thalamus; on the other hand, IgG deposits in recipient mice are confined to neurons in the hippocampus and periventricular area.

3. Active immunization of male Lewis rats with GAS.

   Changes observed:

   ✓ Motor disturbances (impaired manipulation of food and inability to traverse a narrow beam).

   ✓ Increase in induced-grooming behavior.

   ✓ Compulsive and obsessive symptoms found in the clinic for PANDAS cases (Hoffman et al. 2004).

Clinical evidence

Various research findings support the involvement of autoimmunity in childhood-onset tic disorders. Studies of immune-mediated characteristics in experimental animal models reveal a wide range of behavioral changes and abnormalities in brain similar to findings reported in human studies of TS.

Other Infections

TS-like behavioral disruptions consistent with effects on cortico-striato-thalamo-cortical circuits may occur in the context of other infections, possibly due to generic mechanisms of innate immune activation.

Methods used:

1. Prenatal exposure of pregnant mice to the viral mimic, poly I:C, or injection of IL-6

   Changes observed:

   Decrease in Treg along with elevated IL-6 and IL-17 production.

2. A single injection of the bacterial mimic, lipopolysaccharide (LPS), to pregnant Wistar rats in mid-gestation will be given.

   Changes observed:

   ✓ Repetitive behaviors in male offspring.

   ✓ Reduced dopamine synthesis in striatum.

   ✓ Underactivation of dopaminergic pathways (Kirsten et al. 2012).

   Advantages of Immune-mediated models:

   ✓ This animal model establishes disease pattern in almost all mammals tested so far.

   ✓ Several species and strains have been utilized including mice, rats, and rhesus macaque monkeys.

   ✓ Research has been extended further to non-human primates using the advantage of genetic variability and extended preclinical studies.

   Disadvantages:

   ✓ Mechanistic aspect of passage of maternally introduced IL-2 and IL-6 across the placenta into the fetal circulation remains unrevealed.

These models cannot elucidate the factors important to the generation of pathogenic antibodies or reveal the mechanism of entry of circulating antibodies into brain and their binding to the specific antigens. These studies found no effects of anti-neuronal antibody infusion into striatum, although the reasons for failure to detect stereotypes are unclear.

4 Conclusion

This review summarizes the need of animal models in elucidating the pathogenesis and treatment of Tourette's syndrome. However, no model is fully satisfying the following: No single model exhibits construct, face, and predictive validity, and no model has scientifically proven itself by revealing new perspectives regarding the disorder or contributing to the development of new potential treatments. Therefore, careful consideration of animal models will be essential for further understanding of TS pathogenesis and identify new drugs for the improvement of future therapies.

Ethical Statement
All institutional guidelines, national guidelines, state and local laws and regulations with professional standards for the care and use of laboratory animals should be followed. Studies involving animals must state that the Institutional Animal Ethical Committee has approved the protocol. For authors using experimental animals, a statement should be made that the animals' care is in accordance with institutional guidelines and animals used have been treated humanely and with regard to the alleviation of suffering. Researchers should treat animals as sentient and must consider their proper care and use and the avoidance or minimization of discomfort, distress, or pain as imperatives. Animal experiments should be designed only after due consideration of animal health. It should be ensured that all researchers who are using animals have received instruction in research methods and in the care, maintenance, and handling of the species being used. All the surgical procedures should be performed under appropriate anesthesia and follow only those procedures which avoid infection and minimize pain during and after surgery.

References

Amodeo DA, Jones JH, Sweeney JA, Ragozzino ME (2012) Differences in BTBR T + tf/J and C57BL/6 J mice on probabilistic reversal learning and stereotyped behaviors. Behav Brain Res 227(1):64–72

Bos-Veneman NG, Olieman R, Tobiasova Z, Hoekstra PJ, Katsovich L, Bothwell AL, Leckman JF, Kawikova I (2011) Altered immunoglobulin profiles in children with Tourette syndrome. Brain Behav Immun 25(3):532–538

Bronfeld M, Israelashvili M, Bar-Gad I (2013) Pharmacological animal models of Tourette syndrome. Neurosci Biobehav Rev 37(6):1101–1119

Godar SC, Mosher LJ, Di Giovanni G, Bortolato M (2014) Animal models of tic disorders: a translational perspective. J Neurosci Methods 238:54–69

Hoekstra PJ, Dietrich A, Edwards MJ, Elamin I, Martino D (2013) Environmental factors in Tourette syndrome. Neurosci Biobehav Rev 37(6):1040–1049

Hoffman KL, Hornig M, Yaddanapudi K, Jabado O, Lipkin WI (2004) A murine model for neuropsychiatric disorders associated with group A β-hemolytic streptococcal infection. J Neurosci 24(7):1780–1791

Kirsten TB, Chaves-Kirsten GP, Chaible LM, Silva AC, Martins DO, Britto LR, Dagli ML, Torrao AS, Palermo-Neto J, Bernardi MM (2012) Hypoactivity of the central dopaminergic system and autistic-like behavior induced by a single early prenatal exposure to lipopolysaccharide. J Neurosci Res 90(10):1903–1912

Macri S, Onori MP, Roessner V, Laviola G (2013) Animal models recapitulating the multifactorial origin of Tourette syndrome. Int Rev Neurobiol 112:211–237

Martin LA, Ashwood P, Braunschweig D, Cabanlit M, Van de Water J, Amaral DG (2008) Stereotypies and hyperactivity in rhesus monkeys exposed to IgG from mothers of children with autism. Brain Behav Immun 22(6):806–816

Ponzio NM, Servatius R, Beck K, Marzouk A, Kreider TIM (2007) Cytokine levels during pregnancy influence immunological profiles and neurobehavioral patterns of the offspring. Ann N Y Acad Sci 1107(1):118–128

Porrino LJ, Lucignani G, Dow-Edwards D, Sokoloff L (1984) Correlation of dose-dependent effects of acute amphetamine administration on behavior and local cerebral metabolism in rats. Brain Res 307(1–2):311–320

Robertson MM (2008) The prevalence and epidemiology of Gilles de la Tourette syndrome: part 1: the epidemiological and prevalence studies. J Psychosom Res 65(5):461–472

Sallinen J, Haapalinna A, Viitamaa T, Kobilka BK, Scheinin M (1998) Adrenergic α2C-receptors modulate the acoustic startle reflex, prepulse inhibition, and aggression in mice. J Neurosci 18 (8):3035–3042

Singer HS, Mink JW, Loiselle CR, Burke KA, Ruchkina I, Morshed S, Parveen S, Leckman JF, Hallett JJ, Lombroso PJ (2005) Microinfusion of antineuronal antibodies into rodent striatum: failure to differentiate between elevated and low titers. J Neuroimmunol 163(1):8–14

Szabo B, Fritz T, Wedzony K (2001) Effects of imidazoline antihypertensive drugs on sympathetic tone and noradrenaline release in the prefrontal cortex. Br J Pharmacol 134(2):295–304

Yaddanapudi K, Hornig M, Serge R, De Miranda J, Baghban A, Villar G, Lipkin WI (2010) Passive transfer of streptococcus-induced antibodies reproduces behavioral disturbances in a mouse model of pediatric autoimmune neuropsychiatric disorders associated with streptococcal infection. Mol Psychiatry 15(7):712–726

Zohar J, Chopra M, Sasson Y, Amiaz R, Amital D (2000) Obsessive compulsive disorder: serotonin and beyond. World J Biol Psychiatry 1(2):92–100

Zalcman SS, Patel A, Mohla R, Zhu Y, Siegel A (2012) Soluble cytokine receptors (sIL-2Rα, sIL-2Rβ) induce subunit-specific behavioral responses and accumulate in the cerebral cortex and basal forebrain. PLoS ONE 7(4):e36316

Animal Models of Multiple Sclerosis (MS)

Shubhangi Gupta, Sanpreet Kour and Rahul Deshmukh

1 Introduction

Multiple sclerosis is the chronic autoimmune disorder of the sensory system coordinated against its own particular myelin antigens. It is the basic incapacitating neurological illness influencing generally youth and grown-ups between the age of 20 and 40 years. Myelin is a protein that makes up the myelin sheath around the nerve fibers (axons). The myelin cells are called as oligodendrocytes which are the main ground for disease initiation. The term multiple sclerosis is related to formation of scar tissues or plaques or lesions in particular areas of the brain, especially the white matter. Seven decades of research, propose cause behind the disease is exposure to environmental pathogenic organism leads to activation of autoreactive T-cells that recognize CNS autoantigens which led to development of inflammatory reactions and demyelination. It is additionally known to be an inflammatory disease of the white matter portrayed by dynamic and broad deterioration of the myelin sheath and axonal points in neuron, leading to progressive paralysis of hind-limb. The myelin in the oligodendrocytes–myelin–axon unit of the CNS protects and nourishes the axon and increases the cross-sectional diameter of the nerve axon which regulates conduction. This integrally coordinated unit is disturbed due to multiple sclerosis. In MS, decreased axonal density and volume in plaque affected areas as well as in normal appearing CNS tissue contribute to atrophy of brain and spinal cord which lead to permanent disability.

S. Gupta · S. Kour · R. Deshmukh (✉)
Department of Pharmacology, ISF College of Pharmacy, Moga 142001, Punjab, India
e-mail: login2rd@gmail.com

S. Gupta
e-mail: shubhangi.gupta5@gmail.com

P.K. Bansal and R. Deshmukh (eds.), *Animal Models of Neurological Disorders*, https://doi.org/10.1007/978-981-10-5981-0_17

Decades of extensive research showed that complex interplay between environmental factors, genetic susceptibility, and ultimately immune system is found to be a key cause behind the progression of multiple sclerosis. The proposed theory with respect to the pathophysiology of MS was separated on the premise of animal model of experimental autoimmune encephalomyelitis (EAE) that was observed to be helpful in mimicking the disease procedure up to a specific level. Demyelinating disease is related with starting scenes of reversible neurological deficits, trailed by incessant neurological impairment after some time. The disease progression was analyzed by procedures, for example, magnetic resonance imaging (MRI) and cerebrospinal fluid (CSF) examinations. According to the research studies conducted by National Multiple Sclerosis Society (NMSS), approximately 2.3 million people are affected with MS worldwide and 2.5 million people in the USA. The high-risk regions for multiple sclerosis are northern European, northern USA, southern Canada, and Scandinavian countries with a ratio of 50–100 per 1,00,000 population, whereas the number of affected people ratio decreased to 5 per 1,00,000 in Japan proving it to be a low-risk region.

1.1 Side Effects and Symptoms

Symptoms in MS are defined by the term exacerbations or flare-ups. In layman language, recovery from symptoms is called as remission phase, whereas return of symptoms is termed as relapsing phase. Symptoms appearance in MS varies from patients, and it may begin from one to several days lasting up to few months, or slow progression in some cases. Clinical studies in MS patients have showed the presence of various symptoms such as weak stiff muscles with painful muscle spasms, burning sensations, sensory disturbances, numbness or tingling in the arms, legs, or face, ataxia, vertigo, bladder control problems, difficulties in coordinated and balanced movements, and fatigueness. Optic neuritis is the most commonly associated with MS characterized by rapid, blurred, partial, or complete loss of vision. Psychiatric disorders such as depression, euphoria, and mood swinging are often associated with MS patients.

2 Pathophysiology

Multiple sclerosis is unconstrained, procured, sporadic, multifaceted inflammatory demyelinating disease of the CNS. Inquire about reviews exhibited that environmental components that are in charge of bringing on MS incorporate vitamin D lack and Epstein–Barr virus. Voltage-dependent sodium channels in the myelinating axonal membrane are responsible for causing depolarization and triggering saltatory conduction. Demyelinating lesions in the axon leads to thinning of myelin sheath causing redistribution of Na^{+} channels in the conduction block in which there is failure of action potential propagation, insufficient electrical signal transmission at a particular point in axon, therefore impeding salutatory conduction. Myelin antigens present within may be recognized by these T-cell lymphocytes, which may trigger inflammatory reaction cascades, resulting in the formation of sclerotic plaques or demyelinating lesions in the white matter, loss of myelin cells and oligodendrocytes. With the help of this proposed hypothesis, further advances were made with the help of suitable animal models in the existing pathogenesis of MS. The credit for establishing classical pathological characteristics of MS, i.e., demyelinating lesions and axonal pathology goes to Charcot over a century ago. Investigations of traditional animal models express that there is incitement of autoresponsive T-cells (CD4+ and CD8+) and B-cells which penetrate the white matter with myelin parts in healthy people. Mechanisms that cause enactment of immune system cells against myelin sheaths in the CNS are bystander activation, epitope spreading, and molecular mimicry. Collecting confirmations has shown that on initiating myelin particular T-cells in the periphery, they cross the blood–brain barrier (BBB) because of interruption in basement membrane by matrix metalloproteinases (MMPs). T-cell there multiply and discharge proinflammatory cytokines and chemokines like IL-β, IL-6, interferons (INF-ϒ) which additionally cause activation of parenchymal components like microglia, macrophages and astrocytes. During the process of restoration of conduction in the axon, there is reversal of process of Na^{+}–Ca^{+2} exchanger pump creating increased gradient of Ca^{+2} inside the cell, leading to excitotoxicity and neuronal death. However, research emphasis is mainly focused on myelin and to a lesser extent to axonal loss, but still latter is widely acknowledged as the key determinant of permanent clinical disability. A set of recent studies by microarray-based gene expression analysis showed that mitochondrial dysfunctioning due to energy failure, increased oligodendrocytes apoptosis, increased concentration of reactive oxygen species (ROS), is also contributing to MS disease etiology (Fig. 1).

2.1 Need for Animal Models

The need for animal model of MS cannot be further exaggerated as statistical studies show that around 2.5 million people worldwide are suffering from this disease. It has been reported to occur at a young age and effects persist throughout life of the patient. MS is a multifaceted complex disease with variable patterns of sensory, motor, and clinical disturbances due to inter-individual variation. The course of the disease is unpredictable, characterized by alternative episodes of remission and relapses with benignant in nature, whereas in some patients, disease progression is severe from the beginning to permanent disability. Due to complexity in MS disease, research studies are restricted to a certain extent. Secondly, similar to all human diseases, moral and pragmatic issues avoid experimental strategies. Thirdly, the pathological event is implanted in the tissues of the sensory system and difficult to reach to experimental studies. MS research relies on upon the accessibility of reasonable experimental animal model to elucidate the pathogenic components and to create viable and targeted therapeutic strategies. Remedial strategies in multiple sclerosis have focused on immunomodulation. There is an urgent requirement for agents that can restrain progressive multiple sclerosis.

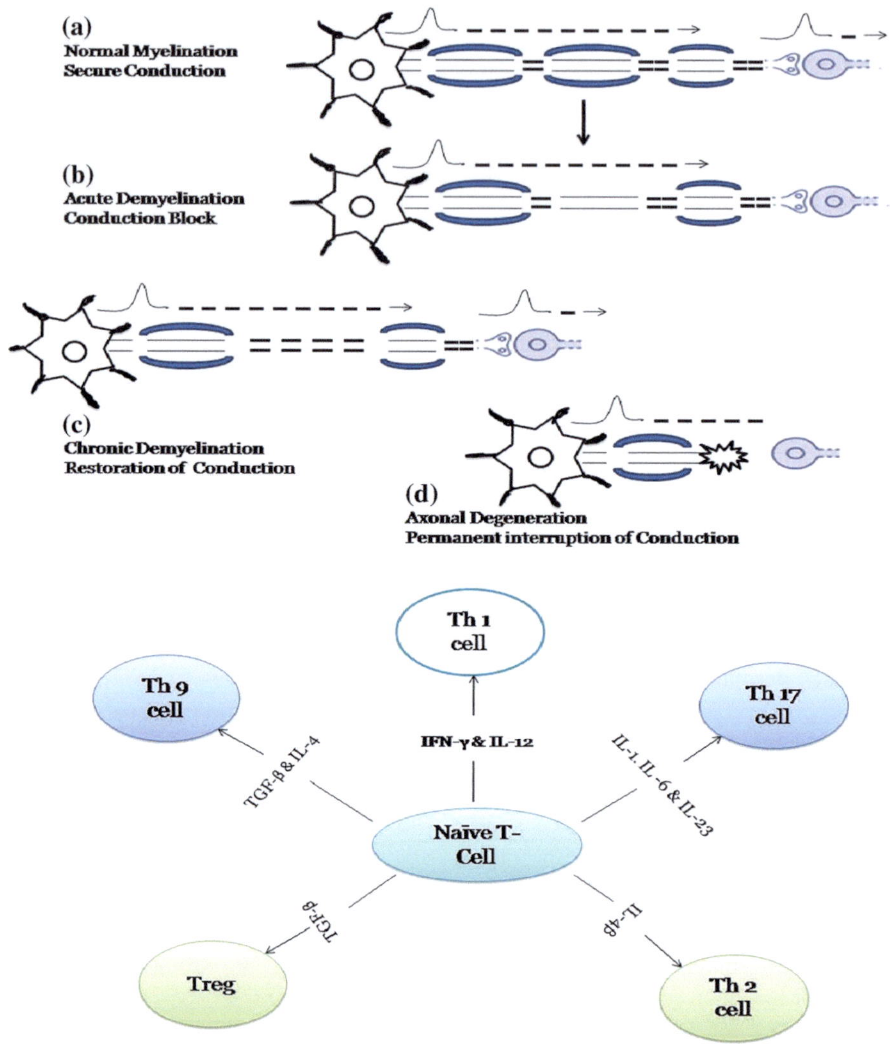

Fig. 1 Pathophysiology of multiple sclerosis

Various animal models have been produced in various species: models induced by inoculation against CNS proteins (EAE) or by infection with neurotropic virus (EBV) or by poison interceded MS by cuprizone and lysophosphatidylcholine (Ethidium Bromide). Although none of the above-mentioned model satisfactorily demonstrates to all phases of human MS disease, however, reflects particular parts of human MS and can be utilized to assess new medicines (Fig. 2).

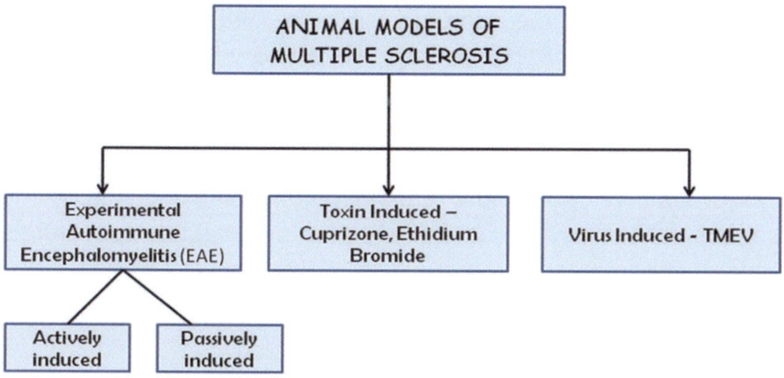

Fig. 2 Classification of animal models of multiple sclerosis

3 Classification of Animal Model of Multiple Sclerosis

3.1 Toxin-Induced Animal Model of Multiple Sclerosis

3.1.1 Cuprizone-Induced Neuro-Intoxication

The neurotoxin effect of copper-chelator cuprizone in mice was first identified in 1966 by Carlton and in the 1970s by Hemm and its co-workers. Cuprizone feeding (bis-cyclohexanone oxaldihydrazone) is a normally utilized model to explore experimental de- and remyelination, with the corpus callosum and the superior cerebellar peduncles being the most much of the time researched white matter tracts. Moreover, examinations have demonstrated that cuprizone feeding likewise prompts demyelination of gray matter structures in the cerebrum and cerebellum. The time course of demyelination is diverse in the white matter and gray matter, demonstrating differentiating tissue reactivity. In the toxin model, grown-up mice fed with the cuprizone prompt reproducible demyelination within weeks. After expulsion of the poison from the food, unconstrained remyelination happens.

3.2 History

Research studies during 1960s proved cuprizone as a neurotoxin for mice, guinea pigs, and rats. Feed containing 0.5% cuprizone resulted in growth retardation, posterior paresis (weakness of the posterior limbs) and severe toxicity, convulsions, and a high mortality rate. Brain edema, demyelination, astrogliosis present in harsh amounts in the white matter of the cerebellum and midbrain are main pathologic characteristics of cuprizone-fed mice. Research work done by Blakemore demonstrated the involvement of toxic cuprizone feeding in inducing oligodendrocyte death with secondary demyelination of the superior cerebellar peduncles. By 5 weeks, many axons were demyelinated, but cells like remyelinating oligodendrocytes were also identified. Widespread remyelination was reported only when the animals were taken off the cuprizone diet. In 1998, Matsushima and his colleagues first portrayed the cuprizone impacts on the white matter of the cerebrum and focused on demyelination in the corpus callosum, where the degree of demyelination could be scored all the more effortlessly and reliably. From that point forward, cuprizone-initiated de- and remyelination forms have been reliably contemplated as productive model in the investigation of MS. In recent years, gray matter structures of the cerebrum and cerebellum have additionally indicated broad de- and remyelination areas.

3.3 Mechanism of Cuprizone-Induced Intoxication

The mechanism capable for cuprizone prompted oligodendrocyte death with consequent demyelination is yet unidentified puzzle. Cuprizone is a copper-chelating compound, which prompts copper lack in serum and brain after nourishing. In any case, copper supplementation at levels up to 100 ppn did not decrease the poisonous impacts of cuprizone, demonstrating that copper lack cannot be the major mechanism of cuprizone activity. Cuprizone feeding prompts the death of well-developed or mature oligodendrocytes, while other cell sorts in the CNS are not influenced. Feeding higher amount of cuprizone prompts changed mitochondrial structure, prompting the formation of giant mitochondria in the liver. Because of cuprizone rich eating routine, movement of mitochondrial-dependent enzyme, for example, monoamine oxidase, cytochrome oxidase, and superoxide dismutase is decreased, while the action of succinate dehydrogenase is expanded in the liver and in cerebrum. One theory expresses that as mitochondrial adenosine-5-triphosphate (ATP) is essential for calcium homeostasis and apoptosis, its dysfunctioning prompts to altered energy metabolism, prompting oligodendrocytes cell death. Studies exhibited that corpus callosum is the most influenced area of the white matter. However, research still failed to demonstrate why just oligodendrocytes are disposed to cuprizone toxication.

3.4 Factors Involved in Cuprizone-Induced Demyelination

Cuprizone is found to induce demyelination in almost all species and strains such as rodents, guinea pigs, hamsters. Dose of cuprizone and the age of the animal sextant decide the intensity of toxication, and the amount of myelin loss varies from various animal experimental models. Mice seem to be especially liable to develop demyelination as a consequence of cuprizone feeding. Cuprizone toxicity was first described for Swiss albino mice and was also tested in BSVS mice, albino mice of the ICI strain, CD1 mice, ddY mice, BALB/cJ mice. Among different mice strains mentioned above, the best experimental model induced demyelination was observed in C57BL/6 mice. Cuprizone toxicity was also seen in male albino rats (0.1–1.5%) and male albino guinea pigs (0.5–1%). Recent studies showed that feeding 0.6% cuprizone for 2- to 3-week-old Wistar rats caused demyelination in the corpus callosum. Further 3% cuprizone feeding to Syrian and Chinese hamsters showed similar pathological symptoms to that of mice like astrocytes hypertrophy, but there was no marked demyelination. To determine demyelination, the histochemical Luxol fast blue (LFB) staining and immune histochemical stainings for different myelin proteins are helpful tools. The cuprizone model offers consistent, anatomically reproducible, and well-detectable processes of degradation of myelin proteins.

Procedure

- In the toxic-induced model of 8- to 12-week-old male C57BL/6 mice, cuprizone (0.2%) feed prompts about entire demyelination of the CNS white matter and gray matter in weeks.
- Oligodendrocyte death and downregulation of myelin genes are available 3–7 days after cuprizone treatment, and weeks before demyelination is visible in the corpus callosum (Carlton et al. 1966). These newly developed mature oligodendrocytes re-show up in the corpus callosum following 5.5 and 6 weeks proceeded with continuous cuprizone feeding.
- When cuprizone feeding proceeds for a more extended period, these oligodendrocytes vanish again after week 6 (Lindner et al. 2009). Microgliosis is a prominent element in cuprizone (0.2%) instigated de- and remyelination in mice (Remington et al. 2007).
- Cuprizone-fed mice do not present obvious neurological deficits, for example, paresis, which is found in other animal models, e.g., in EAE (Miller et al. 2010). An increase of motor coordination deficits upon cuprizone feeding appeared in a various behavioral tests.

Advantages

- The cuprizone model offers the chance to examine the CNS procedures of de- and remyelination without the impact of the peripheral immune system since there is no breakdown of the blood–brain barrier (BBB).
- The cuprizone model of toxic demyelination offers the benefit of reliable, reproducible, and well-detectable de- and remyelination forms.
- Another advantage of the cuprizone model is its lack of BBB damage, thus allowing investigation of the processes of de- and remyelination in the CNS without major influence of the complex peripheral immune system.
- This model provides information more about oligodendrocytes cell death.

Disadvantage

- Insufficient explanations regarding mechanisms of action responsible for de- and remyelination are still not known. Certainly lesion induction is quite different to lesion development in multiple sclerosis (MS), and this aspect cannot be mimicked in this model.

- Since demyelination is induced in an artificial way, the cuprizone model only moderately mimics the complex processes of de- and remyelination in MS.

Clinical Relevance

From four fundamentally diverse examples of demyelination as characterized by two patterns (I and II) demonstrated close resemblance to T-cell-interceded antibody-mediated autoimmune system encephalomyelitis, respectively. Other patterns (III and IV) were focused on oligodendrocyte dystrophy, looked like to the toxin instigated demyelination instead of autoimmunity. Pathological similitudes of pattern 3 lesions of MS with CPZ-actuated lesion are effectively demyelinating lesions with the little association of T-cells, however with the presence of numerous microglia/macrophages, hypoxia-like tissue injury with indications of metabolic anxiety and mitochondrial deficits, which prompts oligodendrocyte apoptosis because of upregulation of p53, BCL-2 components. Also there is axonal swelling with hyperphosphorylated neurofilaments. Motor-related symptoms (exhaustion, weakness, cognitive impairment, loss of sight) and even seizures have been accounted for in MS patients. In the CPZ model demonstrate, demyelination prompts disabled motor coordination among the left and right paws. The level of hippocampal injuries in MS associated with cognitive dysfunction and serious demyelination of the hippocampal structure was seen after CPZ intoxication. Demyelination of the basal ganglia has been connected to MS manifestations, for example, weakness, memory, and motor skill impairment, and extensive demyelination of the basal ganglia was additionally found in the CPZ model. In this manner, CPZ intoxication shapes an amazing clinical model to study pathology and also treatment for various sclerosis diseases with less accentuation to immune system-mediated disease pattern.

3.4.1 Ethidium Bromide Induced Animal Model of MS

Microinjection of lysophosphatidylcholine (or ethidium bromide, which is currently utilized in less amount) into white matter tracts causes immediate demyelination, trailed by remyelination. Also called lysolecithin, it is an activator of phospholipase A.

Procedure

Infusion of 2 microliters of 1% lysolecithin into the spinal cord is an entrenched technique for fast induction of focal areas of demyelination.

Advantages

- To look at cellular and molecular determinants of remyelination, this model has been utilized effectively. The notable components of these straightforwardly applied toxin lie in isolating demyelination and remyelination as distinct process.
- Furthermore, the cuprizone model can likely be adjusted for contemplating hippocampal demyelination or for deciding mechanism of degeneration or survival by continually demyelinated axons, for example, the ones seen

following a 12-week course of cuprizone. Each of these components is important for expanding the treatment choices for MS patients.

Disadvantages

Their shortcomings are the non-appearance of ongoing immune activity as found in MS. There is an extraordinary guarantee of utilizing the injection-based and cuprizone toxic models to distinguish methodologies to enhance remyelination.

3.5 Virus-Induced Animal Model of MS

Epidemiological reviews have proposed a supposition that a viral infection in the cerebrum may bring about an immune-mediated attack against CNS tissue. However, till date, there is no exact virus that has been perceived as an imminent cause or supporter of MS. A conceivable viral and MS connection is proposed by epidemiological reviews. The participation of viral antigens (Ag) and their particular antibodies (Ab's) in the almost every case of MS patient has additionally focused around the virus-mediated theory. The most usually examined virus prompted demyelination animal models of MS are murine hepatitis virus (MHV) and Theiler's murine encephalomyelitis virus (TMEV) models. TMEV is a mouse enteric pathogen which related with the single-stranded RNA picornaviruses. Viruses having a place with this family are exceptionally small in size; on considering electron microscopy, they are identical to the size of a ribosome. In 1937, Max Theiler first watched and expressed that virus which causes flaccid paralysis in mice, however not in monkeys. The TMEV model recreates an incessant progressive disease as a result of the staying power of the virus for the whole life expectancy in suspectable mice. Viral models of axonal damage and unending demyelination have been utilized to study the capacity of viruses in human MS, and they have prompted significant improvement in our comprehension of MS pathology. Pathological elements of virus incited demyelinating disease are ordinarily instigated by the immune system and not by exact viral cytopathy, and the clinical presentation is fundamentally the same as that seen in chronic progressive MS.

Procedure

- Experimentally, a demyelinating disease in subject mouse strains is triggered by intracerebral infusion with TMEV that brought about two stages of the disease.
- Initially on infection, neurons are mainly affected. Chronic phase that starts after 1 month of infection results in slow progressive disability and then de- and remyelination.
- This chronic stage is sometimes alluded to as TMEV-instigated demyelinating disease (TMEV-IDD) (Lipton et al. 1977).

Advantages

The TMEV model can be viewed as appealing for two primary reasons: (1) its virus-induced pathology looks like human MS; (2) it displays an autoimmune response created by a viral infection in the CNS. Epitope creating from virus to self-epitopes has been recognized in the TMEV model.

Mechanisms

In this way, a different mechanism is proposed with plausibility that diverse combination of these mechanisms could be in charge of myelin demolition all through the course of the disease. Direct viral cytopathic impacts on oligodendrocytes, autoimmune obliteration of infected oligodendrocytes, TMEV-particular, epitope spreading, bystander demyelination because of lethal metabolites from activated macrophages and molecular mimicry are a portion of the proposed mechanism that causes MS. In all mice, intracerebral infusion of TMEV prompts intense non-deadly encephalitis, with the biggest viral antigen stack in the midbrain, hypothalamus, brain stem, thalamus, and spinal cord gray matter. The initial period of the disease is exceptionally asymptomatic in majority of the animals. In suspectable mice, the host immune system generally clears the infection from the cerebrum however not from the spinal cord, which brings about endless demyelination with viral persistence in oligodendrocytes and macrophages of the spinal cord white matter demyelination, axonal damage first, bringing about demyelination as a secondary consequence.

Clinical Relevance

The Theiler's murine encephalomyelitis virus (TMEV) model is the most usually utilized MS models, despite the fact that other viral models are likewise utilized, including neurotropic strains of mouse hepatitis virus (MHV) that referee chronic inflammatory demyelination with comparative pathological attributes clinically seen in patients with MS.

Advantages

In TMEV-initiated demyelinating disease:

- Chronic disease course design that goes on for the entire life expectancy of the examined rodents.
- Pathological deviations are restricted to the CNS.
- In spinal cord of some mouse strains, axonal and demyelination injury are in vast scale.
- Numerous viruses are affirmed to bring about demyelination in human (JC viruses, measles, rubella, and HTLV-1).

Disadvantages

- Practically, virus-induced MS models are complex to operate for experimental animal drug discovery.
- Viral models in MS leads to variable outcomes as the experimental setup is complex that involves different mechanisms including epitope spreading and bystander activation.
- Viral models of MS are under strict genetic control, thus the experiment is restricted to fewer genetic strains.

3.6 Experimental Autoimmune Encephalomyelitis (EAE)

Experimental autoimmune encephalomyelitis (EAE) prior called as experimental allergic encephalomyelitis is a CD4+ T-cell-interceded, demyelinating immune system animal model of multiple sclerosis that is portrayed by mononuclear cell penetration through the blood–brain barrier (BBB). EAE is the most broadly studied, most established, and clinically favored animal model for demyelinating disease multiple sclerosis. This model demonstrates the autoimmune origin of the demyelinating disease MS as it can be produced in an experimental animal by inoculation with proteins gotten from the CNS myelin. Inflammation, neurodegeneration and demyelination, axonal loss are the basic pathological components of MS. Studies have demonstrated the participation of myelin-specific T-cells in the early stage of MS. Occurrence of demyelinating lesions was first observed in humans by immunization with rabbit vaccine derived from rabbit brain and attenuate rabies virus. Demyelinating EAE model was firstly successfully established by monkey that was immunized with normal CNS autoantigens homogenate. Following a time of months, neurological signs and CNS lesion were prompted experimentally and demonstrated that loss of motion was related with perivascular infiltrates and demyelination in cerebrum and spinal cord. At first the disease was named as acute disseminated encephalomyelitis which was later changed to EAE, from these reviews, the diseased features shown in monkey's are like one found in human MS. An immune system response prompting EAE in a susceptible individual can be acquired by means of active induction by immunization with CNS myelin protein and by passive (or adoptive) transfer of myelin epitope particular T-lymphocytes responding against myelin antigens to naive recipients. The part of T-lymphocytes was first demonstrated by Paterson who was prevailing with regard to transferring disease by myelin-specific T-cells from inoculated animal. From that point forward, numerous scientists have endeavored to exemplify the part of T-cells in EAE.

EAE serves a valuable clinical animal model for multiple sclerosis (MS) since a large portion of the pathologies seen in the CNS of mice with EAE bears a solid similitude to those found in the CNS of MS patients. In both EAE and MS, the white matter of the CNS was available more with demyelinating lesions related with infiltrating mononuclear cells when compared with gray matter. In addition, foam cell-like macrophages containing phagocytosed hydrophobic myelin debris have been exhibited in plaque. Ascending hind-limb loss of motion is connected with inflammation and demyelinating plaques. Finally, oligoclonal IgG was found in the CSF of both EAE vaccinated mice and MS patients.

Procedure

- The degree of demyelination sickness relies on the dosage and nature of vaccination and encephalitogen, and the hereditary makeup of every species and strain. Myelin proteins generally utilized as encephalitogen are myelin basic protein (MBP), myelin oligodendrocyte glycoprotein (MOG), and proteolipid protein (PLP), myelin-associated oligodendrocytic basic protein and 2'3'- cyclic nucleotide 3'- phosphodiesterase have been utilized to prompt EAE in various species (Sospedra et al. 2005). PLP184-209 is a solid candidate CNS autoantigen in MS since it is the most bounteous CNS myelin protein and also because PLP184-209 is encephalitogenic in mice, it is additionally immunodominant in human and present on the external surface of the myelin sheath, in this manner likewise being a potential target for antibody-mediated demyelination.

- The most commonly opted method that has been utilized for inducing EAE in C57BL/6 is immunization with MOG peptide that is emulsified in Freund's adjuvant.
- The mixture is supplemented along with mycobacterium tuberculosis extract to form an antigenic depot. Further, mice are injected with pertussis toxin on the first day of immunization and on 2 days after which helps in opening of BBB (Baxter et al. 2007).
- This methodology has an advantage of reproducibility of results, but the disease pattern is not relapsing, and it follows a chronic progressive clinical course. After an interval of 10–20 days, neurological and paralytic signs begin to develop. In some strains, neurological symptoms spontaneously build up like in Lewis rats that are immunized against myelin basic protein (MBP).
- In SJL/J mouse, disease pattern is relapsing–remitting MS immunized with $MOG_{92\text{-}106}$ with fewer plaques formation in brain and spinal cord and mononuclear filtrate of lymphocytes, monocytes (Sun et al. 2001).
- The typically susceptible strains, which respond to immunization with MBP and PLP, include Lewis and DA rats and SJL/J and PL/J mice. Immunization against MOG causes EAE in several additional strains, most notably in BN rats and in C57BL/6 and BALB/c mice, which resist MBP or PLP immunization (Gold et al. 2006).
- In MBP-induced EAE of Lewis rats, the inflammatory lesions are primarily located in the lower spinal cord. Whereas MOG-induced EAE in C57BL/6 mice commonly starts in the optic nerve before reaching the spinal cord (Gold et al. 2006). As in the MBP/Lewis rat model, EAE lesions may be purely inflammatory with relatively sparse myelin destruction. By contrast, immunization of BN or DA rats with recombinant MOG protein produces plaque-like lesions with inflammatory foci bordered by large areas of demyelination.
- Actively actuated EAE is in no way confined to rodents but at the same time is delivered in primates, specifically rhesus and squirrel monkeys (the marmoset Callithrix jacchus). As in rodents, the dose, the nature of autoantigen, and the adjuvant utilized as a part of these primates decide the quality of the disease. At present, most reports on primate EAE depend on inoculation with recombinant human MOG in CFA, a treatment that prompts a disease neurologically and histologically near to human MS. This class of models is especially imperative in light of the fact that the sub-human primate immune system intently looks like its human counterpart in structures and reactivity, not at all like the rodent immune system, which varies fundamentally. Impediments of primate models are famously costly, hard to deal with, require specific primate facilities and are

completed utilizing extremely restricted quantities of an experimental animal, and they are basic for preclinical medication (Rivers et al. 1933).

Advantages

- EAE animal model is that it can establish disease pattern in almost all mammals tested so far. Several species and strains have been utilized including mice, rats, and guinea pigs, and research has been extended further to non-human primates using the advantage of genetic variability and extended preclinical studies.
- Proper handling of this model has led to the discovery of new therapeutic strategies in the treatment of MS. Implementation of neural precursor cells (NPCs) in the treatment of MS has been validated in this model, characterizing its widespread variability. It is used as the most invaluable tool for finding out mechanisms responsible for immune response against self-antigens within the CNS.

Disadvantages

- EAE gives no understanding into MS progression. The achievability of constantly demyelinated axons in EAE tissues is diminished, yet mechanism is not known.
- Using C57BL/6 mice loss, the opportunity to study relapses.
- In EAE, remyelination cannot be all around examined: lesion happens arbitrarily concerning timing and limitation.
- EAE is essentially a disease of sub-pial spinal cord white matter, though MS is predominantly a brain disease with prevailing demyelination of the cerebral and cerebellar cortex.
- Participation of the cortex in EAE has been hard to separate and in this way not all around contemplated.
- Most types of EAE are brought about by inoculation strategies that elicit CD4+ T-cell response. The parts of CD8+ T-cells, which dominate in MS lesion and show clonal expansion, have been hard to capture utilizing traditional EAE models.

4 Conclusion

In this given chapter, we have summarized the available animal models for investigating the variable aspects of multiple sclerosis. In the past few decades, substantial advances have occurred in the understanding the central mechanism that underlies the inflammation, demyelination, oligodendrocytes depletion, and neurodegeneration which form the core base of MS. Due to inadequacy in animal models, inconsistency in symptoms, inter-individual variability, evidences are still insufficient regarding pathology and much more is yet to be discovered.

Ethical Statement
All institutional guidelines, national guidelines, local and state laws, and regulation with the professional standard for the care and utilization of research animal ought to be taken followed. Studies including animal must express that the institutional animal ethical board has endorsed the protocol. For authors utilizing experimental animals, a statement ought to be made that the animals' care has been taken as per institutional ethical guidelines and animal utilized have been dealt humanely with respect for the alleviation of suffering. Specialists ought to treat the animal as sentiment and must consider their legitimate care and utilize and the ignorance or minimization of uneasiness, trouble, or agony as goals. Animal experiment ought to be composed simply after due thought of animal well-being. It ought to be guaranteed that all analysts who are utilizing animal have received guideline in research techniques and in the care, maintenance, and handling of the species being utilized. All the surgical methods ought to be performed under right anesthesia and take after just those methodologies which avoid infection and reduce pain during and after surgery.

References

Baxter AG et al (2007) The origin and application of experimental autoimmune encephalomyelitis. Nat Rev Immunol 7(11):904–912

Carlton WW et al (1966) Response of mice to the chelating agents sodium diethyldithiocarbamate, α-benzoinoxime, and biscyclohexanoneoxaldihydrazone. Toxicol Appl Pharmacol 8(3):512–521

Gold R, Linington C et al (2006) Understanding pathogenesis and therapy of multiple sclerosis via animal models: 70 years of merits and culprits in experimental autoimmune encephalomyelitis research. Brain 129(8):1953–1971

Lindner M, Fokuhl J et al (2009) Chronic toxic demyelination in the central nervous system leads to axonal damage despite remyelination. Neurosci Lett 453(2):120–125

Lipton HL, Canto C et al (1977) Contrasting effects of immunosuppression on Theiler's virus infection in mice. Infect Immun 15(3):903–909

Miller RH, Fyffe-Maricich SL et al (2010) Restoring the balance between disease and repair in multiple sclerosis: insights from mouse models. Di Models Mech 3(9–10):535–539

Remington LT, Babcock AA et al (2007) Microglial recruitment, activation, and proliferation in response to primary demyelination. Am J Pathol 170(5):1713–1724

Rivers TM, Sprunt D et al (1933) Observations on attempts to produce acute disseminated encephalomyelitis in monkeys. J Exp Med 58(1):39–53

Sospedra M, Martin R et al (2005) Immunology of multiple sclerosis*. Annu Rev Immunol 23:683–747

Sun D, Whitaker JN et al (2001) Myelin antigen-specific CD8+ T cells are encephalitogenic and produce severe disease in C57BL/6 mice. J Immunol 166(12):7579–7587

Animal Model of Amyotrophic Lateral Sclerosis

Prashant Koshal, Mandeep Kumar, Sumit Jamwal
and Puneet Kumar Bansal

1 Introduction

Amyotrophic lateral sclerosis (ALS) is a mortal neurodegenerative disorder characterized by preferential death of motor neurons. Only 10–20% cases of ALS are familial, and rest are sporadic. ALS is also known as Lou Gehrig's disease and was first described by Charcot in 1874. Behavioral analysis during disease progression has revealed symptoms like muscle weakness, muscle atrophy, altered muscle performance, paralysis, and death after 3–5 years after beginning of symptoms. Altered muscle performance, paralysis, and death occur due to degeneration of upper motor neurons in cerebral cortex and lower motor neurons in brain stem and spinal cord. The prevalence of ALS is about 0.6–2.4/100,000 population with the average age of onset 55 years. More than 50% patients of ALS show cognitive impairment, 5–10% show symptoms of dementia, and these symptoms are higher in man (3.0/100000) as compared to women (2.4/100000). Approximately 30% patients of familial ALS have mutation at C9 or f72, ALS2 (Alsin2 protein), and

P. Koshal · M. Kumar · S. Jamwal · P.K. Bansal (✉)
Department of Pharmacology, ISF College of Pharmacy, Moga 142001, Punjab, India
e-mail: punnubansal79@gmail.com

P. Koshal
e-mail: prashantkoshal@gmail.com

M. Kumar
e-mail: moneyarora8@yahoo.com

S. Jamwal
e-mail: jamwalisf@gmail.com

P.K. Bansal
Department of Pharmaceutical Sciences and Technology, Maharaja Ranjit Singh Punjab Technical University (State Govt. University), Bathinda, Punjab, India

P.K. Bansal and R. Deshmukh (eds.), *Animal Models of Neurological Disorders*, https://doi.org/10.1007/978-981-10-5981-0_18

SOD1. Study has revealed involvement of neurons other than motor neurons, such as ascending spinocerebellar pathways degeneration, depletion of neurons in substantia nigra in case of ALS—dementia.

It is well established that motor neurons are more venerable to AMPA receptor-mediated excite-toxicity due to high density of these receptors on the motor neurons. AMPA receptor complex is made up of the combination of GluR 1-3-4, and it is permeable to Ca^{2+}; but it is impermeable when only one subunit of GluR2 is present. AMPA receptors require week stimulus for Ca^{2+} influx compared to the NMDAR, and during strong stimulus AMPA receptor-mediated depolarization removes the Mg^{2+} block on NMDA receptor, resulting in massive influx of Ca^{2+}, mitochondrial dysfunctioning, and ROS production. Downregulation of EAAT2 glutamate transporter and increase in production pro-inflammatory cytokines such as IL-1β, TNF-α, and iNOS by astrocytes dysfunctioning might be contributing factor for ALS (Fig. 1).

1.1 Why Animal Model Is Needed?

The exact causes of ALS are more or less unknown. This lack of knowledge has motivated the search for animal models to explore the pathogenesis and biochemical mechanisms underlying ALS. A mutation in the superoxide dismutase (SOD1) gene is one of the major factors for ALS in humans. Transgenic mice which carry SOD1 mutation gene reflect similar symptoms as seen in patients with ALS, making them useful models for understanding the pathogenesis of disease and testing new therapies (Fig. 2).

2 Pathophysiology of ALS

Various efforts have been made to understand the basic mechanism underlying motor neurons degeneration in ALS. Mutation in five Mendelian genes such as SOD1, alsin, senataxin (SETX), synaptobrevin/VAMP (vesicle-associated membrane protein)-associated protein B (VAPB), and dynactin has been reported to cause ALS. The first relation between ALS and chromosome 21 was made in 1991, and mutation in SOD1 was recognized as causative factor for 20% of familial cases and 2–7% of sporadic cases. Copper/zinc superoxide dysmutase (SOD1) is abundantly found in the cytosole nucleus, mitochondrial intermembrane space, and peroxisomes of human cells. Mutant SOD1 causes aberrant redox reaction, such as hydrogen peroxide (H_2O_2) or nitronium ion (ONOO−) capable of reacting with reduced SOD1 (SOD1−Cu+). Molecular oxygen (O_2) can react strangely with Zn-deficient SOD1 to produce an excess of superoxide anion (O^{2-}), thus causing release of Cu/Zn that leads to the toxicity.

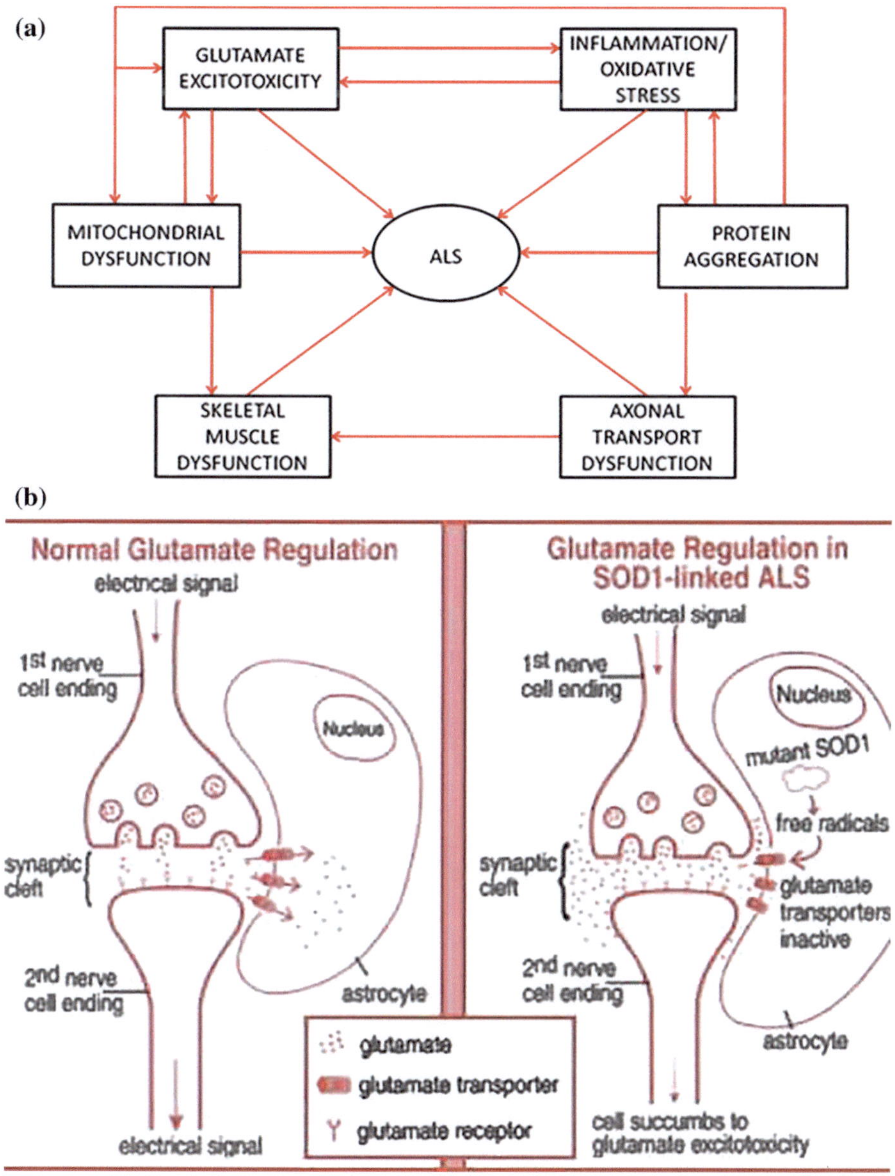

Fig. 1 Pathophysiology of ALS

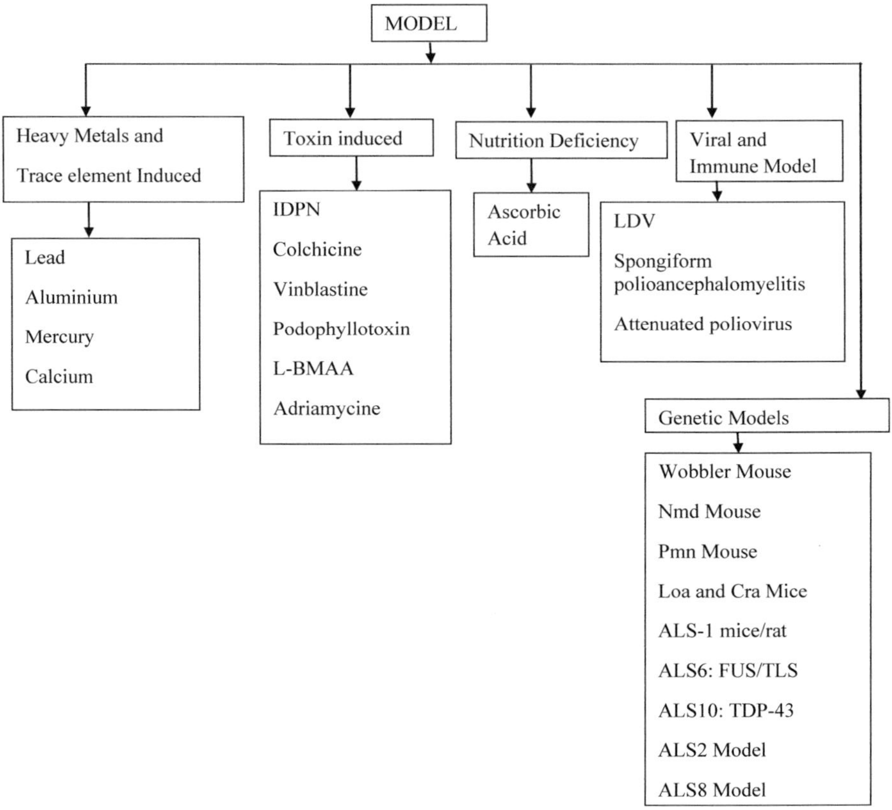

Fig. 2 Animal models of ALS

$$H_2O_2 \rightarrow OH \text{ (Peroxidation reaction)}$$
$$ONOO^- - SOD1 - Cu^+ \rightarrow NO - Tyr \text{ (Tyrosine nitration)}$$
$$O_2 - Zn \text{ deficit } SOD1 \rightarrow O_2 \text{ (Reverse catalysis)}$$

It is now well established that aggregates of misfolded and mutant SOD1 are directly linked with ALS. Improper metalation of the protein, genetic mutations, loss of disulfide bound, and post-translational modifications are some of the reasons that underly misfolding of SOD1 enzyme and its aggregation. These aggregates further produce neurite mitochondrial fragmentation, impaired mitochondrial dynamics, and neuronal toxicity. Altered astrocytes and microglia functions are also responsible for the pathologic condition in ALS, which causes glutamate-induced excitotoxic damage to the neurons and increase neuroinflammation, respectively. Mutant SOD1 can also act as the pro-apoptotic factor that causes release of cytochrome-c, and altered axonal transport is another factor for ALS.

3 Classification of Animal Models of ALS (Fig. 3)

3.1 Lead-Induced Animal Model of ALS

Principle: It is well accepted that lead exerts its toxic effect to the nervous system. In 1907, Kinnier Wilson described associated neurological disease with chronic lead toxicity, in which symmetrical wasting and weakness of muscles were present, and there was indication of pyramidal tract interruption in some, and fasciculation was recorded. Delta amino levulinic aciduria, coproporphyrinuria, increased erythrocyte protoporphyrin, and sideroachrestic anemia are some metabolic changes which are associated with lead intoxication, in which the entry of iron into the porphyrin ring is blocked. In 1940s, it was recognized that children who had been treated for lead poisoning suffered permanent neurological damage. Kostial and Vouk in 1957 demonstrated that lead ions, in concentrations of 5–40 μM/l,

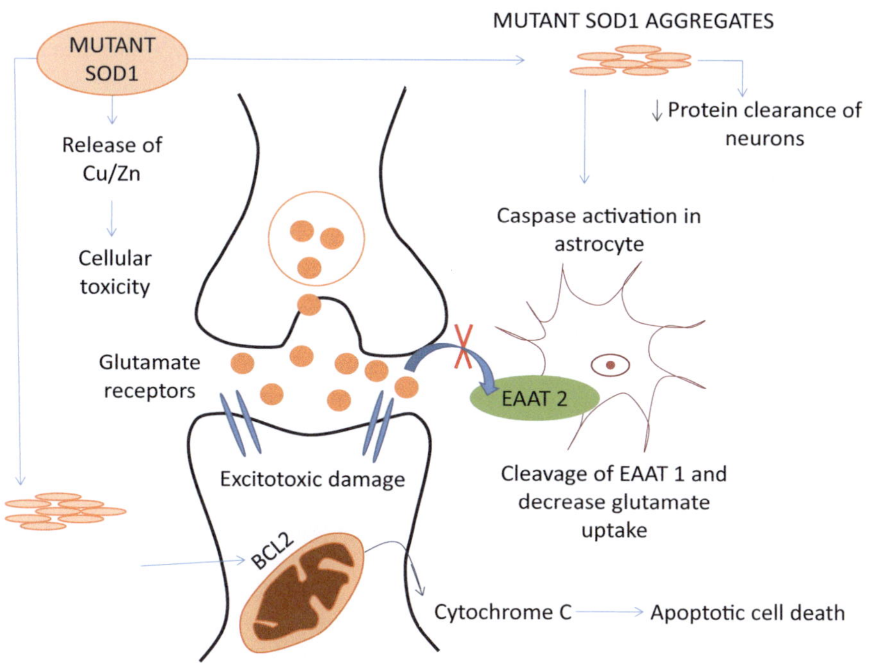

Fig. 3 Zn-/Cu-deficit mutant, SOD1 is a toxic factor for the cells as it has been lost its dysmutase activity. Mutant SOD1 aggregates causes misfolding of chaperons, thus causing decrease in protein clearance from the neuron that leads to the massive protein–protein interaction. Activation of caspase-3 in the astrocytes due to mutant SOD1 leads to breakdown of EEAT2 and causes excitotoxic damage to the motor neuron of spinal cord, and increased permeability of mitochondria due to this mutant SOD1 causes release of cytochrome-c, leading to activation of apoptotic factors. Loss of BCL2 function is also found in the neurons due to increased interaction with mutant SOD1 aggregates

decrease the ganglionic transmission due to decreased output of acetyl choline (Ach) from the superior cervical ganglion. Study reveals that administration of lead in rodent model decreases the locomotor activity.

Nervous system is the primary target for lead. Lead causes activation of the calmoduline and PKc, which in turn causes activation of various signaling cascade and protein function and produces neurotoxic effect. Lead mainly disrupts mitochondria, decreases the calcium uptake, and increases calcium release from mitochondria, as well as releases cytochrome-c and causes apoptotic cell death.

Procedure:

- Male and female SD rats are taken, and lead acetate (5 and 50 ppm) is administered to the animal in drinking water.
- Pre-treatment is given till the 40th day to the animal by lead acetate, and then pre-treated animal is mated.
- Lead acetate treatment is given to the female throughout the gestation period and continued till the lactation.
- The offsprings are exposed in the same way up to 180 days.
- Study reveals that treatment with lead acetate causes decrease in righting reflex, delay in eye-opening, and hypoactivity in rats.

3.2 Mercury-Induced Animal Model of ALS

Principle: Among all environmental neurotoxin, methyl mercury (MeHg) is well-recognized neurotoxin present in the environment, which easily accumulates in the brain tissue and causes irritability, poor memory, depression, anxiety, disturbances of coordination and balance, and motor signs similar to ALS. Pathological reactive oxygen species (ROS) generation due to imbalance in antioxidant defense mechanism has been implicated in various types of neurological disorders like Alzheimer's disease (AD), Parkinson's disease (PD), ALS. Mercury forms complex with the glutathione (GSH) after accumulation in the CNS and causes excessive ROS production. Oxidative stress is able of causing damage to motor neuron populations and causes degeneration of motor neurons by other mechanism also. The uptake of systemically administered mercury is carried out by neuromuscular junction and transported to the cell bodies of motor neuron, and it is thought to be a major cause of sporadic ALS. Mercury is generally found in sea food, and prenatal contact is normally more privileged than early postnatal exposures.

For maintaining calcium (Ca^{2+}) homeostasis, glutamatergic system plays important role in the CNS. Massive release of glutamate causes prolonged activation of the NMDA, thereby increasing calcium influx for a long period and leading to neuronal death. This excitotoxic damage to the neuron is implicated in Alzheimer's disease (AD), Parkinson's disease (PD), and amyotrophic lateral sclerosis (ALS). Study reveals that methyl mercury directly produces damage to the astrocyte glutamate transporter and causes excitotoxic damage to the neurons. Many mechanisms for methyl mercury have been given, like formation of ROS and reactive nitrogen species. Formation of free radicals mediated by MeHg causes many changes to DNA bases, increases lipid peroxidation, and changes calcium and sulfhydryl homeostasis. MeHg toxicity also depletes glutathione and bonding to sulfhydryl groups of proteins.

Procedure:

- Methyl mercury (2.0 mg/kg p.o) is given to the male SD rats for 14 days.
- After administrating MeHg (2 mg/kg) for 14 days, the Hg level of the blood reaches up to 0.8 ppm and remains as high as 0.25 ppm in blood after termination of mercury administration.
- This indicates Hg accumulation to 0.5 ppm in blood, resulting in neurotoxicity which is linked with the dysfunction of the nervous system.

3.3 Aluminum-Induced Animal Model of ALS

Principle: It has been demonstrated that aluminum get deposited in neurofibrillary tangles (NFT) in hippocampal neurons of Guamanian patients who died from ALS or parkinsonism–dementia complex. Increased aluminum content has been found in the cervical anterior horn of the spinal cord in patients who died from ALS. Aluminum interacts with enzymes which are involved in free radical scavenging activity. Aluminum is a light toxic metal and long-term exposure at very low concentrations increases the susceptibility of the central nervous system to neuronal cell death which may be implicated in the pathogenesis of different neuronal disorders such as AD, PD, and ALS.

Neurodegeneration caused by aluminum overload was associated with an imbalance in metal ion levels in the brain. Aluminum alters the function of different parts of the cells, like mitochondria, lysosomes, and nucleus. It inhibits DNA repair system, hampers the activity of antioxidant enzymes, and causes disturbance in cellular metal homeostasis specially that of iron (Fe), increasing ROS generation and causing apoptotic cell death.

Procedure:

- Female white New Zealand rabbits are anesthetized with ketamine 40 mg/kg and acetylpromazine 2 mg/kg, and rabbits are inoculated intracisternally once monthly with 0.1 ml containing 100 μg aluminum chloride in 0.9% NaCl for 14 months.

Evaluation score

- Spontaneous behavior,
- Reaction to handling,
- Posture,
- Paraspinal and limb tone,
- Deep tendon reflex,
- Tonic immobility,
- Focal central nervous system deficit,
- Nystagmus,
- Feeding behavior,
- Retention of urine, and
- Myoclonus and epilepsy.

3.4 Low Calcium and Magnesium

Principle: According to the environmental studies, low calcium and magnesium and high aluminum content in river, soil, and drinking water are responsible for ALS. The prevalence of ALS and parkinsonism–dementia complex was found in the populations of Kii Peninsula of Japan and Guam. The higher prevalence of PD and ALS is due to marked deficiency of calcium and magnesium. The intestinal absorption of aluminum, a well-known environmental toxin, is increased if aluminum is administered chronically under low Ca/Mg, and it is thought to be that it accumulates in the brain and spinal cord and produces neurotoxic effects.

Procedure

- Prepared diet is given to the six cynomolgus monkeys with 0.32% Ca, till 41–46 months, and deionized drinking water is provided in the cage.
- Another four monkeys receive 150 mg of aluminum and 50 mg of manganese daily for 41–42 months, and also deionized water is provided in the cage.
- Two monkeys are provided normal calcium diet and used as control.
- Baked biscuit diet is blended to a powder form, and deionized water is added to it to make dough. After this, prepared dough is pressed and rolled on top of stainless steel cookie sheet; 5 gm of rectangles is made by cutting and dried in autoclave for 2 h at 20 psi jacket pressure at 42.5 °C and stored at −70 °C.
- The diet is prepared in every 6 months.

3.5 IDPN (β,β′-Lminodipropionitrile)

Principle: Study has revealed that normal slow axonal transport in rat sciatic motor axons is interrupted after the administration of IDPN. It has been found that neurofilament protein is retained in the proximal axon after the administration of IDPN prior to microinjection of labeled precursor such as [3H] leucine or [35S] methionine into the cord. The accumulated neurofilament in the proximal axon causes swelling of the proximal axonal which is known as "spheroids." These neurofilamentous giant axonal swellings are found in a number of human disorders. Similarly, neurofilamentous giant axonal swellings are also found in some patients of ALS. Electrophysiological changes are also found in the motor neuron with the IDPN-induced toxicity. Muscle weakness followed by the muscle atrophy, a sign of ALS, is also found in the rodent motor neuropathic model of IDPN.

Neurofilaments are known as cytoskeletal polymers which increase axonal cross-sectional area in a way that propagation of action potentials is increased by neurofilament. Axonal diameter is a main constituent of motor neuron function, which determines the velocity of signal conduction (Barry et al. 2012). The axonal transport of neurofilament proteins are selective impaired due the IDPN, thus causing proximal axonal swelling. Alteration in axonal diameter due to the axonal swelling causes changes in potential impedance. Study revealed that swellings may slow down or ultimately block the impulse conduction. Myelin alteration and remodeling at the site of the swelling worsen the condition when impulse conduction is slow due to changes in axonal diameter.

Procedure:

Mongrel Cats:

- IDPN 50 mg/kg, i.p., is administered in a 3% (v/v) solution of 5 mM phosphate buffer once in a week to Mongrel cats and examined at 7th day, 14th day, or 35th day.

Rats

- Male Sprague Dawley rats are kept in stainless steel cages having 100 cm/rotation and 15 cm width running wheel.
- The animal is allowed to run freely over this wheel.
- Animal is housed under laboratory condition, and motor activity in running wheel was measured until it reached to the steady state, approximately 2–3 weeks.
- IDPN in a concentration of 0.13% was given in drinking water. Deionized water is used to dilute the IDPN for administration.

3.6 Vinblastine, Podophyllotoxin, Colchicine

Principle: Intrathecal administration of vincristine, vinblastine, podophyllotoxin, and colchicine in the rabbit led to marked alteration in neuron, such as neurofilaments with few normal neurotubules which were found in the perikarya of the anterior horn. Similar results were also found with the intrathecal administration of maytansine, maytanprine, and oncodazole in rabbits, in which axonal swellings were found with the maytansine-intoxicated rabbits. All these agents provoke the growth of 10-nm neurofilaments and have high affinity for the tubulin.

Procedure

- In this three series experiment (vinblastine, podophyllotoxin, colchicine) and 18 rabbits of either sex are used.
- First injection of 100 μg colchicine in 100 μl distilled water is given in the subarachnoid space of the lumbosacral spinal cord.
- Second injection of 200 mg vinblastine in 100 μl of distilled water is given in the subarachnoid space at the cisterna magna.
- Podophyllotoxin 1 mg is dissolved in the two drops of 75% ethanol, and 2 ml volume is made by adding distilled water.
- Further, this podophyllotoxin suspension was given into the subarachnoid space at the cisterna magna (Wisniewski et al. 1968).

3.7 L-BMAA (β-N-Methylamino-L-Alanine)

The high prevalence of ALS and parkinsonism–dementia complex in the native Chamorro population of Guam island is probably due to β-N-methylamino-l-alanine (l-BMAA) which is a cyanobacterial excitotoxin. Moreover, L-BMAA was also found in the patients of sporadic Alzheimer's disease in US–Canada region. Additionally, higher prevalence of ALS was also found in some subpopulations such as Italian and American football players, Gulf War veterans who stayed in Qatar desert, and in both cases cyanobacteria and L-BMAA were found in the pitch and desert soil. L-BMAA is structurally similar to alanine, and it was thought that L-BMAA causes the misfolding of proteins. It is hypothesized that L-BMAA causes excitotoxic damage to the neurons by activating NMDA and mGluR5 receptors, but higher concentration is required (>100 mg/kg) in rodents.

Principle: It is proposed that L-BMAA can act by the following mechanism such as direct activation of NMDA receptor, mGluR5 and mGluR1 receptors, and generation of reactive oxygen species by inhibiting the cysteine-glutamate exchange transporter, which causes depletion of GSH. It has been proposed by Weiss and Choi that a ternary receptor/bicarbonate/L-BMAA complex is formed by the interaction of L-BMAA with bicarbonate/carbonate at the β-methylamino function, which is similar in shape that is formed by the glutamate and NMDA receptor.

Procedure

- Male Wistar rats (21 day old) are used in the study.
- L-BMAA in a dose ranging 100–350 mg/kg i.p is given daily for 5 days.
- Lower dose, i.e., 100 mg/kg/day, i.p., produces mild neurological damage, where as 200–350/kg/day i.p produces similar neurological damage as observed clinically.
- Neurological evaluation is done by ambulation, tail lift test, and strength test.

3.8 Adriamycin

It is a potent anticancer drug and used for the treatment of hematologic and solid tumors. Adriamycin produces potent oxidative stress in the treated patients, which is the cause of cardiomyopathy and heart failure. Adriamycin was restricted for the clinical use due to the nephropathy and cardiac myopathy (Balakumar et al. 2008). Study reveals that adriamycin is placed on tongue, it reached to the hypoglossal neurons in a retrograde manner, and this retrograde transport of drug may exert harmful effect to the motor neuron of spinal cord and brain stem.

Principle: Adriamycin is transported to the somatic motor neurons and dorsal root ganglion cells and causes the degeneration of the target motor neurons within a week, when it is injected into rat sciatic nerve or distal hindlimb muscle. It was found that nerve cells left from the corresponding motor neuron columns, and after 2 weeks of treatment most dorsal root ganglion cells had also disappeared.

Procedure:

- Adriamycin is dissolved in a concentration of 1 or 2 mg/ml and administered to the mice in the following way.
- At first, adriamycin is administered to the intact sciatic nerve and injection is given by using 35 μl Hamilton syringe.
- Endoneurial microinjection of adriamycin in a dose of 12.5 μg is given to the 2 animal and survival period is 15 min, dose of 25 μg is given to 4 animal and survival time is 15 min, then dose of 25 μg is given to 2 animal and survival time is 60 min, intramuscular injection of adriamycin is given at a dose of 25 μg to the 3 animals and survival time is 15 min.
- The injection of adriamycin is given to the ligated animal in the following way, epineurial injection is given after 7 days of ligation in a dose of 25 μg in 4 animals, and intravenous injection is given immediately after 24 h in a dose of 10 mg/kg to 10 animals.
- Neurological evaluation is done by ambulation, tail lift test, and strength test.

3.9 Ascorbic Acid Deficiency in Guinea Pig

Numerous studies have investigated the possible role of dietary factors in death of motor cells in the guinea pig. Den Hartog Jager in 1985 said that chronic deficiency of ascorbic acid (vitamin C) leads to the death of motor neuron, but further Sillevis Smitt et al. in 1991 had made correction on this proposal and said that ascorbate deficiency causes the myopathy. It was first explained by the Hoist and Frolich in 1907 that scorbutic guinea pigs possess the pathological changes in skeletal muscle, and several authors has given the conformation to these skeletal muscle abnormalities. Moreover, hemorrhages in the brain and spinal co-regeneration of large motor neuron in spinal cord were found by the Meyer and McKormick in 1928. The detailed account on animal model of ALS due to chronic deficiency of Ascorbic acid in Guinea pig was given in 1985 by Den Hartog Jager (Sillevis Smitt et al. 1991).

Principle: Ascorbic acid competes with the superoxide dismutase (SOD) and eliminates the superoxide radicals, thereby provides prevention to the SOD. It also prevents the generation of toxic molecules such as hydroxyl radicals and peroxynilrite anions, derived from O_2^- free radicals. It increases the regeneration of vitamin E (α-tocopherol), and study reveals that vitamin E delays the onset of ALS in the transgenic animal. Moreover, glutamatergic excitation is also maintained by the ascorbic acid and its metabolite known as dehydroxyascorbate (DHA), which prevents the binding of glutamate to the NMDA receptor (Kok 1997).

Procedure:

- Male Guinea pigs are employed in this study.
- The semisynthetic diet is prepared with rice, casein, cod-liver oil, carrot, salt mixture, riboflavin, thiamine, pyridoxin, magnesium oxide, ascorbic acid, nicotinic acid, vitamin B_{12}, vitamin E, vitamin K, inositol, folic acid, L-hydroxyproline, L-mithionine, and L-cysteine.
- The amount of ascorbic acid in concentration of 40, 100, 2, 100, 500–250 mg is used in all diet, and higher amount of ascorbic acid is added in the third experimental group.
- The ascorbic acid deficiency is not a cause of ALS, but ascorbic acid-deficient guinea pig shows degeneration in the anterior horn motor cells with neuronophagia and demyelination of the pyramidal tract.

Ethical Statement

All institutional guidelines, national guidelines, state and local laws and regulations with professional standards for the care and use of laboratory animals should be followed. Studies involving animals must state that the institutional animal ethical committee has approved the protocol. For authors using experimental animals, a statement should be made that the animals' care is in accordance with institutional guidelines and animals used have been treated humanely and with regard to the alleviation of suffering. Researchers should treat animals as sentient and must consider their proper care and use and the avoidance or minimization of discomfort, distress, or pain as imperatives. Animal experiments should be designed only after due consideration of animal health. It should be ensured that all researchers who are using animals have received instruction in research methods and in the care, maintenance, and handling of the species being used. All the surgical procedures should be performed under appropriate anesthesia and follow only those procedures which avoid infection and minimize pain during and after surgery.

References

Balkumar P, Chakkerwar V, Kumar V et al (2008) Experimental models for nephropathy. J Renin-Angiotensin-Aldosterone Syst 9:189–195

Barry D, Stevenson W, Bober B et al (2012) Expansion of neurofilament medium C terminus increases axonal diameter independent of increases in conduction velocity or myelin thickness. J Neurol Sci 32:6209–6219

Kok A (1997) Ascorbate availability and neurodegeneration in amyotrophic lateral sclerosis. J Med Hypotheses 48:281–296

Sillevis Smitt P, De Jong J, Troost D et al (1991) Muscular changes in the guinea pig caused by chronic ascorbic acid deficiency. J Neurol Sci 102:4–10

Wisniewski H, Shelanski M (1968) Effects of mitotic spindle inhibitors on neurotubules and neurofilaments in anterior horn cells. J Cell Biol 38(1):224–229

Commonly Used Laboratory Anaesthetics

Shubhangi Gupta, Rajat Bhardwaj and Rahul Deshmukh

1 Introduction

Anaesthetics are the drug that induces temporary state of unconsciousness, loss of senses in order to carry out surgical procedures. Anaesthetics are used during tests and surgical operations to prevent pain and discomfort and enable a wide range of medical procedures to be performed on patients. Analgesia is the absence of pain in response to stimulation that would normally be painful. An analgesic drugs act at the level of the central nervous system or at the peripheral site of inflammation to diminish or block pain signals.

At cellular level, the main function of anaesthetics is to inhibit synaptic transmission by reducing transmitter release and inhibition of transmitter action or reduction of the excitability of the post-synaptic cell or by blocking axonal conduction. For a drug to be useful as an anaesthetic, it should be readily controllable, so that induction and recovery are rapid, allowing the level of anaesthesia to be adjusted as required during the course of operation. Anaesthetic drugs are usually given in combination with analgesics or sedatives in order to achieve full anaesthesia. For anaesthetic drugs, the duration of action has not been provided. Duration of anaesthesia is influenced by the drugs used, strain, age, sex, body weight, procedure performed and the amount of stimulus during the procedure

S. Gupta · R. Bhardwaj · R. Deshmukh (✉)
Department of Pharmacology, ISF College of Pharmacy,
Moga 142001, Punjab, India
e-mail: login2rd@gmail.com

S. Gupta
e-mail: shubhangi.gupta5@gmail.com

R. Bhardwaj
e-mail: 9rajatbhardwaj9@gmail.com

P.K. Bansal and R. Deshmukh (eds.), *Animal Models of Neurological Disorders*, https://doi.org/10.1007/978-981-10-5981-0_19

2 Classification of Anaesthetics

Anaesthetics are broadly classified into two categories: local and general anaesthetics depending on goals and type of medical intervention (Fig. 1). The anaesthetics are widely being used for carrying out surgical procedures in experimental animals. This chapter discusses the general pharmacology of anaesthetics and their use in laboratory animals.

2.1 General Anaesthetics

General anaesthetics (GA) are used to render patients unaware of and unresponsive to painful stimulation during surgical procedures for almost 170 years. Prior to their discovery, surgery was a traumatic and barbaric affair, yet today, it is accepted as a routine and essential part of modern medicine. They are given systemically and exert their main effects on central nervous system. These drugs produce reversible loss of all sensation and consciousness associated with immobilization, analgesic, amnesic effect. GA acts primarily by depressing synaptic transmission. The main target for action of GA appears to be thalamic nuclei or reticular activating system (RAS), hypothalamus for memory related affects or blocking of muscular reflexes at spinal level. As the concentration of GA is increased, all the brain functions are affected including motor control, reflex activity, respiratory depression and autonomic regulation. Therefore, doses should be carefully monitored and adjusted within therapeutic levels by the anaesthetist before using it for surgical procedures on animals.

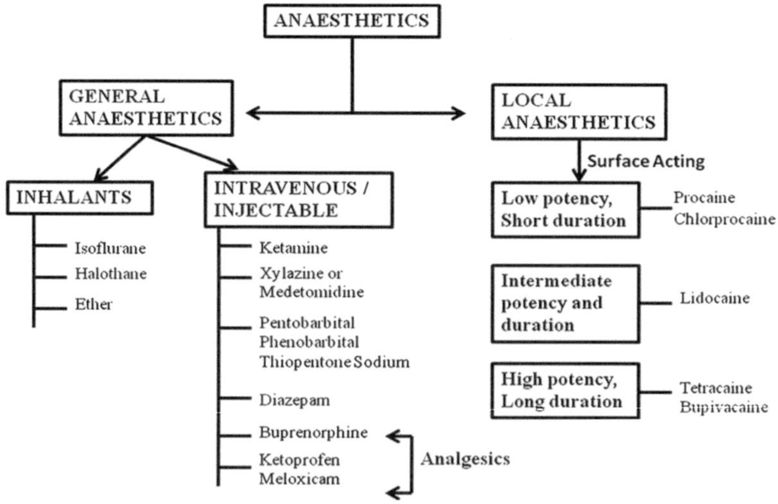

Fig. 1 Classification of commonly used laboratory anaesthetics

2.1.1 Mechanism of Action

In general, some GA may selectively inhibit excitatory ion channels (NMDA type of glutamate receptor which gates Ca^{+2} ion channel (for ex: N20, ketamine). Whereas, few of them potentiate the action of inhibitory transmitter GABA to open Cl^- ion channels (barbiturates, BZDs, inhalational) (Fig. 2).

Lipid Theory

Overton and Meyer suggested that general anaesthetics act on the plasma membrane and exert their action. This is supported by proof that the potency of the drug has immediate, affirmative association with the lipid solubility of the blood. The mechanism of action was recommended to be enhanced fluidity of the membrane. The elucidation of the Overton and Meyer finding has been challenged and discredited. In 1978, a workgroup processed conformational model pathological situations which alter lipid membrane composition could decrease nervous response to anaesthetics.

Ion Channels

General anaesthetics inhibit excitatory functions of some CNS receptors such as glutamate or 5-HT receptors. Some general anaesthetics also excite inhibitory receptors, notably $GABA_A$ receptors and TREK (Temperature sensitive,

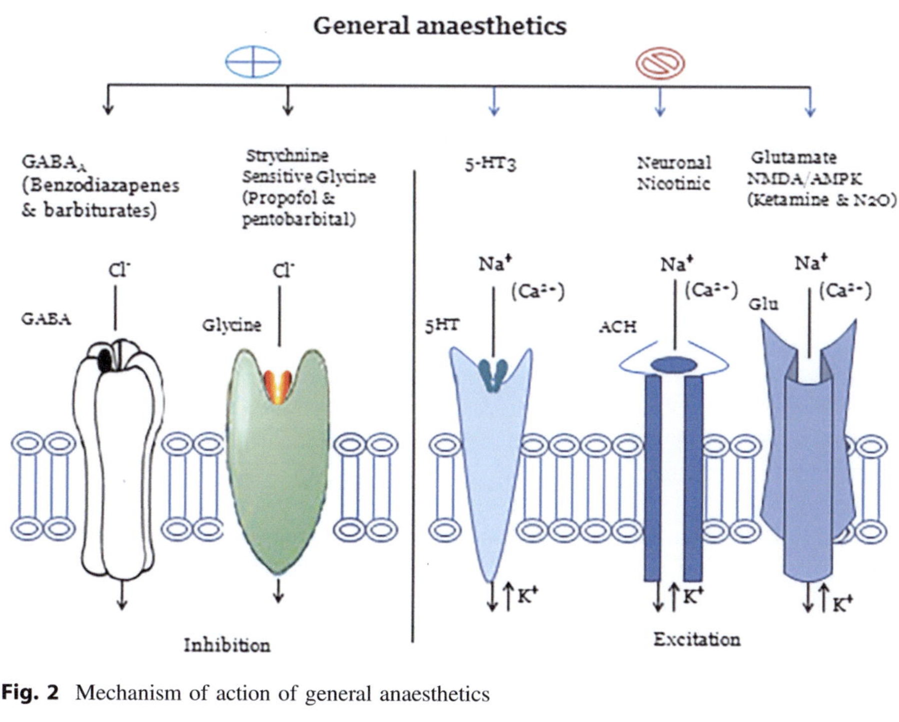

Fig. 2 Mechanism of action of general anaesthetics

osmosensitive and mechanogated K^+ channel). General anaesthetics may decrease transmitter release pre-synaptically or decrease excitability of post-synaptic neuron.

The exact mechanism of action of general anaesthetics is not known. The most accepted mechanisms are:

- Inhaled and some intravenous anaesthetics bind to specific sites on GABA receptor chloride channels and activate these receptors and increase the inhibitory neurotransmission which leads to CNS depression.
- Inhalational anaesthetics also enhance the sensitivity of glycine—gated chloride channels to glycine. These glycine receptors bring about inhibitory neurotransmission in the brainstem.
- Some anaesthetics like ketamine and nitrous oxide bind to and inhibit the N-methyl D-aspartate (NMDA) receptors.
- Inhalational and intravenous agents act at multiple sites in the nervous system and depress the neuronal activity at many sites in the brain.
- Some GA may activate the potassium channels in the brain leading to hyperpolarization of the membranes and thereby inhibitory effect. This in addition to their effects on the $GABA_B$ receptors.

2.1.2 Classification of General Anaesthetics

General Anaesthetics are broadly classified into two categories:

Intravenous and inhalational anaesthetics

Inhalational Anaesthetics

Inhalational anaesthetics are in the form of gases or volatile liquids and most preferred method in terms of safety and efficacy. It is easy to maintain and adjust the anaesthetic depth in case of inhalational anaesthetics. Because the exhalation, a process by which anaesthetics are eliminated from the blood and with less dependence on drug metabolism to eradicate the drug from the body, there is less possibility for drug-induced toxicity and allowing rapid induction and smooth recovery.

The disadvantages to inhalant anaesthesia are the intricacy and cost of the equipment required for administering the anaesthesia and possible hazards to personnel involved. As all inhalant drugs are volatile liquids, they should be stored in seperate rooms because the vapours are either flammable or toxic to inhale over extended period of time.

- Most inhalational anaesthetics that are widely used are ether, chloroform, cyclopropane which has been now replaced in developed countries particularly by halothane, enflurane, nitrous oxide (N_2O) and isoflurane.
- Inhalant anaesthetics offer a secure, consistent, reversible and reproducible means of depiction rodents unconscious in order to perform surgeries and other

complex or potentially painful procedures. Inhalant anaesthesia of small rodents is commonly maintained and makes use of face masks or nose cones.

Ether (Diethyl Ether):

- Ether is a highly volatile liquid and a potent anaesthetic which produces good analgesia and marked muscle relaxation.
- However, it is currently not employed for surgical procedures because of its unpleasant and inflammable properties.
- Low cost and easy dispensing by open drop method makes it use still in animal laboratories by inexperienced personnel.

Cyclopropane:

- It is a colourless gas with sweet odour and taste, available as liquid under pressure. It produces analgesia without loss of consciousness in 1–2% concentration, in 6–8% it produces loss of consciousness while 20–25% is required to produce surgical anaesthesia.
- It has low blood solubility. The induction and recovery are rapid and smooth. Blood pressure and cardiac contractility are well maintained with cyclopropane even on prolonged administration.
- Muscle relaxant activity is fairly good. Because of its highly inflammable and explosive nature, the close circuit has to be used to conserve the drug and to keep its concentration in the operating room low.
- Cyclopropane also sensitizes the myocardium to adrenaline and may produce a variety of cardiac irregularities such as tachycardia and fibrillation.

Chloroform:

- Chloroform is a commonly used laboratory inhalational anaesthetic.
- It is a colourless, volatile, liquid derivative of trichloromethane with an ether-like odour.
- One possible mechanism of action for chloroform is that it increases movement of potassium ion through certain types of potassium channels in nerve cells.
- Due to its highly inflammable and toxicity characteristics, it has been withdrawn from the market for surgical procedures and is no more used.
- Chances of cardiac disarrhythmia.

Halothane:

- It is widely used inhalational anaesthetic as it is non-explosive, non-irritant; induction and recovery are relatively fast.

- However, it is not a good analgesic or muscle relaxant which limits its use for surgical procedures.
- It is highly potent and can easily produce respiratory and CVS failure partly due to myocardial depression and vasodilation, so precise control of administered concentration is important. Two rare but serious adverse reactions associated with halothane are hepatotoxicity and malignant hyperthermia.

Enflurane:

- Halogenated anaesthetic is similar to halothane but less soluble in blood and, less metabolized, therefore less risk of toxicity.
- Enflurane cause faster induction and recovery compared to halothane probably due to its ability to cause less accumulation of fat.
- High doses are associated with risk of epileptic seizures.

Isoflurane:

- Isoflurane is the first choice of anaesthetic used for animal restraint or surgical procedures in laboratory animal species. Isoflurane is delivered via a nose cone and inhaled in rodents or provided through an intratracheal tube in larger species. Maintenance anaesthesia is typically between 1.5 and 3% isoflurane. Induction of anaesthesia with gas is typically achieved with <2 min exposure to 3–5% isoflurane.
- It is similar to enflurane, but lacks epileptogenic property.
- It is more potent and more volatile and less soluble in blood.

Advantages: It has better adjustment of depth of anaesthesia and low toxicity.

Disadvantages: It is of high cost and is a respiratory irritant.

Desflurane:

Desflurane is a congener of isoflurane. It has all the advantages of isoflurane. In addition, it has low solubility in blood and tissues because of which it rapidly attains therapeutic concentrations in the alveoli. Therefore, induction and recovery are very rapid and smooth.

Disadvantage:

- It is pungent which may induce coughing and sometimes laryngospasm. It is therefore used for maintenance of anaesthetia and is not preferred for induction.
- Because of low volatility, a special vaporizer is required for administration.
- It can cause transient sympathetic stimulation and tachycardia.

Sevoflurane:

Sevoflurane is the latest introduction to inhalational anaesthetics. It has the benefits of desflurane but is not pungent. It is a good bronchodilator and has rapid and smooth induction and recovery because of low solubility in blood and tissues. This also makes it suitable for day—care surgeries.

Disavantages:

- Sevoflurane is chemically unstable and is degraded by carbon dioxide absorbents (soda lime) to a metabolite that can cause nephrotoxicity. Postoperative restlessness is avoided by premedication with midazolam.
- It undergoes biotransformation (about 3%) in the liver to release fluoride ions which can cause nephrotoxicity.
- It can precipitate malignant hyperthermia in genetically susceptible individuals.

Nitrous Oxide (N_2O):

Nitrous oxide (N_2O) was discovered in 1793 by the English scientist Joseph Priestley, who also discovered oxygen (O_2). In 1799, Sir Humphrey Davy administered N_2O to visitors at the Pneumatic Institute and gave it for the first time the term "laughing gas". He astutely noted the analgesic effects of the gas and even predicted its application in suppression of pain during surgical procedures.

- Nitrous oxide is given usually as an adjuvant to other anaesthetic due to its low potency, negligible toxicity, rapid induction and recovery and good analgesic properties.
- It may pose risk of bone marrow depression with long-term administration and accumulation in gaseous activities.

Newer anaesthetic:

Xenon:

Xenon is an inert gas that has properties very close to an ideal anaesthetic. It is available in selected centres in some countries.

Advantages:

- Rapid induction, insoluble in blood and tissues.
- Rapid recovery.
- Potent anaesthetic.
- No effect on hepatic, renal or pulmonary function.
- Not metabolized in the body.

Disadvantages:

Xenon cannot be manufactured but can only be extracted from air which makes it very expensive and largely unaffordable. If this problem is taken care of, xenon will top the list of anaesthetics.

Intravenous Anaesthetics

These are drugs which on i.v. injection produces rapid unconsciousness in about 20 s, as soon as drug reaches the brain from its sight of injection. At the same time, larger number of animals is maintained under anaesthesia. Repeated injections and constant rate infusions (CRI) can be helpful for maintaining anaesthesia for longer period. When using parenteral anaesthetics, it is important to consider accurate dosing with correct multidrug use ratios, storage conditions and feasibility of immediate use following reconstitution. It is critical to weigh each animal accurately prior to administration of a calculated dose of anaesthesia to avoid either over-dosing or under-dosing.

Barbiturates:

- The principal effect of a barbiturate is depression of the CNS by hindrance with passage of impulses to the cerebral cortex. Barbiturates act directly on CNS neurons in a manner similar to that of the inhibitory transmitter GABA, therefore acts as GABA-A agonists. Barbiturates are considered to be good anaesthetic agents but with less analgesic and sedative effect.
- Phenobarbital sodium is used for prolonged experiments.
- Pentobarbital sodium (Nembutal), the most commonly used drug of this class that causes rapid and long-acting anaesthetic action (6% dissolved in 10% alcohol). In prolonged anaesthesia, the drug has also been shown to cause respiratory depression and impaired myocardial contractility.
- Methohexital and Thiopental are considered short and ultra-short acting anaesthetics and are more commonly used as induction agents in large animal species.
- Sodium pentobarbital (Nembutal) and sodium thiopental (Pentothal) are currently the two most commonly used barbiturates. The duration of action of pentobarbital is considerably longer than that of thiopental.

Advantages: Rapid anaesthetic onset provides a prolonged duration of surgical anaesthesia.

Disadvantages: Prolonged recovery time; inadequate analgesic properties; extremely expensive; narrow margin of safety; produces respiratory depression at higher dosages; non-rodent species may experience a distressful anaesthetic recovery.

Thiopentane sodium:

- It is an only ultra-short acting barbiturate used for animal surgical operations. On i.v. injection it causes unconsciousness within 20 s that lasts for 20 min due to its high-lipid solubility. It is highly soluble in water, so it should be prepared freshly before injection.
- It is widely used as an induction agent for routine purposes.
- It is not a good analgesic and is a weak muscle relaxant. It is used as a sole anaesthetic for short operations that is not painful.
- Risk of CVS and respiratory depression

Ketamine:

- Ketamine is a dissociative anaesthesia characterized by profound analgesia, immobility, amnesia with light sleep and feeling of dissociation from one's own body and the surroundings. The target area for ketamine action is in the cortical and subcortical areas, not in the reticular activating system.
- It inhibits the entry of Ca^{+2} ions by blocking NMDA type of excitatory glutamate receptor; onset of action is 2–5 min.
- Ketamine is the most commonly used injectable anaesthetic used in a variety of species. However, ketamine used as the sole anaesthetic is not recommended. In most cases, ketamine is used in combination with other injectable agents such as α-2 agonists (xylazine or medetomidine) or benzodiazepines to reduce or eliminate many of the less desirable side effects if used alone.

Advantages: Ketamine has a wide margin of safety in most species; residual analgesic effect following anaesthetic recovery. In combination with other drugs, it can provide a surgical plane of anaesthesia for about one-half hour.

Disadvantages: Ketamine alone does not provide muscle relaxation and muscle spasms may be observed.

Ketamine combinations:

(A) **Ketamine + Xylazine**: A single syringe must be used for the administration of both drugs after mixed well to produce a deep level of sedation. This combination is the most common injectable anaesthetic used in rodent species.

- Advantages of ketamine/alpha2-agonist combinations are that they may produce short-term anaesthesia and also shows good analgesia, and the healing can be hasten by reversing the alpha2-agonist with atipamezole or yohimbine. Whereas, in some cases, the combination can cause profound cardiac depression.

(B) **Ketamine + Xylazine + Acepromazine**: These drugs can be mixed well before administration because the drugs are incompatible and once mixed, will decrease over time. The acepromazine is added to the ketamine/xylazine mixture consequences in deeper and/or longer plane of anaesthesia in small rodents, especially rats, and probably in mice.

(C) **Ketamine + Diazepam**: Before administration of these drugs, it can be mixed well in a single syringe. Advantages of ketamine and diazepam have advantage as compared to ketamine/xylazine combinations include it shows limited cardiovascular effects like minimal hypotension. However, in rodents, ketamine/diazepam is only combination which provides light anaesthesia so it may be suitable for chemical restraint. This combination is favoured for non-painful procedures and imaging since it is safer than the ketamine/alpha2-agonist combinations.

Alpha2-agonists (Xylazine or Medetomidine):

The alpha2-agonists (xylazine or medetomidine) are hypnotic analgesics. It is able to produce sufficient depth of anaesthesia for even minor surgical procedures. But xylazine and ketamine combinations produce good analgesics effect that may be useful during surgery.

Benzodiazepines:

It includes diazepam, midazolam and zolazepam. All drugs of this class have anti-anxiety and anticonvulsant drugs with excellent muscle relaxation activity. They have negligible cardiovascular and respiratory effects. Sedation is minimal in most species, except for swine and non-human primates with no analgesic effect. The primary use of these drugs in anaesthesia is in combination with other drugs. Ketamine—diazepam, midazolam—narcotic combinations can be very useful for induction of general anaesthesia and for short procedures.

Chloralose:

- It is a mixture of chloral and glucose prepared by heating equal parts of anhydrous glucose and chloral, when both α-chloralose (active form) and β-chloralose (inactive form) are formed.
- Chloralose is the active form (α-chloralose) freely soluble in hot water, alcohol and ethanol and slightly soluble in cold water. It is prepared by heating 1% solution by boiling in 0.9% NaCl (saline) or in distilled water, and administered i.v. or i.p. at a temperature of 30–40 °C before the chloralose comes out of the solution. Duration of action for α-chloralose is 8–10 h.
- It is suitable only for acute experiments, usually in dogs or cats and induce consistant depth of anaesthesia. The respiration and CVS depression does not occur at optimum doses.
- Main problem is associated with low-water solubility of chloralose but can make up 10% solution in PEG. It is not a suitable anaesthetic for rabbits since they get narcotized, large volume needed and may produce convulsions on slight stimulation.

Urethane (Ethyl Carbamate):

- Readily soluble in water (25%) giving a neutral solution.
- It is suitable only for acute experiments since it has delayed toxic effect on liver and can cause agranulocytosis and pulmonary adenomata. It produces little or no effect on nerve transmission and little reflex depression.
- It is suitable for rabbits or rats; duration of anaesthesia is 3–4 h or more. Mice develop an exceptionally high incidence of lung tumour regardless of route of administration.

Paraldehyde:

- It has a wide range of safety because it depresses only the cerebrum and not the medullary centres.
- On i.v. injection it produces cardiac dilation and pulmonary congestion and oedema.
- Bilateral carotid occlusion produces poor pressor response or even a depressor response.

Magnesium Sulphate:

- A 20% magnesium sulphate solution 5 ml/kg i.v. produces anaesthesia for about an hour. Its principal use is in producing euthanasia.

Chloral hydrate:

- Chloral hydrate (trichloroacetaldehyde monohydrate) has been used in physiological experiments in laboratory animals (particularly rodents) because of their hypnotic property. Chloral hydrate is generally considered to be a good hypnotic but a poor analgesic even at anaesthetic doses. For beginning/continuation of surgical anaesthesia (and for a smoother initiation/recovery period), it must be commonly administered with some other anaesthetic or tranquilizing agent.
- It is widely used in physiological experiments because at hypnotic doses, the drug has minimal effects on cardiorespiratory function. Larger doses may cause dangerous respiratory and thermoregulatory depression, cardiac arrhythmias and severe hypotension. At hypnotic doses, the drug has minimal effects on cardiorespiratory function.

Analgesics:

Improved pain management is the main objective in the use of experimental animals. Analgesics are those agents which decrease or relieve pain without loss of consciousness during surgical procedures. The distress, suffering and pain in laboratory animals is an ethical duty for all individuals who work with these species in biomedical research. Systemic and/or local analgesics may also cut the anaesthetic necessity and have a preventive effect on pain perception which continues into the period of recovery. Analgesic administration should be given both pre-operatively and post-operatively for sufficient pain relief in rodents.

In experimental animal species, two most commonly used analgesics are opioids and nonsteroidal anti-inflammatory drugs (NSAIDs). But type of experimental model in the study is also very important for the choice of drug as anaesthetics (Fig. 2).

Opioids (Buprenorphine):

- Opioids drugs are firstly bind with three different receptors [mu (μ), kappa (κ), and delta (δ)] as agonists, partial agonists or antagonists and produce their effect. The expression of these receptors is mainly within the brain and spinal cord.

Advantages: It is a potent analgesia; during surgery, simultaneous administration opioids can lower the dose of barbiturate as general anaesthetic; receptor mediated mechanism in the brain and spinal cord; long history of use in research; the effect is antagonised by naloxone.

Disadvantages: Highly probable for human abuse and addiction; duration of action is very short; repetitive use may consequence in tolerance development.

Nonsteroidal Anti-Inflammatory Drugs (NSAIDs):

- Carprofen, Meloxicam, Ketoprofen, Ibuprofen, Acetaminophen.
- All classes of drugs belong to this group which shows inhibitory activity of the cyclooxygenase (COX) enzyme. The COX enzyme assists the production of Prostoglandin G2 (PGG2) which stimulates the range of enzymatic processes that are involved in normal physiological processes and production of Prostoglandin E2 (PGE2). PGE2 particularly plays an important role to produce pain in the periphery and within the central nervous system. Thus, inhibition of COX blocks the action of PGE2 that is effective in managing the pain, discomfort in periphery and within the CNS. COX-1 and COX-2 are the main two forms of the COX enzyme. Additionally, COX inhibitors are categorized as non-selective COX inhibitors and selective COX-2 inhibitors. This peculiarity has been made because COX-2 inhibition is supposed to be chief mechanism of NSAID to give analgesia and anti-inflammatory action even though this "consensus" is still under debate.
- Blockage of both COX-1 and COX-2 by NSAIDs is important in peri-operative pain practice.
  In the peri-operative situation, coxibs give an additional advantage over NSAIDs by avoiding the production of platelet thromboxane and clotting.

Advantages: Newer drugs include carprofen; meloxicam shows an analgesic activity for prolonged time; it reveals analgesic quality that equals to some opioids.

Disadvantages: Caution for inflammation models, infectious disease and coagulation research because of anti-inflammatory properties; COX-1 has side effects such as gastrointestinal complications, prolonged coagulation times and changes in kidney function with non-COX-2 selective forms.

2.2 Local Anaesthetics

Local anaesthetics (LA) are drugs that block nerve conduction when applied locally to nerve tissue in appropriate concentrations. Their action is completely reversible. They act on every type of nerve fibre and can cause both sensory and motor paralysis in the innervated area. They act on axons, cell body, dendrites, synapses and other excitable membranes that utilize sodium channels as the primary means of action potential generation.

Lidocaine, Bupivacaine, Proparacaine:

- Local anaesthetics particularly bind with voltage-gated Na + channel in the nerve cell membrane which blocks nerve impulses. Sufficient concentration of local anaesthetics is able to inhibit the conduction of nerves when applied locally.
- Routes of administration of local anaesthetic are either topical to mucus membranes of nose, eye, etc. or directly injected into tissues and around nerve bundles. Local anaesthetics would be considered as an additional analgesic to opioid and NSAID analgesics and is administred before starting surgery.
- The two most commonly used local anaesthetics in veterinary patients are **lidocaine** (xylocaine or novocaine) and **bupivacaine** (marcaine or sensocaine). Lidocaine has 1–2 min onset and 1½–2 h of duration of action. Bupivacaine shows slower onset (5–10 min) and a much longer duration of action (4–12 h, site dependent).

Advantages: Administration of local anaesthetics can provide a good adjunct to general anaesthesia for pain relief in pre and intra-operatively procedure.

Disadvantages: Avoid intramuscular and intravenous administration because both routes reach systemic circulation very rapidly. Signs and symptoms of overdose or systemic toxicity include CVS effects, seizures and death.

3 Anaesthetics Used Alone in Laboratory

Anaesthetics	Species	Dosage (per kg of body weight)	Route of administration
Isoflurane	Mice, rats	Induction—3–4%	Inhalation, nose cone method
	Guinea pigs, rabbits	Maintenance dose–1–2%	
Pentobarbital	Mice	35 mg	I.V
		50–90 mg	I.P
	Rats	30–40 mg	I.V

(continued)

(continued)

Anaesthetics	Species	Dosage (per kg of body weight)	Route of administration
		30–60 mg	I.P
	Guinea pigs	30–50 mg	I.P
	Rabbits	50–60 mg	I.V
Alpha-chloralose	Mice	114 of 5% solution	I.P
	Rats	31–65 mg	I.P
	Rabbits	80–100 mg	I.V.
Urethane (ethyl carbamate)	Rats	1000–1500 mg	I.P
	Guinea pigs	1.5 g	I.P
	Rabbits	1000 mg	I.V, I.P
Chloral hydrate	Mice	370–400 mg	I.P
	Rats	300–450 mg	I.P
	Guinea pigs	400–600 mg	I.P
Buprenorphine	Mice	2 mg	I.P, S.C, P.O
	Rats	0.01–0.05 mg	I.P, S.C, P.O
	Guinea pigs	0.05 mg	S.C, I.M, I.P
	Rabbits	0.01–0.05 mg	S.C, I.M, I.V
Ketoprofen	Mice	2 mg	I.P, S.C, P.O
	Rats	0.01–0.05 mg	I.P, S.C, P.O
	Guinea pigs	0.05 mg	S.C, I.M, I.P
	Rabbits	0.01–0.05 mg	S.C, I.M, I.V
Meloxicam	Mice	1–2 mg	I.P, S.C
	Rats	1–2 mg	I.P, S.C, P.O
	Guinea pigs	0.5 mg	S.C, P.O
	Rabbits	0.3 mg	S.C
Lidocaine	Mice, rats, guinea pigs, rabbits	2–4 mg	Topical (surface), S.C, intra-incisional
Bupivacaine	Mice, rats, guinea pigs, rabbits	1–2 mg	Topical (surface), S.C,
Thiopentone sodium	Mice, rats	80–100 mg	I.P
Diazepam	Mice, rats, guinea pigs, rabbits	2.5–5 mg	I.P

4 Anaesthetics Used in Combination in Laboratory

Anaesthetics	Species	Dosage (per kg of body weight)	Route of administration
Ketamine + Xylazine	Mice	80–100 mg + 7.5–16 mg	I.P
	Rats	40–80 mg + 5–10 mg	I.P, I.M
	Guinea pigs	40 mg + 5 mg	I.P
	Rabbits	25–35 mg + 5 mg	I.M
Ketamine + Xylazine + Acepromazine	Mice	80–100 mg + 20 mg + 3 mg	I.P
	Rats	75–90 mg + 5–10 mg + 1–2 mg	I.P
Chloralose + Pentobarbitone	Rats	100 mg + 30 mg	I.P
	Guinea pigs	31–65 mg	
	Rabbits	100 mg + 30 mg	I.P.
Chloralose + Urethane	Rabbits	60–72 g + 1–1.2 g	I.V
Pentobarbitone Sodium + Urethane	Rats	400–500 mg + 5–20 mg	I.P
	Guinea pigs	1.5 gm + 6 mg	I.P
	Rabbits	700 mg + 40 mg	I.P

5 Ethical Statement

All institutional guidelines, national guidelines, state and local laws and regulations with professional standards for the care and use of laboratory animals should be followed. Studies involving animals must state that the institutional animal ethical committee has approved the protocol. For authors using experimental animals, a statement should be made that the animals' care is in accordance with institutional guidelines and animals used have been treated humanely and with regard for the alleviation of suffering. Researchers should treat animals as sentient and must consider their proper care and use and the avoidance or minimization of discomfort, distress or pain as imperatives. Animal experiments should be designed only after due consideration of animal health. It should be ensured that all researchers who are using animals have received instruction in research methods and in the care, maintenance and handling of the species being used. All the surgical procedures should be performed under appropriate anaesthesia and follow only those procedures which avoid infection and minimize pain during and after surgery.

Euthanasia Procedure Used in Experimental Laboratory

Mandeep Kumar and Puneet Kumar Bansal

1 Introduction

The word euthanasia is basically derived from the two ancient Greek words: "eu" means good, and "thanatos" means death. This word is generally used to demonstrate killing animal in a way that minimizes or eliminates pain and distress. A good death is equivalent to the humane termination of an animal's life. It is a well-accepted and recommended procedure in all aspects of veterinary medicine and scientific experiments involving animals.

Principles for euthanasia

- Animal should be killed with the highest respect.
- Suitable techniques should be selected to kill the animals according to the type of species, age, and health of the animal.
- Euthanasia techniques should be given importance to make the animal's death painless and distress-free.
- Techniques should need minimum restraint of the animal and should minimize distress and anxiety felt by the animal, before loss of consciousness. Signs of fear and anxiety include distress vocalizations, freezing, salivation, struggling,

M. Kumar · P.K. Bansal (✉)
Department of Pharmacology, ISF College of Pharmacy, Moga 142001, Punjab, India
e-mail: punnubansal79@gmail.com

M. Kumar
e-mail: moneyarora8@yahoo.com

P.K. Bansal
Department of Pharmaceutical Sciences and Technology, Maharaja Ranjit Singh Punjab Technical University (State Govt. University), Bathinda, Punjab, India

P.K. Bansal and R. Deshmukh (eds.), *Animal Models of Neurological Disorders*, https://doi.org/10.1007/978-981-10-5981-0_20

defensive aggression, escaping attempts, urination, defecation, shivering, tremors, and sweating, etc.
- Euthanasia techniques should lead to speedy loss of consciousness, followed by respiratory or cardiac arrest and loss of brain function.
- Training programmes should be mandatory for personnel responsible for carrying out the euthanasia techniques in a successful and humane way, to identify signs of pain, distress, and fear in different species, and to confirm death in various species.
- Restraint by an experienced handler should be kind but firm. During euthanasia, careful handling, stroking, and talking often have a calming effect on many animals.
- Death must be confirmed following euthanasia. Cessation of heart beat and respiration, loss of reflexes, and the lowering of the body temperature are the most important aspects in verification of death.
- Equipments used during euthanasia of animals should be regularly checked, in good working condition, and cleaned properly for traces of urine, blood, and faces after killing one animal to minimize the anxiety to other animals.
- While selecting euthanasia technique, human psychological reactions should be taken into consideration, but this should not overcome animal welfare considerations. Regular exposure to euthanasia can have both negative and positive impacts on personnel's as some become more perfect in conducting the procedure and some might lose efficiency.
- Effective decisions and care should be taken while disposing off carcasses and waste as animals might have been treated with radioactive or other toxic chemicals and this waste or carcass can enter food chain in one or other way.

2 Recommended Chemical Methods for Euthanasia

2.1 Inhalational

Overdose of an inhalation anesthetic is well documented as efficient method of euthanasia. Chambers or anesthetic circuits are widely used to deliver inhalational agents either in vapors form or as a gas. Anesthetic can be used in combination with a sedative in situations where administration of the sedative causes minimal stress.

Carbon dioxide—Carbon dioxide acts as an anesthetic agent when used at concentrations above 60% and causes rapid loss of consciousness. CO_2 should be used above 70% for euthanasia of small animals, because it is effective and humanitarian. Recommended source for using CO_2 is compressed CO_2 gas so that gas inflow to the chamber in a precisely controlled. The most favorable flow rate for CO_2 euthanasia systems should displace 10–30% of the chamber air (Fig. 1).

Volatile inhalational agents include halothane, enflurane, and isoflurane. These agents in combination with or without N_2O are suitable for euthanasia of laboratory rodents. These agents are usually given in vapor form to animals because when applied

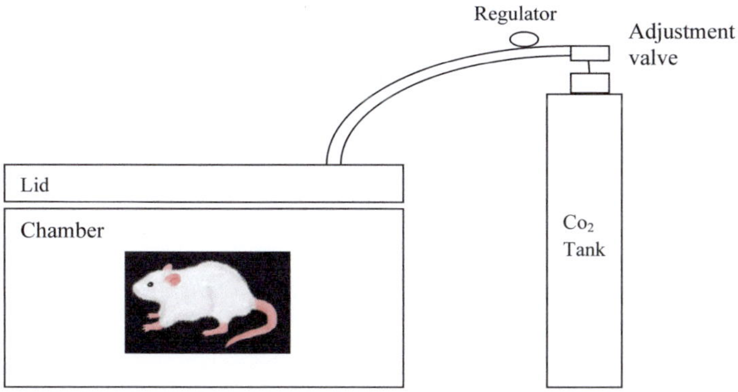

Fig. 1 Inhalation method for euthanasia

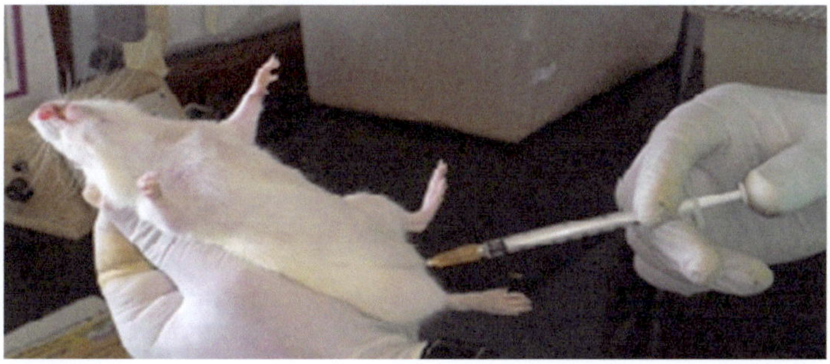

Fig. 2 Injectable method for euthanasia

topically in liquid form causes high level of irritation to animals. For euthanasia, these agents should be delivered via vaporizer or anesthetic chamber (open-drop technique), and exposure period of animals needs to be longer to ensure death.

2.2 Injectables

Injectable anesthetics should be given by intravenous (*i.v.*) route because it is rapid and reliable, but intraperitoneal (*i.p.*) route is mostly preferred in case of small animals. As an alternative, intracardiac injection may be considered in cases of already sedated or anesthetized animals (Fig. 2).

Barbiturates are the most commonly used agents for performing euthanasia of animals. They cause depression of CNS in descending order, starting from the cerebral cortex, then leading to loss of consciousness and finally resulting in anesthesia. Barbituric acid derivatives, oxybarbiturates (sodium pentobarbitone, secobarbital), thiobarbiturates (thiopentone), and a variety of barbiturate mixtures come under the category of barbiturates. Among all the barbiturates, sodium pentobarbitone is the most widely used agent for euthanasia, either alone or as commercially existing mixtures. The dose of sodium pentobarbitone for euthanasia is usually three times greater than the dose required for anesthesia and varies from 85 mg/kg (larger species) to 200 mg/kg (some rodents).

T-61 is an injectable, non-barbiturate agent. It consists of a mixture of embutramide, mebozonium iodide, and tetracaine hydrochloride and is given by i.v. route. The prerequisite for using T-61 is that animal must be sedated before administrating T-6. Mebozonium causes paralysis, and embutramide results in unconsciousness. The controversy has been raised that the paralytic effect of mebozonium occurs before embutramide-induced unconsciousness, which results in high distress to animals. On the other hand, in a study conducted by Hellebrekers and colleagues, the loss of consciousness and muscle activity occurred at the same time in rabbits and dogs. In spite of this positive report, T-61 is now no longer used and manufactured in countries like USA, but it is still used and available in some countries.

3 Recommended Physical Methods for Euthanasia

Physical methods of euthanasia include techniques like captive bolt, concussion, electrical stunning, cervical dislocation, decapitation, and microwave irradiation. These methods cause trauma to the brain which immediately lead to loss of consciousness.

3.1 Captive Bolt

Penetrating captive bolts are used for euthanizing ruminants, horses, swine, laboratory rabbits, and dogs. Captive bolts should be placed properly to provide sufficient restraint. The projectile should effectively disrupt the cerebral hemisphere and the brainstem to induce speedy unconsciousness followed by death of animals (Fig. 3).

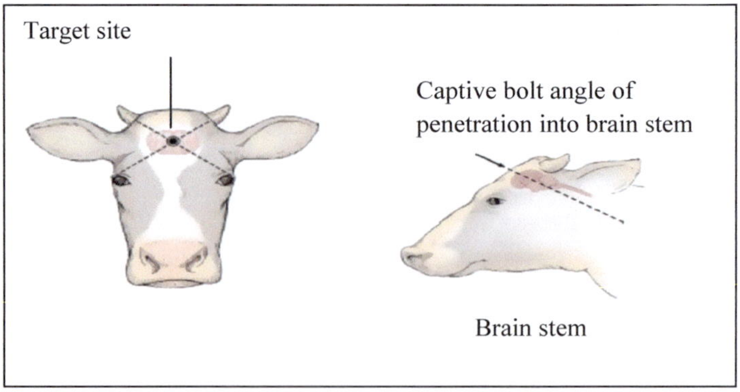

Fig. 3 Captive bolt method for euthanasia

3.2 Concussion

This is carried out by crushing a blow on the head of the animal, resulting in rapid CNS depression. This technique is used in case of animals having thin cranium like small rabbits, newborn kittens, newborn puppies, young guinea pigs, hamsters, rats, mice, birds, small reptiles, amphibians, and fish.

3.3 Electrical Stunning

This method is widely performed in finfish, amphibians, birds, dogs, poultry, pigs, sheep, calves, fox, goats, and rabbits. Specific equipments should be used to perform this kind of euthanasia. The maximum power voltage of this equipment ranges from 1.3 to 10 KW. This method results in death of animals by cardiac fibrillation and cerebral hypoxia.

3.4 Cervical Dislocation

This method is used to euthanize small birds, poultry, mice, immature rats (<200 g), and rabbits (<1 g). For rats and mice, the thumb and index finger are placed at the base of the skull on either side of the neck, and the base of the tail or the hind limbs is quickly pulled with the other hand, causing disconnection of the cervical vertebrae from the skull. It is essential to confirm that the neck is broken after performing the procedure by palpation of the vertebrae. If carried out properly, it should cause widespread damage to the brainstem and immediate unconsciousness (Fig. 4).

3.5 Decapitation

This procedure has been used for killing rodents, small rabbits, fish, amphibians, and birds. In this procedure, a sharp instrument is used for the severing of the neck

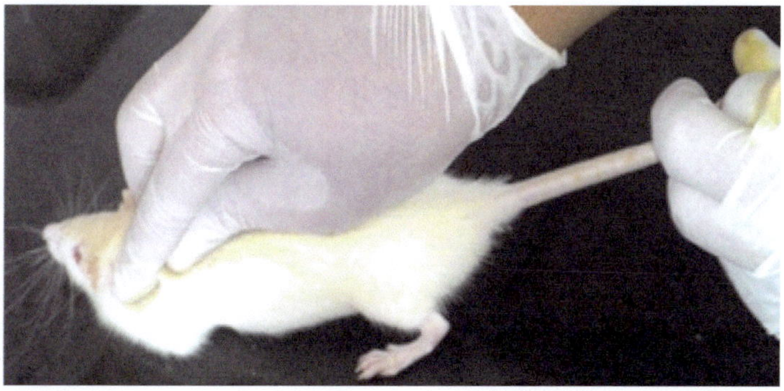

Fig. 4 Cervical dislocation method for euthanasia

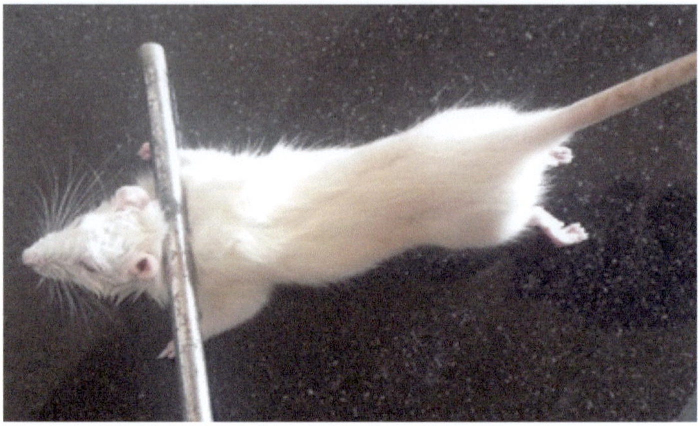

Fig. 5 Decapitation method for euthanasia

of the animal, close to the head (Fig. 5). Decapitation results in an instant interruption of the blood supply to the brain, causing fall in blood pressure in the brain and ultimately leads to loss of consciousness. This is valid only for warm-blooded animals. In case of cold-blooded animals, the animals must be stunned or make insensible before performing decapitation as they are very tolerant of anoxia.

3.6 Microwave Irradiation

This method involves the use microwave beam directed precisely at a specific part of the brain. Neurobiologists make wide use of this as it fixes brain metabolites without the loss of anatomical integrity. Only specially designed apparatus should be used for this purpose. The use of domestic microwave ovens is strictly not allowed. This procedure should only to be carried out on small animals such as rats, small rabbits (<300 g), amphibians, birds, and mice.

4 Methods and Agents Not Recommended for Euthanasia

Physical methods that should not be used for euthanasia includes pithing, rapid freezing, exsanguination, decompression, hypothermia, hyperthermia, drowning, neck crushing, and strangulation. Chemical agents like nitrogen/argon, ethanol, chloral hydrate, potassium chloride, N_2O, cyclopropane, ether, chloroform, methoxyflurane, trichloroethylene, hydrogen cyanide, 2-phenoxyethanol, urethane, neuromuscular blocking agents, ketamine, sedatives, and magnesium sulfate should not be used for euthanasia.

Table 1 Euthanasia methods for particular species

Name of species	Physical methods	Chemical methods
Rats and mice	Cervical dislocation with exsanguinations, decapitation, concussion	CO_2, halothane, sodium pentobarbitone
Guinea pigs	Stunning with exsanguinations, cervical dislocation	CO_2, halothane, sodium pentobarbitone
Reptiles	Captive bolt, concussion	Sodium pentobarbitone
Birds	Cervical dislocation, microwave irradiation, concussion, electrical stunning	CO_2, halothane, enflurane, isoflurane, sodium pentobarbitone, and T-61
Rabbits	Cervical dislocation, concussion, captive bolt, electrical stunning	Halothane, enflurane, isoflurane, sodium pentobabrbitone, and T-61
Fish	Concussion, cervical dislocation	Halothane and barbiturates
Amphibians	Concussion, microwave irradiation electrical stunning	Sodium pentobarbitone and T-61
Dogs, cats, and ferrets	Captive bolt, electrical stunning	Halothane, enflurane, isoflurane, sodium pentobarbitone, secobarbital, and T-61
Large mammals (pigs, sheep, goats, cattle, horses)	Captive bolt, concussion, electrical stunning	CO_2, halothane, enflurane, isoflurane, sodium pentobarbitone, and T-61
Non-human primates	Captive bolt, concussion, electrical stunning	CO_2, halothane, enflurane, isoflurane, sodium pentobarbitone, and T-61

5 Euthanasia Methods for Particular Species

See Table 1.

6 Ethical Statement

All institutional guidelines, national guidelines, state and local laws and regulations with professional standards for the care and use of laboratory animals should be followed. Studies involving animals must state that the institutional animal ethical committee has approved the protocol. For authors using experimental animals, a statement should be made that the animals' care is in accordance with institutional guidelines and animals used have been treated humanely and with regard to the alleviation of suffering. Researchers should treat animals as sentient and must consider their proper care and use and the avoidance or minimization of discomfort, distress, or pain as imperatives. Animal experiments should be designed only after due consideration of animal health. It should be ensured that all researchers who are using animals have received instruction in research methods and in the care, maintenance, and handling of the species being used. All the surgical procedures should be performed under appropriate anesthesia and follow only those procedures which avoid infection and minimize pain during and after surgery.

Behavioral Tests for Rodent Models

Sumit Jamwal, Shamsher Singh and Puneet Kumar Bansal

1 Behavioral Parameters for Motor Coordination Evaluation

1.1 Rotarod

Introduction

The "rotarod" technique was originated by N.W. Dunham and T.S. Miya and has proved to be of great value in preclinical research in testing of drugs which affect motor coordination. The rotarod test is a performance-based test on long cylindrical rotating rod with forced motor activity being applied, usually by rodents. This test measures parameters such as latency to fall (in seconds). This test is used to evaluate balance, grip strength, and motor coordination of the rodents, especially during testing of experimental drugs in animal models of movement disorders or after traumatic brain injury.

S. Jamwal · S. Singh · P.K. Bansal (✉)
Department of Pharmacology, ISF College of Pharmacy, Moga 142001, Punjab, India
e-mail: punnubansal79@gmail.com

S. Jamwal
e-mail: jamwalisf@gmail.com

S. Singh
e-mail: shamshersinghbajwa@gmail.com

P.K. Bansal
Department of Pharmaceutical Sciences and Technology, Maharaja Ranjit Singh Punjab Technical University (State Govt. University), Bathinda, Punjab, India

P.K. Bansal and R. Deshmukh (eds.), *Animal Models of Neurological Disorders*, https://doi.org/10.1007/978-981-10-5981-0_21

Fig. 1 Rotarod apparatus

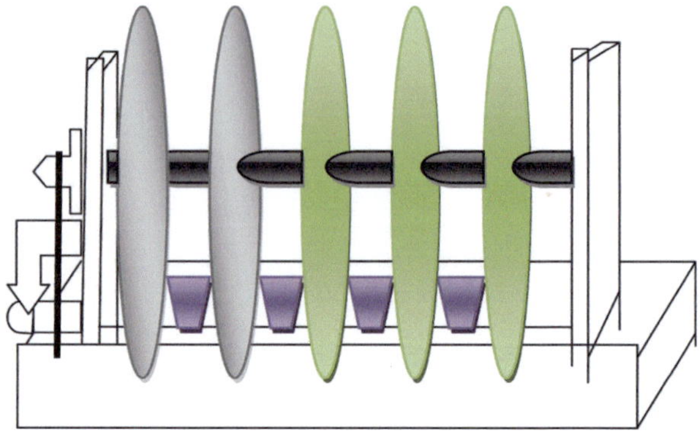

Principle

This apparatus is used to determine balance, motor coordination, and muscle strength in rats and mice. In this test, a rodent is placed on a long cylindrical rod which rotates along its long axis. When rodent fell off from the rod onto the plate placed below, the animal latency to fall (in seconds) is recorded.

Procedure

- Rotarod usually consists of two, three, or four compartments of 75 mm width each having a long cylindrical rotating rod (25 mm diameter) which rotates at speed of 5, 10, 15, 20, and 25 rpm with time interval counter in each compartment.
- The rotarod apparatus is turned on and animals are provided with training session consisted of 3 trials and time 60 s revolve at constant speed 25 rpm (Fig. 1).
- Rats (maximum of four at one time) are placed on the rod as it is being rotated at 25 rpm constant speed. When placement of animals is finished, acceleration button is pressed.
- The latency to fall from rod for each rat is recorded.
- Any other activity or observations during the test like occurrence of jumping, passive rotations are also recorded on data sheet.

1.2 Open-Field Activity

Developed by Calvin S. Hall, the open-field test provides unique opportunity to simultaneously measure the exploratory behavior, total locomotor activity (number of square crossed, rearing, and grooming), and anxiety-like behavior in rodents. The open field is usually an enclosed space, which may be square, rectangular, or circular in shape with surrounding walls that prevent animal from escaping.

Principle

This test is based on tendency of animal to explore novel area and avoiding bright light. When animals are placed into open field for the first time, they tend to remain near the walls. Open-field activity is a valid measure of "anxiety-like" behaviors and locomotion in animals. Increased exploratory behavior or locomotion in the center of arena indicates low anxiety, whereas preference for residing near walls and decreased locomotion/exploratory behavior indicates high level of anxiety in animals.

Apparatus

Different shapes for open-field apparatus have been designed, but the most commonly used design for mice is a large square chamber ranging in size from 28 × 28 cm to 56 × 56 cm and for rats chamber size ranges from 72 × 72 cm to 100 × 100 cm. One of the walls is clear Plexiglas, so mice could be visible in the apparatus. The square chamber is divided into number of small squares by drawing lines. The apparatus is located in a 1.8 × 4.6 m test room and lit by a 60-watt lamp for background lighting. The measures of number of square crossed can be obtained manually or with an automated camera-based computer tracking system (Limelight, Actimetrics) fixed to the ceiling, 2.1 m above the apparatus (Fig. 2).

Procedure

- The animals are carried to the test room in their home cages.
- Animals are placed into the center or one of the four corners of the open field and allowed to explore the apparatus for 5 min.

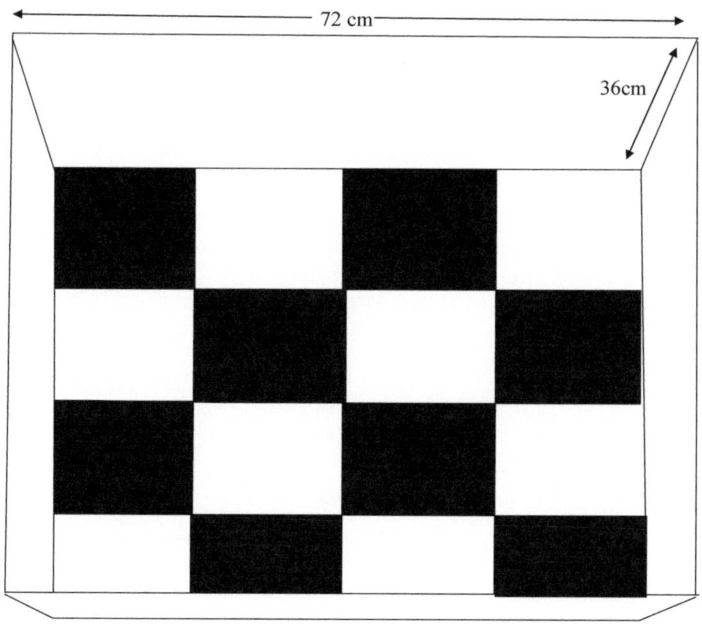

Fig. 2 Open-field apparatus

- After the 5-min test, animals are returned to their home cages and the open-field arena is cleaned using 70% ethyl alcohol and allowed to dry for 2 min.
- To assess the process of habituation to the novelty of the arena, the rat is exposed to the apparatus for 5 min on 2 consecutive days.
- During the actual experimentation, animals are placed in the center for 12 min and total locomotor activity during the last 10 min is recorded.

Behavioral Assessment

1. Line crossing: Rate with which the animals cross one of the lines with all four paws.
2. Central arena duration: Total duration of time for which animal remains in the central area.
3. Rearing: Number of times the animals stood on their hind legs.
4. Stretch attend postures: Number of times the animal showed forward elongation of the head and forelimbs and then pulling back to the original position.
5. Grooming: Duration of time the animal spent licking or scratching his body while motionless.
6. Freezing: Duration of time for which the animal remains complete stationary.
7. Urination: Number of puddles or streaks of urine.
8. Defecation: Number of fecal boli produced.

Each animal is awarded with a score for total locomotor activity and is the sum of total number of squares crossed, number of rearing, grooming and other activities.

1.3 Grip Strength

Grip strength meters are generally used in scientific research to assess neuromuscular and striatal functioning by measuring the amount of force applied by animal to grasp steel grid with forelimbs/hindlimbs. The amount of force applied is either recorded manually or automatically with the included software. Forelimb and hindlimb assessments can be measured simultaneously using dual sensor models or in separate trials using single stand model. The values are expressed in pounds, kilograms, or newtons.

Principle

The grip strength test is a simple and rapid noninvasive method widely used to evaluate the muscle force of forelimbs/hindlimbs in vivo. The animal is allowed to grasp a horizontal metal bar or grid with its forelimb/hindlimbs and then pulled backward. The bar or grid is attached to a transducer, and the force produced during the animal pulling is repeatedly measured at intervals (e.g., weekly) during experimental protocol. This method is highly economical; therefore, grip bar strength test is the most commonly used in vivo test for assessing impaired limb strength (forelimb and/or hindlimb) caused by pathology progression or during therapeutic interventions.

Procedure

Forelimb grip strength measurement.

- Set the meter to 0, and use KgF as a unit for values.
- Gently lift the rat/mice from the tail to the certain height where forelimbs/hindlimbs are just above the grip bar.
- Allow the rat/mice to gently grab the bar with both forepaws. This must be clearly seen.
- Then, gently pull the rat/mice backward at a constant rate until its grip is lost (Fig. 3).
- Peak tension (grams of force) is recorded on a digital force gauge as rat/mice release their grip. The transducer will save the value at the time of the release. The transducer is then again set to 0 before starting new measurement.
- Repeat the same three or four times, and calculate the mean value.

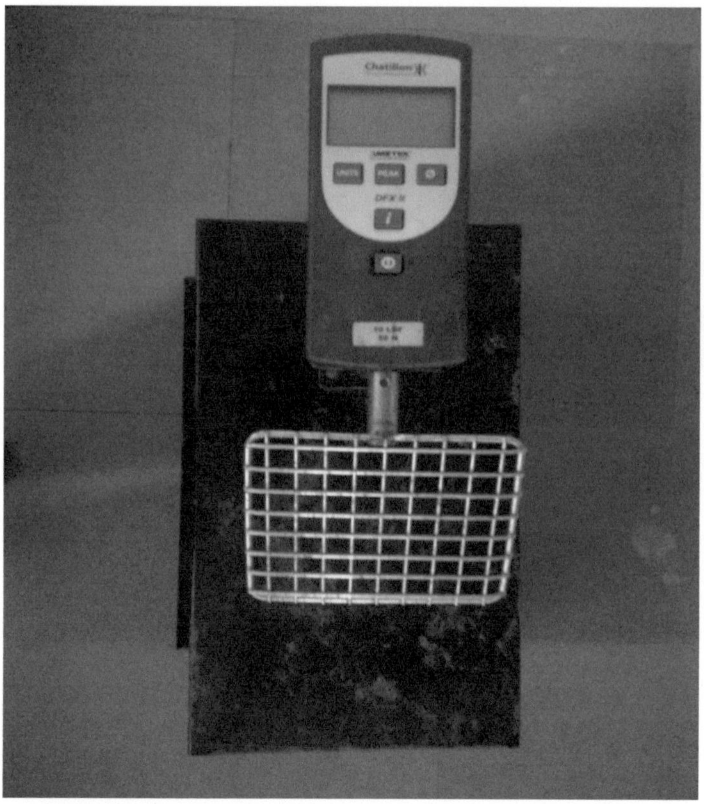

Fig. 3 Grip strength meter

Advantages

- This test is noninvasive and does not produce any damage to the muscle. Therefore, it is a useful tool to evaluate the effect of toxins and therapeutic treatment.
- When the test is performed in a continuous manner, the animals do not lose interest in the test over time period.

Disadvantages

- Duration: This test has to be performed for 3–4 times during single trial, with 1 min time lapse between each determination. The testing requires several minutes for single animal. This leads to a considerable time investment, especially when several experimental groups are being concurrently assessed.
- Variability: This is an in vivo test and shows a certain level of variability and requires good experience to achieve the best results.

1.4 Actophotometer

The majority of the CNS depressants (barbiturates and diazepam) and CNS stimulants (caffeine and amphetamines) drugs influence the locomotor and ambulatory activities in man and rodents. The locomotor activity is measure of index of wakefulness of brain activity.

Principle

The locomotor activity of rodents is easily measured using an actophotometer which operates on photoelectric cells connected in electric circuit to a counter. When the beam of infrared light falls on the photoelectric cell is cut off due to movement of animal in a closed chamber, a count is recorded. An actophotometer may be circular or square arena, in which the animal moves freely.

Apparatus It consists of six in-built photosensors and digital counter to measure spontaneous ambulatory activity. It is integrated with facility of activating rats by providing electric shock (100 volts) for very small duration. The stimulus is given in range of 1–100 V.

Procedure

- Weigh the animals and number them.
- Turn on the equipment, and all the photocells are checked for accurate working. Animals are placed individually for period of 10 min in the activity cage, and the basal locomotor activity or score is recorded.

Fig. 4 Actophotometer

- The test drug is injected, and after 30 min, each animal is evaluated for activity scores for 10 min.
- The difference in the activity before and after test drug administration is noted down.
- In the end, calculate percentage decrease or increase in motor activity (Fig. 4).

1.5 Narrow Beam Walking

Principle

It is used for the assessment of motor coordination, particularly of hindlimb. This apparatus is used for measuring gait abnormalities. Gait abnormalities and foot slips count are measured by narrow beam walk apparatus.

Apparatus

A narrow beam comprises of two platforms (8 cm in diameter) connected by a wooden beam (0.5 mm in thickness, 2.0 cm in width, and 120 cm in length) is used. The beam is elevated 1 m above the ground. A sawdust-filled box was placed below the beam to cushion the fall of the rat. A black box is placed at the end of the beam as the finish point. Nesting material from home cages is placed in the black box to attract the rat to the finish point. A white lamp (60 W) is placed near the start point and serves as an aversive stimulus. The time to cross the narrow beam is measured. During testing, the rats should be given 1 min to traverse the beam. The latency or time taken to cross the beam and number of foot slips encountered on narrow beam are recorded. If the animals fell off from the narrow beam, or freeze in between, the rats are given a maximum latency of 60 s to cross the beam and maximum six foot slips (Fig. 5).

Procedure

- Animal is allowed to acclimatize to the environment for 5 min before training.

Fig. 5 Plexi glass cage with mirror

- Animals are placed on one corner of narrow beam and allowed to walk across the narrow beam from one end to the other for at least 3 times.
- The time interval between two trials is 2 min.
- The number of foot slips encountered and time taken to cross the beam in each trial were recorded.
- Average of three trials is done to yield a final value.

1.6 Wire Suspension Test/String Test

Principle

The principle of this test is the degeneration of neurons in basal ganglia circuits that cause motor deficit. This wire suspension task is used to check muscle strength and prehensile reflex (capacity of the animal to hold a tightly stretched horizontal wire (1 mm in diameter, 35 cm in length, and 50 cm above the table top) with its forepaws and to remain suspended on the wire).

Procedure

- Before start training, rat is permitted to habituate to the environment for 5 min.
- Forelimbs of rat are put into the apparatus to grasp a tight horizontal wire and remain then suspended in air.
- Check the latency (time to drop from the wire) of the animal in seconds, with a maximum latency of 60 s.

1.7 Limb Withdrawal Test

Principle

This test is considered to be an important parameter to measure hindlimbs abnormalities due to striatal degeneration. In this test, the animal is put on a Perspex platform (20-cm-high 30 cm × 30 cm) containing four holes of 5 cm and 4 cm diameter for the hindlimbs and forelimbs, respectively.

Procedure

- The animal is positioned on the platform by locating first the hindlimbs and then the forelimbs into the holes.
- The time taken by the animal to retract its first hindlimb and the contra lateral hindlimb is recorded.
- The retraction time difference of both hindlimbs is determined.
- At every 30-min interval, the test is performed thrice and the average values were reported.

1.8 Classic Rotation Test

Principle

In classic rotation test, the dopamine receptor agonist and amphetamine are used to induce contralateral rotations or ipsilateral rotations in a rodent with a unilateral dopamine lesion. The number of rotations is counted by an automated rotation tracker or counted manually in a set time (usually 90 min). This test has been reported the gold standard for 6-hydroxydopamine (6-OHDA) lesions, because it causes striatal dopamine receptor sensitivity by producing complete unilateral loss of nigrostriatal dopamine.

Procedures

- Different procedures are available in the literature for classical rotation test.
- Generally, first 6-OHDA is administered to the brain of the animal by stereotaxic surgery, and then after 2–3 weeks, classical rotation test is performed.
- Rats are administered with dopamine receptor agonist and presynaptic stimulator amphetamine to produce contralateral and ipsilateral rotations.
- The total numbers of both contralateral and ipsilateral rotations are counted for 90 min.

1.9 Neurological Deficit Score

Scoring scale with a score of 0–8 is used to test neurological dysfunction. This test scores recumbency, dystonia of hind legs, gait abnormality, imbalance on a platform, and grasping problems. Total score of 8 points denotes maximal neurological deficit (animal showing near-death recumbency), and a score of 0 points indicates normal performance.

1.10 Vacuous Chewing Movements (VCMs), Grooming, Rearing, and Tongue Protrusions Measurement

Principle

Extrapyramidal side effect like orofacial dyskinesia is produced by longtime administration of neuroleptics. VCMs are characterized by symptoms like purposeless (non-directed) mouth openings, with or without tongue swelling. VCMs are induced secondary to various pharmacological treatments and during surgical lesions.

Procedure

- Rats are individually placed in a small cylindrical Plexiglas cage (30 × 20 30 cm) for the observation of orofacial dyskinesia.
- Before behavioral assessments, rats are allowed to habituate to the cage for 2 min. To determine the presence of oral dyskinesia, manually scored the VCMs, facial jerking, tongue protrusion, sniffing and grooming by using scoring scale.
- If during a period of grooming, VCMs, facial jerking, and tongue protrusion are occurred, they were not counted.
- Mirrors are placed under the floor and behind the back wall of the cage to help the observer to carry out the observation of oral dyskinesia when the animal is not directly visible to him.
- The behavioral parameters of oral dyskinesia are measured continuously for a period of 10 min.

1.11 Catalepsy

This test is defined by failure to correct an externally imposed posture over a prolonged period of time. The block test and bar test (commonly used in mice and rats, respectively) are two common tests of catalepsy. In this method, front paws of animals are placed on a bar over 6 cm and then 9 cm. The time retention of rats in this imposed posture is measured as a bar test elapsed time. Both forepaws of the rats are put on the bar in a half-rearing position. Latency of the animal to removal of the paw is recorded.

Apparatus

Small wooden block (4 cm high) and bar (1 cm in diameter and 10 cm long) are used.

Procedure

- During the training period, rat is placed on the wooden bar.
- Now after training, test drug is administered to the animals for inducing catalepsy.
- Rat is placed again on the apparatus for measuring the duration to move both forelimbs or hindlimbs from block to ground.
- The end point of catalepsy was considered to occur when both front paws were removed from the bar.
- A cutoff time of 1100 s is applied. Alternatively, if animal does not move from its place when kept on apparatus, then score is considered to be zero.

2 Behavior Parameters for Memory Evaluation

2.1 Elevated Plus Maze

Principle

Elevated plus maze is the simplest equipment used to study the drugs affecting learning and memory. It can be explored using transfer latency (TL) as parameter for acquisition and retention of memory process. This apparatus is also used to assess the drugs which produce acquisition deficit.

Apparatus and Procedure

The elevated plus maze comprises of two open arms (16 cm × 5 cm) and two covered arms (16 cm × 5 cm × 15 cm) expanded from a central platform (5 cm × 5 cm) and to a height of 25 cm the maze was raised from the floor. On the day one, each animal is located at the end of an open arm, opposite from the central platform. Transfer latency (TL) was described as the time taken by the animal to travel from the open arm into one of the covered arms with all its four legs. TL of each animal is accounted on the first day (i.e., 10th day of test drug administration). Push the animal gently into one of the two covered arms if the animal did not enter into one of the covered arms within 90 s, and TL was assigned as 90 s. The animal is permitted to walk around the maze for another 2 min and then returned to its home cage. Retention of this learned-task (memory) was examined 24 h (11th day) after the first day trial (Fig. 6).

Application

1. It helps in estimation of different other drugs which affect the memory like sedative (e.g., diazepam), scopolamine etc.
2. It is a easy, convenient, and time-saving method for studying memory deficit in animals.

Fig. 6 Elevated plus maze

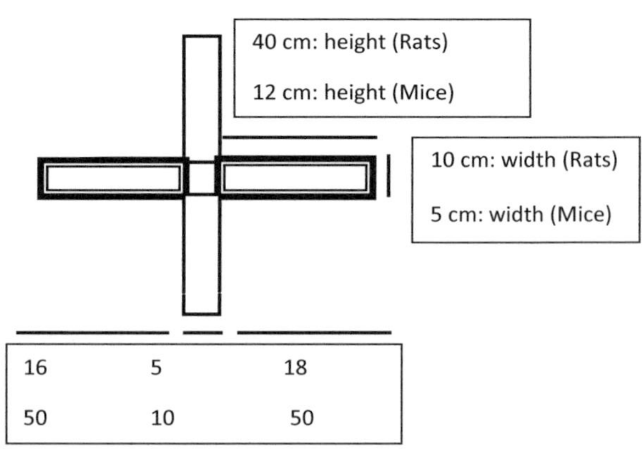

2.2 Morris Water Maze

Principle

Morris water maze is the most familiar test used to assess cognitive functions associated with memory. Morris water maze (MWM) is a task widely used and established by behavioral physiologist and pharmacologist to evaluate and compare learning and memory in rodents. In principle, rodents (rats or mice) can avoid swimming by climbing onto the platform and in the end the rodent apparently learns the spatial position of the platform from any starting point at the boundaries of the pool. Thus, the platform offers no local cues to direct the escape behavior of the animal. The only spatial cues are those exterior of the tank mainly the visual cues. The MWM procedure offers several advantages as a means of evaluating cognitive function in rodents:

- Both learning and retrieval processes can be evaluated Through the use of training as well as probe or transfer trials,
- Non-mnemonic behaviors or strategies can be described, and motor or motivational insufficiencies can be recognized Through the use of video tracking strategies.
- Visible platform tests can recognize gross visual deficiencies that might confound interpretation of results obtained from standard (MWM) testing.
- Both learning and re-learning experiments can be done by changing the position of platform.

Apparatus

Morris water maze comprises of a circular water tank (180 cm diameter, 60 cm height) made of black dense polyvinyl chloride or hardboard layered with fiberglass and resin and then surface pointed white (1.8–2.0 m in diameter and 0.4–0.6 m in height). The pool is filled with water (25 ± 1 °C) to a depth of 40 cm. A non-toxic water-dispersible emulsion was used to make the water cloudy. Four equally spaced locations just around border of the pool (north, south, east, and west) were used as start points, which divided the pool into four quadrants. An escape platform (10 cm in diameter) was located in the pool 2 cm beneath the surface of water. The escape platform was located in the center of one of the randomly selected quadrants of the pool and placed in the same location during the whole experiment (northeast for this study). The rats were permitted to swim freely into the pool for 120 s without platform prior to the training started (Fig. 7).

Procedure

Spatial acquisition task:

- Put the animal in the desired start location in the maze, in front of the tank wall. Then, animal is released into the water. During the release of animal, a timer or computer tracking program is started.
- Timer is started when the animal reaches (touches) the platform. A trial limit of 1 or 2 min per trial is average, usually, 2 min for rats and 1 min for mice.

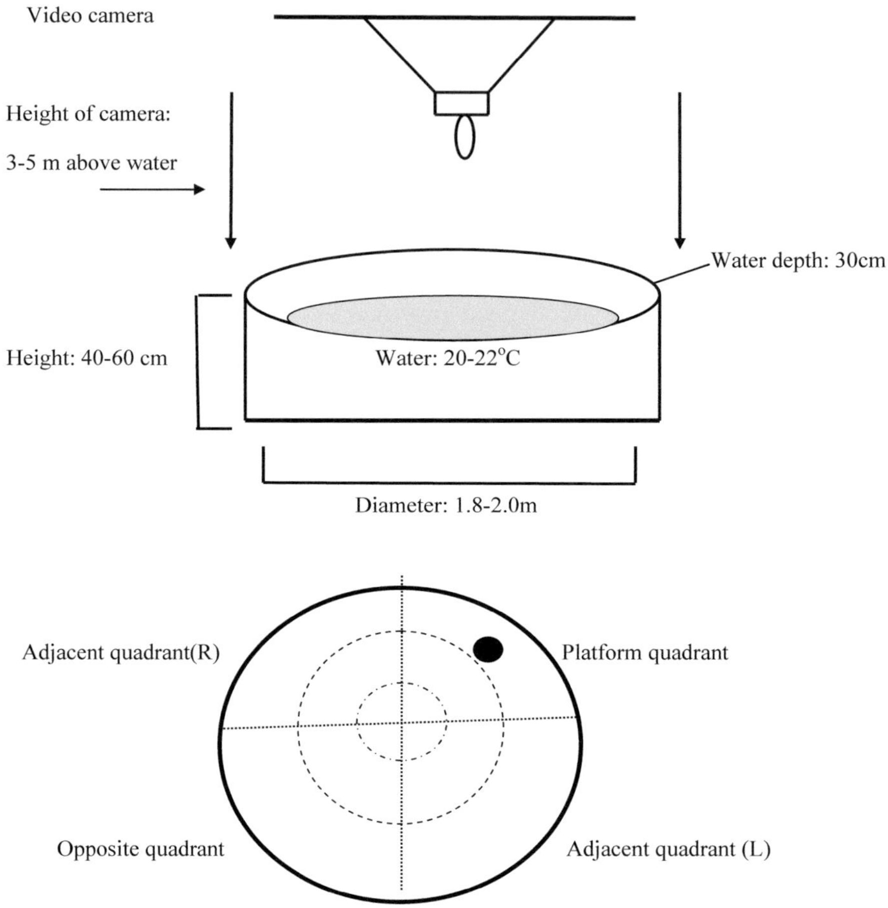

Fig. 7 Morris water maze depicting the basic assembly and the four quadrants

Animals are guided to the platform if they are not able to find the platform within this time limit.

- During the inter-trial interval (ITI), place the animal on the platform. Inter-trial intervals of 30 or 60 s were once used, but 15 s has become usual and produces better learning. However, longer ITIs are often used for mice, particularly during the first test session, and this enhances learning. This may be because mice are more susceptible to hypothermia-induced performance effects. The goal behind leaving the animal on the platform is to permit it to familiarize to its position in space and remember the location of the objective in relation to nearby cues.
- Situate the animal in the maze at a new start position, and repeat the trial (Steps 1–3) until the animal has the preferred number of trials for that day. Animals are normally given numerous trials per day. The most frequent number is 4. As there are four primary start positions, this keeps the start locations balanced each

day. With four trials per day, it takes 5–10 min per animal on the first day and progressively less time per animal each day thereafter. An another approach is to permit the animal for the time period of 15 s on the platform, then eliminate it to its cage and test the second animal on Trial 1 and do again this rotation until all animals have finished Trial 1, then repeat the process for consequent trials.

Reference memory: Probe trial (retention)

- The aim of the probe trial is to find out whether or not the animal remembers where the platform was situated.

Procedure:

- Eliminate the platform.
- Put the animal in a new start location in the maze, in front of the tank wall—for example, 180° from the original platform location.
- Rats were permitted to swim freely in the pool for 60 s, and the % time spent in target quadrant was recorded. The time spent in the target quadrant shows the degree of memory consolidation which had taken place after the acquisition trial. We use a fresh start location during the probe trial to make sure that its spatial preference is a reflection of the memory of the objective position rather than for a definite swim path.
- Indications of such memory comprise numeral platform-site crossovers, time and distance spent in the target quadrant as compared to the other quadrants, time in a pre-defined annulus surrounding the target that is greater than the target itself, average distance to the target site, angle (bearing) to the target site, and latency to first target-site crossover.

Evaluation:

- Escape latency (time to find the hidden platform from the starting point).
- Time spent in target quadrant.

2.3 Step-Down-Type Passive Avoidance Test

Principle

The step-down type of passive avoidance task has been used to examine the long-term memory based on negative reinforcement.

Apparatus

It consists of an acrylic cage (30 cm × 30 cm × 40 cm high) with a grid floor, inserted in a semi-soundproof outer box (35 cm × 35 cm × 90 cm high). The cage is illuminated with a 15 W lamp during the experimental period. A wooden platform (4 cm × 4 cm × 4 cm) is fixed in the center of the grid floor. Electric shocks (1 Hz, 500 ms, 45 V DC) are delivered to the grid floor with an isolated pulse stimulator.

Procedure

The training is carried out in two sessions:

- Each mouse is placed on the platform set in the center of the cage, and when the mouse steps down and placed all its paws on the grid floor, shocks were allowed for 15 s. When the mouse was placed in the test cage, the current resistance varied from 100 to 250 kΩ comparable to an electric shock ranging from 0.15 to 0.38 mA per mouse.
- Step-down latency (SDL), frequency of vocalizations, and flinching reactions are recorded.
- Animals showing SDL in the criterion range (3–30 s) during the first training session were used for the second session and the retention test.
- The second session is carried out after 90 min of the first, with an upper cutoff time of 60 s. Animals which did not step down within 60 s are considered as remembering the task and taken off, without receiving electric shocks any more.
- The retention test is carried out 24 h after training in a similar manner, except that the electric shocks were not applied to the grid floor.
- Each mouse is again placed on the platform, and the SDL was recorded, with an upper cutoff time of 300 s.
- Test drugs are administered only once, 30 min before the first training, injections are not given before the second training, or during the retention test.
- During post-training experiments, animals were trained following the same procedure and retention is examined after 7 days.

Evaluation

- Step-down latency (SDL) and the number of vocalizations and flinching reactions are measured.

2.4 Radial Arm Maze

Principle

Radial arm maze is used to evaluate the working memory in animals. The rat uses spatial information provided by the distal visual cues in the room to efficiently locate the baited arms. The radial arm maze allows to learn spatial learning and memory processes in the rats.

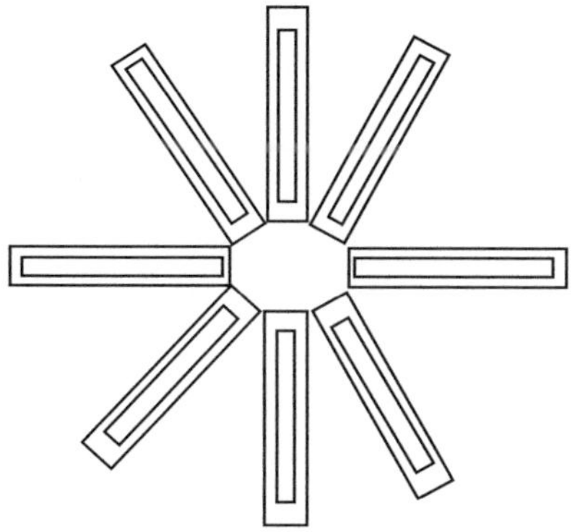

Fig. 8 Radial arm maze apparatus

Apparatus

Radial arm maze consists of eight arms. Each arm (50 cm × 12 cm) of the eight-arm radial maze extends from an octagonal-shaped center of 30 cm diameter. The platform is elevated 40 cm above the floor. Black plastic cups (3 cm in diameter and 1 cm deep) can be mounted at the end of each arm as containers for reinforces (food). Guillotine doors surround the hub (Fig. 8).

- Arm length: 50 cm
- Platform height: 40 cm
- Arm width: 12 cm
- Food cups diameter: 3 cm
- Central hub diameter: 30 cm
- Food cups depth: 1 cm
- Hub doors: guillotine

Procedure

- A rat is placed on the center hub with all guillotine doors lowered. Then, all the doors are simultaneously opened to allow the rat to choose arm freely (at the beginning of trial, two food pellets are placed in each receptacle).
- When the rat enters one of the arms, the doors to the seven remaining arms are closed.
- The open door is closed after the animal returns to the center hub. Then, all the doors are raised again simultaneously.
- The trial is considered complete when the rat visits all eight arms or spends 10 min in the maze.
- The number of correct responses before committing the first error is calculated as the index of radial maze performance.
- The session is terminated after the rat makes eight choices. The rat has to obtain the maximum number of rewards with minimum number of errors.

Evaluation

- The number of errors is counted during the session.

2.5 Y-maze

Principle

The Y-maze is widely used to assess behavioral task in preclinical research for studying spatial learning and memory. This test is based on the principle that rodents are encouraged to explore the maze arena in search of food. This maze provides animal with only two options: to enter into left arm or to the right arm, each containing a food item. Once a food is removed from one of the arms, the animals have natural tendency to search food and obtain the food reward from the opposite arm. This ability of animals to remember spatial locations is used to test cognitive function. It involves the use of hippocampal-dependent spatial reference memory, and this ability to remember arm is affected by the administration of drugs or disease models. The Y-maze is widely accepted as behavior task to evaluate short-term memory, general locomotor activity, and stereotypic behavior.

Apparatus

The maze consists of wood or plastic having black or grayish appearance. Each arm has dimension of 40 cm length, 13 cm height, 3 cm wide at the bottom, 10 cm wide at the top, and joins at an equal angle (Fig. 9).

Procedure

- Animal is placed at the end of one arm away from center and allowed to move freely through the maze for a 5- or 8-min session (small or large Y-maze, respectively) while being monitored by automated tracking system. Animal tends to explore the maze systemically entering each arm in turn.

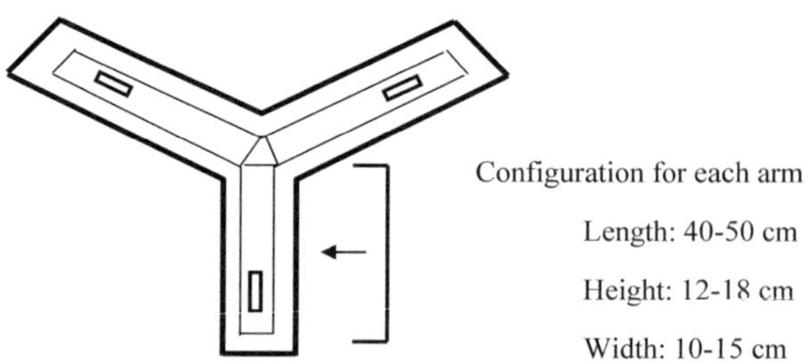

Fig. 9 Y-maze

- The number of entries into each arm is recorded (one entry means all four paws inside entering arm).
- Animal is returned to home cage, and the number of fecal pellets is recorded.
- The percentage of alteration (number of successive entries into the three arms, on overlapping triplet sets) is calculated as the ratio of actual alterations to possible alterations, defined as the total number of arm entries minus two, multiplied by two.

2.6 Radial Water Maze

Principle

The elevated radial water maze is a modified version, which is used to assess both spatial memory and working memory, i.e., the capability to memorize the list of already visited maze arms and to use this information on subsequent choices.

Apparatus

Radial water maze is made opaque by the addition of milk or a non-toxic white color and consists of a circular pool (120 cm in diameter, 60 cm high) filled 30 cm deep with water (25 °C). Eight 12-cm-wide channels formed by a 40 × 40 cm plastic wall project radically from the central area (30 cm in diameter). In the middle of the central area a circular platform (15 cm in diameter) can be moved vertically by a pneumatic device between the upper positions 1 cm above the water surface, to the bottom of pool by an electromechanical device a horizontal bench (10 × 11 cm) is fixed which can be collapsed. The maze is situated in a fixed position in a room with a large window, shelves, and racks visible from the maze (spatial cues).

Procedure

- One trial per day is given to the animals.
- The rat is placed on the raised central platform of the maze, at the beginning of each trial.
- The platform is lowered, and the rat is forced to swim until it reaches a bench at the far end of one of the eight arms and climbs upon it, after every 10 s.
- The bench is collapsed, and the rat must go back to the simultaneously raised central platform after every 10 s.
- At the end of the trial, the bench remains collapsed, and the sequence is repeated by lowering the central platform after 10 s.
- To assess the channel (traversing three quarters of the channel length) which has already been assessed in the same trial is determined as an error and until a correct choice is made the animal remains allowed to choose again. The central platform is not raised after an inaccurate choice.

- The animal visits all eight channels and the trial is terminated, after 16 choices are made, or after 20 min elapse.
- The animal is taken back from the raised central platform and dries off in a cage.
- Training continues 5 days a week; for 3 weeks, the training continues.

Evaluation: The evaluation of performance is done by the following:

The total number of choices required for visiting all channels and the time required for completing the trail, and number of errors in the first eight choices.

2.7 Novel Object Recognition Task

Principle

The novel object recognition (NOR) is widely used animal model for the investigation of memory alterations. This task is useful to check short-term memory, intermediate-term memory, and long-term memory, by assessment of the retention interval, that is, amount of time for which animals must retain memory of the sample objects placed during the recognition phase before the test phase, when one of the familiar objects is replaced by a novel one.

Apparatus

The NOR is performed in a wooden box (80 × 60 × 40 cm). The objects to be differentiated are of two different shapes (triangle and cylindrical), which are made up of painted wood, having 10 cm height.

Procedure

- Animals are allowed to the habituation period of 15 min to explore the apparatus before one day of beginning of experiment.
- Then the sample trial test is performed after 24 h for first 3 min with two identical objects of shape and color (i.e., FO1 and FO2) placed at the corners of the box, is started.
- After T1, the animals are placed back in their home cage and interval of 60 min is given before the second trial (T2).
- In the second trial (T2), one of the objects placed in T1 is replaced by a novel object (NO).
- The apparatus and the objects are cleaned properly after each trial to avoid the presence of olfactory trail.
- Exploration is considered as directing the nose to the objects at a distance less than 2 cm from the objects or touching it with the nose. Sitting or standing on the object is not recognized as exploration.
- The times spent by animal in exploring two objects in T1 and T2 are recorded separately. Now calculate the series of variables and total time spent in exploring two identical objects in T1.

- The differentiation between the familiar and the novel object during T2 is measured by making out the comparison between the time spent in exploring the familiar object with that of the novel object.
- The discrimination (D) index can be calculated by the formula: $D = N - F/N + F$ (F indicates familiar and N-novel).

3 Behavioral Parameters for Depression and Anxiety

3.1 Elevated Plus Maze

Principle

The principle behind elevated plus maze is based on the fact that animals have selective preference toward the safe areas (closed arm) and anxiety toward the open and high space (open arm). It is a simple behavioral assay that reveals anxiety-like behavior in rodents and used for the evaluation of anxiolytic actions of drugs. Several reports suggest that rodents enter closed arm more frequently, spending more time in them, consequently, revealing the anxious behavior as a measure of number of entries into each arm.

Apparatus For rats, the plus maze consists of two open arms (50×10 cm) and two closed arms ($50 \times 10 \times 40$ cm), and an open roof with the entire maze elevated 50 cm from the floor. For mice, the apparatus consists of two open arms (16×5 cm) and two closed arms ($16 \times 5 \times 12$ cm) elevated at a height of 25 cm.

Procedure

1. The animal is placed at the center of the maze with their head toward the open arm.
2. The test is carried out for 5 min. During the test, the number of entries into the open and closed arms, the time spent in each arm, and transfer latency are measured.
3. Transfer latency is defined as the time taken by the animal to move from open to the closed arm within 90 s cutoff time.

3.2 Mirror Chamber Apparatus

The mirror chamber consists of a wooden compartment ($40 \times 40 \times 30.5$ cm) having a mirror chamber ($30 \times 30 \times 30$ cm) enclosed within it. The mirror chamber is placed in the center of wooden box in such a way that a 5-cm corridor is formed which completely surrounds the mirrored chamber. The animal is placed inside the chamber of mirrors at a fixed corner (Fig. 5). During the 5-min test session, the following parameters are assessed:

(a) Latency to enter the mirror chamber
(b) Number of entries
(c) Total time spent by animal in mirror chamber
(d) Time spent per entry in mirror chamber.

3.3 Elevated Zero Maze

Principle:

The elevated zero maze is a modification of elevated plus maze and works on the same principle. This test is based on the natural tendency of animal to avoid open and elevated areas and preference to spend time in closed areas. Elevation of maze causes more anxiety and greater avoidance conflict as in elevated plus maze. It is used to evaluate the anxiogenic or anxiolytic effects of various pharmacological agents in rodents.

Apparatus:

This maze is white or black, circular apparatus having outer diameter of 45 cm and inner diameter of 30 cm. The annular platform where the animal can explore is 6 cm wide, and is divided into four quadrants, two "open" quadrants without walls and two opposing "closed" quadrants having 12-cm-high walls. The walls have thickness of 0.75 cm.

Procedure:

- Each animal is placed at the entry of the closed quadrant with the animal facing toward the closed arm.
- The behavior of rodent is recorded by using video tracking software using a camera fixed above the maze and analyzed for a period of 5 min.
- An entry in open arm is considered if all four paws of the animal are in the open arm.
- The parameters to be assessed are as follows:

  (a) Latency to first open area entry
  (b) Number of open area entries
  (c) Percent of time spent in open/closed areas
  (d) No. of stretched postures into the open quadrants.

3.4 Black and White Box Test

Principle:

This method is based on the natural aversion of rodents to highly bright areas. The change in exploratory behavior for assessment of anxiety is done by variation of illumination within the test box.

Apparatus:

The metal test box (45 × 27 × 27 cm) is open at the top, and the base is divided into 9-cm squares. The apparatus consists of two compartments, one is painted black and illuminated and the second is painted white and brightly illuminated with a 1 × 60 W light source. The compartments are connected by an opening 7.5 × 7.5 cm located at floor level in the center of partition.

Procedure:

- Animals are placed in a dark testing room and allowed to adapt for 1-h period to the new environment, after which they are placed into the test box.
- After exploration, animals are placed in the center of the white compartment and are observed for five behaviors:

  (a) Number of exploratory rearing in the white and black sections
  (b) Number of line crossings in the white and black sections
  (c) Number of transitions between the two compartments
  (d) Time spent in the white and black areas
  (e) The latency of the initial movement from the white to the black area.

4 Behavioral Parameters for Neuropathic Pain

4.1 Assessment of Thermal Hyperalgesia

Principle

The Plantar test apparatus is used to detect a peripherally mediated response to thermal stimulation (caused by drugs) in rodents.

Procedure

- Hyperalgesia to thermal stimulation is determined using a Plantar test apparatus (Ugo Basile, Comerio, Italy)
- Rats are placed individually in Plexiglas cubicles mounted on a glass surface maintained at 25 ± 0 °C.

- During this time, the rats initially established exploratory behavior, but subsequently stopped exploring and stood quietly with occasional sessions of grooming.
- A thermal stimulus, in the form of radiant heat has been emitted from a focused projection bulb, which is located under the glass floor, is focused onto the plantar surface of the left hind paw, and paw withdrawal latencies (PWLs) are recorded at interval of 15 min and the mean of the three values is used for analysis.
- The intensity of radiant heat was adjusted to give 17–18 s withdrawal latency in rats.
- A cutoff latency of 30 s was set to avoid tissue damage. The response latency is determined using a timer linked to the photodiode motion sensors in the plantar reflex device.

Evaluation

- Paw withdrawal latency.

4.2 Assessment of Mechanical Allodynia

Principle

The Dynamic Plantar Aesthesiometer has been designed to assess "touch sensitivity" on the plantar surface of the rodents. This test is used to assess the threshold for touch-evoked sensations and also used to measure the continuation of pain through different behaviors such as withdrawal, licking, immobility, and vocalization.

Procedure

- The threshold for touch sensitivity is measured in both hind paws, using an automated apparatus for applying reproducible light touch (Dynamic Plantar Aesthesiometer 37400-002; Ugo Basile, Comerio, Italy).
- Animals are placed in their compartments on the surface made up of metal mesh.
- After a small period, in which animals show exploratory behavior, they remain still in a resting position and at this time the test begins.
- With the help of an adjustable angled mirror, the touch stimulator unit is placed beneath the selected hind paw to position the filament below the plantar surface of the animal.
- When the unit is started, the electrodynamics actuator lifts the stainless steel filament, which touches the plantar surface and starts to apply an upward force below the threshold of feeling.

- The force increases, until the animal moves its paw or the point at which greatest present force is met. The maximum value of force in grams (50 g) is previously fixed.

Evaluation

- Paw withdrawal threshold.

4.3 Assessment of Cold Allodynia in Rats

Principle

Pain is elicited by cold, and a major feature of many neuropathic pain states is that normally innocuous cool stimuli initiate to produce pain (cold allodynia). Cold allodynia is measured as the number of foot withdrawal responses after use of cold stimuli at the planter surface of the paw.

Procedure

- Animals restrained gently, and both the hind paws are immersed on cold water (4 °C) for a period of 15 s (cutoff time).
- Paw withdrawal latency for each hind paw is measured, and the experiment is repeated 3 times for each rat.
- Paw withdrawal latency was expressed as threshold levels in seconds.

Evaluation

- Paw withdrawal latency.

4.4 Randall–Selitto Test

Principle

The Randall–Selitto test or paw pressure test is a basic pain research technique used for the measurement of the pain response in animals. This test is used to check the efficiency of analgesics by distinguishing the reaction to progressively increasing pressure on an inflamed paw. *Randall*–Selitto *test* method has been used to detect and quantify neuropathic pain responses in rats.

Procedure

- The nociceptive withdrawal threshold is assessed by using the Randall–Selitto electronic algesiometer.

- Before the test, each animal receives 5 min of handling to get used to manipulation; then, it is placed into a soft cotton cloth and carefully immobilized with the same hand used to hold the tested paw.
- The test consists of the application of an increasing mechanical force, in which the tip of the device is applied onto the dorsal surfaces of both fore and hind paws or the medial portion of the plantar until a withdrawal response resulted.
- The point of application is marked with ink in order to retain the location over repetitive trials. The maximum force which has to be applied should be kept limited to 250 g to avoid any skin damage.
- Measurements in the skin of the dorsal and lateral parts of the trunk are also performed to assess at-level neuropathic pain after SCI, with a maximum force of 350 g (400 g is the maximum reliable measurement).

Evaluation

- Paw withdrawal threshold.

4.5 Hot Plate Method

Principle

In this method, heat is used as a source of pain. Animals are placed individually on a hot plate which is maintained at constant temperature of 55 °C, and the response of animals, such as paw licking or jump response, is taken as the end point. Analgesics increase the reaction time.

Procedure

- Animal is placed on the heated surface of the plate.
- The temperature (55 °C) of the hot plate is maintained.
- The basal reaction time is taken by observing licking of hind paw or jump response (whichever appearing first) in animals. Normally, animals show such response in 6–8 s.
- A cutoff period of 15 s is observed to avoid any damage to the paws of animals.
- The time of latency is defined as the time period between the zero point, when the animal is kept on the hot plate surface, and the time when the animal licks its paw or jumps off to avoid thermal pain.

Evaluation

- Hind paw licking and jump response.

4.6 Tail Flick Method

Principle

The method uses radiant heat for evaluating the effect of analgesics in experimental animals. The analgesic activity of the test drug can be studied by measuring drug-induced changes in the sensitivity of the prescreened animals (the intensity of the light beam has been experimentally defined such that naive animals will withdraw their tails within 2–4 s) to heat stress applied to their tails by using analgesiometer.

Apparatus and Procedure

- Apparatus is consisted of one heating wire and a galvanometer which deflect the amount of current passing into wire (at constant rate of 5 A).
- Firstly, put on the apparatus and make sure that wire is fully heated up and after that place the tail of animal gently over heating wire.
- The distance between heat source and the tail was 1.5 cm and site of application of heat on the tail was kept within 2 cm, measuring from the root of the tail.
- Note down the tail flick latency before and after test drug administration in the animals.
- Cutoff reaction time was +10 s to avoid any tissue injury during the process.

Application: It is used to analyze the analgesic effect of opioid and all other analgesics.

5 Ethical Statement

All institutional guidelines, national guidelines, state and local laws, and regulations with professional standards for the care and use of laboratory animals should be followed. Studies involving animals must state that the Institutional Animal Ethical Committee has approved the protocol. For authors using experimental animals, a statement should be made that the animals' care is in accordance with institutional guidelines and animals used have been treated humanely and with regard to the alleviation of suffering. Researchers should treat animals as sentient and must consider their proper care and use and the avoidance or minimization of discomfort, distress, or pain as imperatives. Animal experiments should be designed only after due consideration of animal health. It should be ensured that all researchers who are using animals have received instruction in research methods and in the care, maintenance, and handling of the species being used. All the surgical procedures should be performed under appropriate anesthesia and follow only those procedures which avoid infection and minimize pain during and after surgery.

Author Index

P.K. Bansal and R. Deshmukh (eds.), *Animal Models of Neurological Disorders*, https://doi.org/10.1007/978-981-10-5981-0

MIX
Papier aus verantwortungsvollen Quellen
Paper from responsible sources
FSC® C105338

If you have any concerns about our products,
you can contact us on
ProductSafety@springernature.com

In case Publisher is established outside the EU,
the EU authorized representative is:
Springer Nature Customer Service Center GmbH
Europaplatz 3, 69115 Heidelberg, Germany

Printed by Libri Plureos GmbH
in Hamburg, Germany